Embedded Artificial Intelligence and the Internet of Things for Photovoltaic Systems

From Monitoring to Maintenance and Security

Embedded Artificial Intelligence and the Internet of Things for Photovoltaic Systems

From Monitoring to Maintenance and Security

Adel Mellit
Faculty of Sciences and Technology,
University of Jijel, Jijel, Algeria

ELSEVIER

Elsevier
Radarweg 29, PO Box 211, 1000 AE Amsterdam, Netherlands
125 London Wall, London EC2Y 5AS, United Kingdom
50 Hampshire Street, 5th Floor, Cambridge, MA 02139, United States

ISBN: 978-0-443-44835-5

For Information on all Elsevier publications
visit our website at https://www.elsevier.com/books-and-journals

Publisher: Priscilla Braglia
Acquisitions Editor: Edward Payne
Editorial Project Manager: Naman Singh
Production Project Manager: Paul Prasad Chandramohan
Cover Designer: Raman Kumar

Typeset by MPS Limited, Chennai, India

Contents

6. Smart monitoring systems for photovoltaic power plants

Adel Mellit, Nassiha Boutana and Soteris Kalogirou

7. Advances in anomaly detection and fault diagnosis for photovoltaic power plants

Adel Mellit and Chourouk Moussaoui

9. Cyber-physical security in photovoltaic power plants

Nassiha Boutana, Adel Mellit and Abdesselam Bouzerdoum

List of contributors

Lotfi Baghli Génie électrique, Université de Lorraine - GREEN – ENSEM, Vandoeuvre-lès-Nancy Cedex, France

Mohamed Benghanem Department of Physics, University of Medinah, Medinah, Saudi Arabia

Maamar Bettayeb Department of Electrical Engineering, University of Sharjah, Sharjah, United Arab Emirates

Nassiha Boutana Department of Electronics, USTHB University, Algiers, Algeria

Abdesselam Bouzerdoum Hamad Bin Khalifa University, Doha, Qatar

Rokia Chekirou Faculty of Sciences and Technology, University of Jijel, Jijel, Algeria

Mahmoud Dhimish Department of Photonics Engineering, Technical University of Denmark, Roskilde, Denmark

Nadia Drir Faculty of Electrical Engineering, USTB University, Algiers, Algeria

Bekkay Hajji Department of Electronics, Computer Science and Telecommunications, National School of Applied Sciences, Mohammed First University, Oujda, Morocco

Soteris Kalogirou Department of Mechanical Engineering and Materials Science and Engineering, Cyprus University of Technology, Cyprus

Adel Mellit Faculty of Sciences and Technology, University of Jijel, Jijel, Algeria

Chourouk Moussaoui Faculty of Sciences and Technology, University of Jijel, Jijel, Algeria

Abdelhamid Rabhi University of Picardie Jules Verne, Amiens Cedex 1, France

Giuseppe M. Tina Department of Electrical, Electronic and Computer Engineering, University of Catania, Catania, Italy

Marco Zennaro ICTP, Trieste, Italy

About the author

Adel Mellit is a professor at the Faculty of Sciences and Technology, University of Jijel, Algeria. He received his MS and PhD degrees in electronics from the University of Science and Technology Houari Boumediene (USTHB), Algeria, in 2002 and 2006, respectively. His research focuses on the development and application of artificial intelligence techniques for solar photovoltaic systems. Prof. Mellit has authored or coauthored more than 200 papers in international peer-reviewed journals, as well as numerous conference publications (many with IEEE) primarily in the field of photovoltaic systems. He has also contributed six book chapters, edited five conference proceedings published by Springer, and authored three books. He founded and directed the Renewable Energy Laboratory at the University of Jijel from 2012 to 2022. With more than 25 years of experience in solar energy systems, he has been an associate member of the Abdus Salam International Centre for Theoretical Physics (ICTP), Italy, since 2007. He has participated in various national and international research projects, mainly related to solar energy, microgrids, and electric vehicles. Prof. Mellit serves on the editorial boards of leading journals such as *Renewable Energy* and *Renewable and Sustainable Energy Reviews*, is an editor of the *IEEE Journal of Photovoltaics*, and is a subject editor for *Energy*.

Preface

Over the past three decades, solar photovoltaic (PV) technology has evolved from a niche renewable solution into a major pillar of the global energy transition. At the same time, artificial intelligence (AI) has emerged as a transformative tool for improving system performance, reliability, and sustainability. This book is a continuation of my previous work, *Handbook of Artificial Intelligence Techniques in Photovoltaic Systems: Modeling, Control, Optimization, Forecasting and Fault Diagnosis* (2022), and reflects many years of experience, research collaboration, and practical applications in the field of AI-based solar PV systems.

The motivation behind this book comes from a long research journey involving numerous experimental studies, collaborations, and more than 200 scientific publications. It represents the fruit of many years of work on the integration of AI, embedded systems, and smart monitoring techniques into real solar PV plants. While the earlier handbook focused mainly on classical AI applications in PV systems, this new book explores the latest developments, such as embedded AI (EAI), explainable AI (XAI), TinyML, large language models (LLMs), Artificial Intelligence of Things (AIoT), digital twins, blockchain, and cybersecurity. These technologies are becoming essential for the next generation of intelligent solar PV plants. Solar PV energy is one of the fastest-growing renewable energy sectors, but it also faces important challenges related to system reliability, fault diagnosis, maintenance, cybersecurity, and the need for scalable intelligent solutions. This book aims to provide researchers, engineers, practitioners, and graduate students with a comprehensive overview of recent advances and future directions in embedded and intelligent solar PV systems. It presents not only theoretical developments but also practical implementations, real case studies, datasets, and technological roadmaps that can help bridge the gap between academic research and industrial deployment.

The topics covered in this book are especially important for advancing the solar PV sector toward smarter, more efficient, and more resilient energy systems. By combining AI, the Internet of Things (IoT), embedded systems, and digital technologies, solar PV plants can move toward autonomous operation, predictive maintenance, and optimized performance. This book is intended to serve as a valuable reference for engineers, researchers, and practitioners in the solar PV sector.

This book is organized into nine chapters:

Chapter 1 is divided into two parts: the first part provides a brief introduction to PV technology, including the basic principles of solar cell operation, a brief description of cell technologies (monocrystalline, polycrystalline, thin-film, and new solar cell technologies), recent trends in PV materials, configurations, and latest efficiencies. The second part describes the various configurations of current PV systems, such as grid-connected and off-grid systems, as well as newer configurations, such as bifacial modules, agrivoltaics, floating PV systems, and PV in buildings and architectural configurations. The chapter also presents global trends in PV installation, costs, decarbonization, and solar energy development. Finally, the chapter presents several emerging issues in the field of PVs, from efficiency and material sustainability limits to system integration and environmental impact.

Chapter 2 first provides a brief introduction to AI, including ML, DL, and GenAI. Then it reviews the concept of embedded AI, including TinyML, and optimization techniques (such as quantization and pruning), which are essential for deploying embedded AI on resource-constrained edge devices (e.g., limited memory size, storage capacity, and processor speed). Various lightweight models, including MobileNets, EfficientNets-lite, and ShuffleNet, among others, will be presented and compared using illustrative tables. This chapter will also discuss the main advantages, limitations, and future trends in the field of embedded AI.

Chapter 3 commences with a brief presentation of cloud, fog, and edge computing. Then covers the concept of Edge AI, where different types of edge devices, including microcontrollers (e.g., Arduino Nano BLE sense, STM, etc.), microprocessors (CPUs), single-board computer (e.g., Raspberry Pi), graphical processors (e.g., Nvidia Jetson Nano), tensor processing unit (e.g., Google Coral), field programmable gate array (FPGA), and Application Specific Integrated Circuit (ASIC) are offered. It also provides platforms (such as Edge Impulse, TensorFlow Lite, uTensor, ONNX, OpenVINO, X-CUBE-AI, Embedded Learning Library, and PyTorch Mobile) for developing and deploying AI models. Comparative tables highlighting various edge devices and available platforms will be presented and discussed.

Chapter 4 provides an overview of the Internet of Things (IoT) technology and its integration with AI, often referred to as AIoT. It begins with a definition of IoT, its history, and its layered architecture and proceeds with a detailed examination of communication protocols spanning the physical, data link, network, transport, and application layers, including criteria for selecting appropriate protocols. The chapter then explores IoT devices and platforms, focusing on microcontroller units (MCUs) as key facilitators of IoT applications. Based on this, the chapter introduces the concept of AIoT, including its definition and framework, and how edge AI facilitates real-time intelligent processing. The chapter discusses the benefits and challenges of AIoT in detail, leading into a discussion of secure IoT that emphasizes the importance of privacy and robustness in smart IoT systems. It also presents the concepts of blockchain and digital twins. The chapter links AIoT fundamentals to emerging trends in the field, providing an understanding of the technological foundations and future paths.

Chapter 5 explores data augmentation and synthetic data generation as key factors for improving PV defect diagnosis using electroluminescent (EL) and infrared (IR) imaging. It initially provides a systematic review of conventional augmentation techniques, geometric and optical transformations, and their impact on addressing dataset imbalances. Building on this foundation, it conducts a comparative analysis of advanced generation approaches, including generative adversarial network (GAN)–based models, feature-guided adversarial frameworks, and conditional variational autoencoders (CVAEs), for producing realistic synthetic PV images. The chapter introduces an evaluation framework that combines histogram-, distribution-, and statistical-based metrics to evaluate both qualitative and quantitative analysis of the generated data. It also highlights the advantages and disadvantages of augmentation strategies, demonstrating how synthetic data can complement real datasets in enhancing model robustness. Free and open-source datasets will also be presented and discussed. These datasets will play a crucial role in the following chapters to address various issues related to PV plants, such as anomaly detection, object detection, and diagnosis. The chapter concludes with perspectives on integrating synthetic data pipelines into scalable PV diagnostic systems and outlines future research directions.

Chapter 6 presents a comprehensive overview of traditional and modern PV monitoring systems, with a particular emphasis on IoT and AIoT techniques. It begins by outlining the fundamentals of PV monitoring and the role of Supervisory Control and Data Acquisition (SCADA) systems in ensuring stable and reliable plant operation. The chapter then explores the integration of IoT sensors, data collection protocols, and AI techniques for advanced data analysis in PV applications. A literature review highlights recent approaches to smart PV monitoring, followed by two detailed case studies: the first demonstrates an IoT-based PV monitoring prototype, while the second presents an AIoT-enabled framework for anomaly detection in PV modules. The discussion also addresses key challenges, including data transmission, interoperability, and system scalability, while offering potential solutions and future research directions. The chapter concludes with insights on the evolution of smart PV monitoring, emphasizing the transition toward autonomous, AI-driven platforms for enhanced efficiency and reliability of solar PV plants.

Chapter 7 presents a comprehensive overview of advanced anomaly detection and diagnosis methods and their real-time implementation for solar PV systems. It covers both traditional electrical techniques and modern image-based approaches, highlighting their potential in identifying and classifying different types of faults. Electrical methods based on I-V curve analysis are examined alongside image processing techniques, including visual, IR, and EL imaging. Hybrid paradigms that combine I-V analysis with image-based methods are also discussed, offering a multiperspective approach for robust fault detection. A major emphasis of the chapter is placed on AI-driven solutions. ML and DL models, such as CNN (Convolutional Neural Network), VGG (Visual Geometry Group), MobileNet, and YOLO (You Only Look Once), are introduced as powerful tools for automated fault classification. These models are trained and tested on real-world datasets comprising visual, IR, and EL images of PV modules and arrays. Their practical application in real-time scenarios is demonstrated through illustrative examples and end-to-end solutions. In addition, the chapter explores embedded AI frameworks for deploying lightweight models on edge devices, enabling real-time, on-site monitoring without relying heavily on cloud infrastructure. Real-world case studies are included to illustrate practical deployments, showcasing the transition from algorithm design to field implementation. Finally, the chapter concludes with an outlook on future trends and a roadmap for integrating AI into PV fault diagnosis. This includes potential improvements in model efficiency, data acquisition strategies, and the fusion of multimodal monitoring techniques.

Chapter 8 examines maintenance strategies for PV power plants, covering both traditional and modern approaches. It discusses corrective, preventive, and predictive maintenance methods and compares their suitability for different sizes of PV power plants. Preventive maintenance is explained in detail through time-based and condition-based practices, while predictive maintenance is examined through AI-enabled techniques. The chapter also discusses key diagnostic and cleaning methods, including visual inspection, drone-based surveying, and automated cleaning, in terms of their effectiveness and practical feasibility. It addresses current challenges such as data limitations, costs, and model interpretability. Emerging technologies based on Edge AI, TinyML, AIoT, and LLMs are presented as promising solutions for intelligent maintenance. Finally, the chapter provides recommendations and future directions for improving the reliability and operational efficiency of PV power plants.

Chapter 9 addresses the security of PV systems, addressing both physical and cyber aspects. It focuses on potential threats that could compromise the cyber domain, including communication links, cloud platforms, and SCADA systems, as well as physical infrastructure such as inverters, sensors, and controllers. The chapter analyses the impacts of common threats, including false data injection, denial of service, data integrity attacks, and spoofing, on the operational reliability, safety, and stability of the grid. To mitigate these threats, the chapter emphasizes the growing importance of DL and ML methods as effective tools for detecting, classifying, and preventing cyberattacks in PV power plants. The concepts of intrusion detection systems and blockchain technology have also been introduced. By combining different types of strategies with real-time monitoring, these technologies offer promising solutions for enhancing resilience and ensuring the safe operation of PV power plants.

Acknowledgments

The author would like to thank Dr. A. Massi Pavan, Dr. N. Blasuttigh, and Prof. V. Lughi from the University of Trieste (Italy), as well as Dr. N. Kellil from UDES (Algeria), for providing valuable datasets used in this book. The author also gratefully acknowledges the support of Abdus Salam International Centre for Theoretical Physics (ICTP) through the Associates Programme (2023–28). Finally, the author would like to sincerely thank all contributors (Prof. M. Benghanem, Prof. M. Bettayeb, Prof. L. Baghli, Prof. B. Hajji, Prof. M. Zennaro, Prof. H. Rabhi, Prof. M. Dhimish, Dr. N. Drir, Dr. N. Boutana, Prof. S. Kalogirou, Ms. C. Moussaoui, Prof. M. Tina, Ms. R. Chekirou, and Prof. S. Bouzerdoum) to this book for their efforts, collaboration, and dedication in preparing the chapters.

Chapter 1

Introduction to photovoltaics

Adel Mellit[1] and Mohamed Benghanem[2]
[1]*Faculty of Sciences and Technology, University of Jijel, Jijel, Algeria,* [2]*Department of Physics, University of Medinah, Medinah, Saudi Arabia*

Introduction

Photovoltaic (PV) technology has emerged as a cornerstone of the global transition toward sustainable energy. It now plays a pivotal role in decarbonizing energy systems, enhancing energy security, and enabling the large-scale integration of clean and distributed power generation (Peters, 2025). Advances in PV materials, manufacturing processes, power electronics, and system design have accelerated its adoption, from large-scale solar farms to rooftop installations and off-grid applications, making it one of the fastest-growing and most impactful renewable energy technologies worldwide (Green, 2019). Globally, the installed capacity of PV solar energy systems has grown rapidly, with widespread deployment in Asia, Europe, and North America, as well as accelerating adoption in emerging markets. Annual PV installations continue to set new records, reflecting advancements in manufacturing efficiency, system performance, and grid integration. This sustained progress positions PV as one of the fastest-growing and most cost-competitive renewable energy technologies, playing a pivotal role in the global energy transition and carbon reduction efforts (Snapshot of Global PV Markets., 2023; Photovoltaics report, 2025).

This chapter provides a comprehensive overview of PV systems, beginning with the fundamental principles of solar energy conversion and the historical evolution of solar cell technologies. It introduces key concepts such as the photovoltaic effect, solar cell modeling, and the development of various cell types, including monocrystalline Si, polycrystalline Si, thin-films (amorphous silicon. Cadmium Telluride, Copper Indium Gallium Selenide, etc), emerging materials (organic, quantum dot, dye-sensitized perovskite, etc), and multi-junction technologies (III–V, Perovskite–Si tandem, etc.) highlighting recent advancements in efficiency and design. The chapter then explores the diverse configurations of modern PV systems, ranging from traditional off-grid and grid-connected setups to novel applications, such as bifacial modules, agrivoltaics, floating PV installations, and building-integrated PVs. These configurations

Embedded Artificial Intelligence and the Internet of Things for Photovoltaic Systems.
DOI: https://doi.org/10.1016/B978-0-443-44835-5.00009-6

reflect the growing versatility of PV technology in meeting energy demands across residential, commercial, and utility-scale sectors. To evaluate system performance, the chapter outlines essential indicators such as reference yield, final yield, array yield, specific yield, and performance ratio. It also examines global trends in PV installation capacity, cost dynamics, and the role of solar energy in decarbonization efforts and environmental sustainability. Finally, the chapter addresses current challenges in the PV sector, including limitations in efficiency, material sustainability, and system integration. It concludes by discussing future development pathways that aim to enhance the scalability, resilience, and environmental impact of PV technologies.

Overview of photovoltaic technology

Brief historical evolution

PV is the technology that directly converts sunlight into electricity (the PV effect), first observed by Becquerel (1839), Anderson and Chai (1976). Major advances have marked the evolution of PV technology (Starowicz et al., 2023):

- 1954: Bell Laboratories developed the first functional silicon solar cell with an efficiency of 6%.
- 1970s: PV technology gained importance in space applications, particularly for satellites.
- 2000s: Economies of scale and technological advances led to grid parity, with costs falling by approximately 20% per doubling of production capacity.
- 2020: Rise of perovskites, tandem cells, and multijunction cells.

Principle of the photovoltaic effect

A PV effect is the production of an electric current between two points of a semiconductor exposed to sunlight (Rappaport, 1959). It is the most fundamental principle of converting solar energy into electricity. Inside a solar cell, there is a p-n junction, which is produced by connecting n-type and p-type semiconductor material. When the sunlight (photon) falls on the solar cell, the photon with appropriate energy promotes an electron from the valence band to the conduction band and creates an electron–hole pair. The electric field at the p-n junction separates the charge carriers to allow the current to flow. This simple principle is shown in Fig. 1.1, which shows how solar energy can be directly converted into electricity (Kaygusuz, 2001).

Solar cell modeling

Solar cells are based on semiconductors doped with engineered p-n junctions that absorb light efficiently and separate charge carriers (Nelson, 2003)

- Incident photons generate a pair of charge carriers (electrons and holes) in the semiconductor.

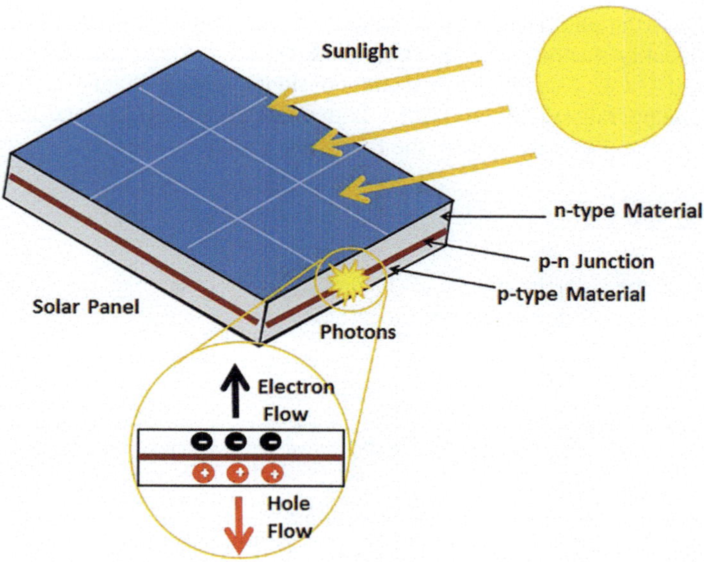

FIGURE 1.1 Principle of the PV effect. The figure illustrates how sunlight striking a photovoltaic cell generates electron–hole pairs within the semiconductor material. These charge carriers are then separated by the internal electric field, producing an electric current that can be extracted for power generation. *PV*, photovoltaic.

- The internal electric field at the p-n junction results in the movement of charge carriers to opposite contacts.
- Separated charges generate an electromotive force between the terminals.
- Electrical energy is delivered to an external load with some losses.

Photocurrent generation has two main physical components: photon absorption and separation of moving charge carriers. Only photons with energy greater than the bandgap of the semiconductor are effective at generating a pair of moving charge carriers. Charge carriers recombining (an electron and a hole recombine) can lower the efficiency by converting the photon energy into heat. The built-in electric field at a p-n junction prevents this from happening because it efficiently sweeps minority charge carriers across the junction before they can recombine. Under short-circuit conditions, the resulting charge flow completes the circuit, and DC is generated.

The basic equation describing the I–V characteristics of a p-n junction solar cell is obtained by inserting the appropriate parameter values into the Shockley diode equation and adding a term for photocurrent generation:

$$I = I_{ph} + I_0\left(e^{\frac{qV}{nkT}} - 1\right) \tag{1.1}$$

where I_{ph} is the photogenerated current, I_0 is the reverse saturation current, n is the ideality factor (≈1 for ideal diodes), q is the elementary charge (1.602×10^{-19} C), k is the Boltzmann constant (1.381×10^{-23}J/K), T is the absolute temperature (Kelvin), and V is the voltage across the solar cells.

The I_{ph} can be expressed by the following equation:

$$I_{ph} = I_{sc}\left(\frac{G}{G_{ref}}\right)[1 + \alpha(T - T_{ref})] \tag{1.2}$$

where G is the solar irradiance (W/m^2), G_{ref} is the reference solar irradiance (1000 W/m^2), α is the temperature coefficient of I_{sc}, and T_{ref} is the reference temperature (25°C).

The I_0 is given as:

$$I_0 = I_{0,ref}\left(\frac{T}{T_{ref}}\right)^3 \exp\left(\frac{-qE_g}{nk}\left(\frac{1}{T} - \frac{1}{T_{ref}}\right)\right) \tag{1.3}$$

where $I_{0,ref}$ is the reverse saturation current at T_{ref}, and E_g is the bandgap energy of the semiconductor.

The fill factor (FF) is a critical parameter in solar cell performance; it represents the ratio of the maximum power output (P_{max}) to the theoretical power ($V_{oc}\times I_{sc}$). It quantifies how square the solar cell's I–V (See Fig. 1.2) curve is.

$$FF = \frac{P_{\max}}{V_{oc} \times I_{sc}} = \frac{V_{mpp} \times I_{mpp}}{V_{oc} \times I_{sc}} \tag{1.4}$$

where V_{oc} represents the open-circuit voltage (V), I_{sc} is the short-circuit current (A), $V_{mpp}$and I_{mpp} represent the voltage and current at the maximum power point.

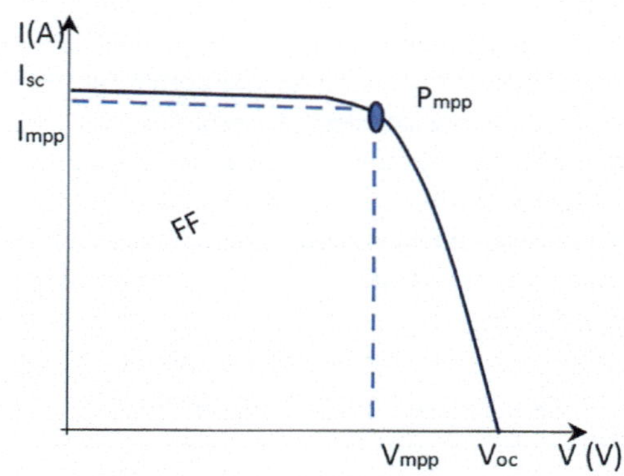

FIGURE 1.2 The I–V characteristic of a solar cell. The I–V characteristic (current–voltage curve) shows how much current a solar cell delivers at different output voltages. It is obtained by varying the voltage from 0 V (short circuit) to maximum voltage (open circuit) and measuring the current.

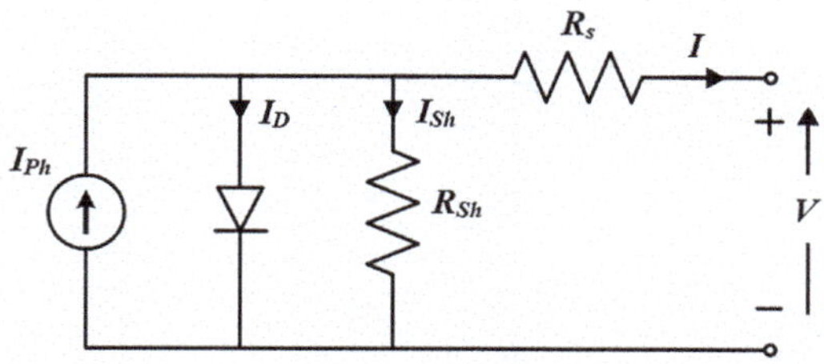

FIGURE 1.3 The equivalent circuit of a single diode model of a solar cell. The figure shows the equivalent circuit of a solar cell using the single-diode model, which includes a current source, a diode, a series resistance, and a shunt resistance. This circuit represents the electrical behavior of a real solar cell by modeling its generation, losses, and nonlinear characteristics.

The efficiency η of a solar cell is given by the following formula.

$$\eta[\%] = \frac{V_{mpp} \times I_{mpp}}{1000 \times Cellarea} \times 100 \quad (1.5)$$

The equivalent circuit model shown in Fig. 1.3 describes the full I–V behavior of a solar cell, module, or array as a continuous function under specific operating conditions. A widely used and physically grounded model is the single-diode model, which captures the essential electrical characteristics of a solar cell.

The mathematical equation is given as follows:

$$I = I_{ph} + I_0\left[\exp\left(\frac{V + IR_s}{nV_T}\right) - 1\right] - \frac{V + IR_s}{R_{sh}} \quad (1.6)$$

where V_T is the kT/q, $R_s$is the serial resistor, R_{sh} is the shunt resistor

Solar cell technologies

Despite their apparent simplicity, solar cells come with sophisticated technology behind them, with several technologies adapted for different applications. Most of the technology deployed in the field (Snapshot of Global PV Markets., 2023) (about 90% of all PV installed worldwide) uses silicon-based technologies with limited performance characteristics. Both monocrystalline and polycrystalline silicon cell performance is largely a function of the atomic-level purity and crystalline quality of the material. The purer the material, the higher the conversion efficiency but also the higher the cost of production. However, there is generally no correlation between pure material and cost, because there is often little impact from the end user on system decisions, which tend to be more a function of cost per watt and space than pure technical performance. Fig. 1.4 illustrates the best research cell efficiencies as a function of technology as reported by NREL (NREL Best Research-Cell Efficiency Chart, 2024).

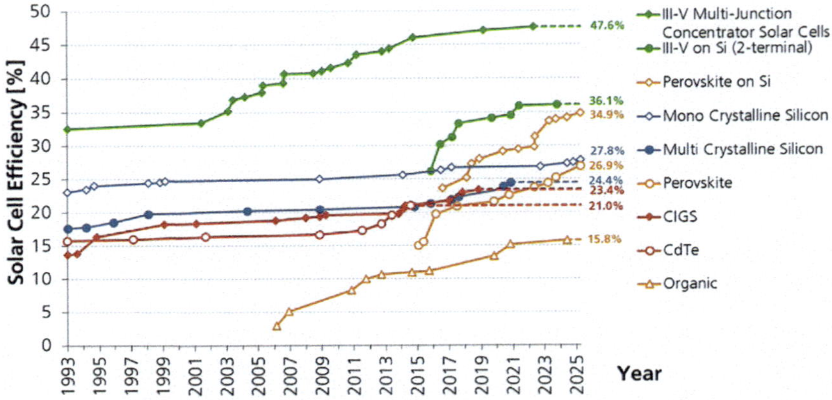

FIGURE 1.4 Best Research-Cell efficiency chart. The figure presents NREL's Best Research-Cell Efficiency Chart, showing how the record efficiencies of various PV technologies have improved over time. Each colored line represents a different PV technology, highlighting the historical progression and current performance limits of leading cell designs. *PV*, photovoltaic.
NREL Best Research-Cell Efficiency Chart. (2024). https://www.nrel.gov/pv/cell-efficiency.

From Fig. 1.4, the following points can be highlighted:

- Rates of improvement are fast at the beginning, with efficiencies gained quickly in mature technology. However, the rate of improvement slows over time. Advancements in silicon solar cells were fast in the 1970s and 1990s; today, improvements are slower.
- Multilayer and tandem technologies have repeatedly broken efficiency limits, suggesting that future solar research will be in layered structures to harvest a wider range.
- New PV technologies, such as perovskite and quantum dots, will completely change the market if they can be stabilized and commercialized.
- The efficiency limits mentioned here reflect the performance of "state-of-the-art research cells," not the efficiency of commercialized components, which are typically slightly lower due to practical losses.

More details about the latest solar cell efficiencies can also be found in (Green et al., 2025).

Every time the cumulative production of solar PV modules doubled, the unit price fell by about 25.7% over the past 44 years. This long-term decline, driven by efficiency gains, large-scale manufacturing, and supply chain optimization, makes PV more competitive and a key pillar of the global energy transition.

This represents a clear shift from the dominance of poly-crystalline silicon before 2016 to the near-complete dominance of mono-crystalline silicon today, with thin-film technology retaining a small share. This reflects the industry's shift toward mono-crystalline silicon, due to its higher efficiency.

Crystalline silicon solar cells

Monocrystalline PV cells are composed of high-efficiency single-crystal silicon ingots (Luque and Hegedus, 2011). They are usually of one consistent dark shade and have an octagonal shape due to the way cylindrical ingots are produced. To gain more efficiency and reduce waste, manufacturers slice off the rounded edges, creating almost a square shape to maximize the useful area. On the other hand, polycrystalline PV modules have been commercially available since 1981 and are produced by melting raw silicon into square molds and letting it cool, resulting in perfectly square cells with identical dimensions. There is very little wasted because the square cells produced from this method have uniform dimensions and are large enough to yield a nearly square panel after trimming off the edges for structure. Crystalline silicon is the most commonly available technology on the market today for solar, with the highest energy conversion efficiencies of any commercially available solar cells or modules (Saga, 2010). Crystalline silicon modules are highly trusted with excellent efficiency and long-term stability, as well as reliable performance.

Thin-film solar cells

Thin-film solar cell manufacturing employs a layered deposition technique, in which numerous ultra-thin photovoltaic layers are deposited one after another to create a full solar module (Chopra et al., 2004). This manufacturing technique allows for the creation of many different kinds of thin-film solar cells, each with a different light-absorbing layer. The main categories of thin film solar cells are a-Sc (amorphous silicon), CdTe (cadmium telluride), copper indium gallium selenide (CIGS), and organic cells. The conversion efficiency of thin film photovoltaics ranges from 10% to 13% depending on the type of cell (Lee and Ebong, 2017). Thin film cells have low efficiency compared to crystalline silicon cells, but they have low material cost, are flexible, lightweight, and have better performance at low light and high temperature.

Emerging solar cells

Silicon, CdTe, and CIGS are traditional materials that have been used to support solar energy production over the years, but limitations of efficiency, cost, and sustainability led to a continuous quest for new materials and advanced cell architectures (Priya and Stonier, 2025). Next-generation PVs: Currently developing PV technologies, organic cells, perovskites, dye-sensitized, and quantum dot cells. However, at this time, these technologies have efficiencies that are lower than traditional silicon cells, with perovskite tandem cells (i.e, combining perovskite with silicon), reaching somewhat higher efficiencies at about 25% (Lim et al., 2024). Notably, recent progress is especially rapid for perovskite tandem cells, which indicates a great potential for future high-efficiency and low-cost solar energy. Hybrid tandems (typically perovskite on silicon or other materials) also hold a lot of promise for increasing efficiency above 30% (Liu et al., 2025). This represents an active area of research to combine the advantages of different materials.

Multijunction solar cells

Multijunction solar cells, represented by purple lines and symbols, achieve the highest efficiency gains among photovoltaic technologies (Yamaguchi et al., 2005). Their efficiency began at less than 30% in the late 1970s and has improved rapidly, reaching approximately 47.6% by 2023 (NREL Best Research-Cell Efficiency Chart, 2024), according to the latest data. These cells achieve this high efficiency by using multiple layers to capture different parts of the solar spectrum. This steep upward curve reflects significant research progress and achievements in multijunction technology. The manufacture of multijunction solar cells is inherently complex, and challenges such as limited availability of key materials and slow production rates result in low yields, hindering large-scale commercialization. The integration of concentrator PV technology offers a potential path to improving performance while reducing production costs. However, to effectively compete with other PV technologies, the manufacturing costs of multijunction solar cells must be significantly reduced. One promising approach is the development of tandem structures that combine perovskite solar cells with silicon-based cells (Raisa et al., 2025).

Photovoltaic system configurations

PV systems are broadly classified based on how they are connected and operated within the electrical grid or independent systems. The main configurations include: off-grid, grid-connected, and hybrid PV systems.

Off-grid photovoltaic systems

Off-grid PV systems operate independently of the utility grid, using solar modules, battery storage, and often a backup generator to supply power continuously. Designed to store excess solar energy during the day and supply electricity during the night or cloudy periods. These types of systems are more suitable for locations that are isolated from the grid (Ghafoor and Munir, 2015). Common applications of an off-grid PV system include powering agricultural operations, electrification of rural sites, water pumping, telecommunications infrastructure, and electrification in remote areas. A charge regulator is always needed to manage the power between the battery and the load. A basic configuration of an off-grid PV system is shown in Fig. 1.5. It comprises a PV array, a charge regulator, batteries, and an inverter. The capacity of such PV systems ranges between 1 kW and 10 kW. These types of systems require careful sizing to avoid power loss during cloudy weather, especially in areas with little sunlight and requiring no special maintenance. With advanced photovoltaic inverters, most system information, including current, voltage,

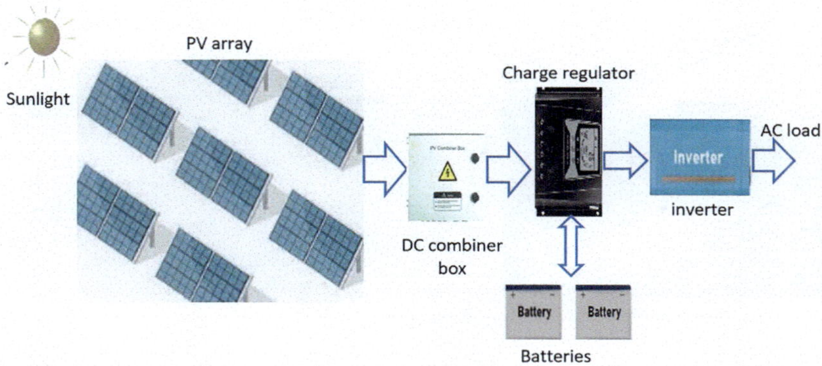

FIGURE 1.5 Basic configuration of an off-grid PV system. The figure presents the basic configuration of an off-grid photovoltaic system, showing how solar panels, a charge controller, and batteries are connected to supply power. It highlights the role of storage and power conditioning components in delivering electricity independently from the utility grid. *PV*, photovoltaic.

and power, can be displayed automatically. The battery bank is the critical component of these systems and must be carefully selected.

Residential photovoltaic systems

Residential PV systems are small-scale off-grid PV power systems installed on individual homes or apartment buildings to generate electricity for domestic use. They are typically grid-connected, although off-grid and hybrid versions are also available. A hybrid inverter is necessary in this type of PV system to manage power and connect to the grid during periods of low irradiance. In such a system, batteries are rarely integrated. The capacity of such PV systems does not exceed 20 kW for domestic applications (Leloux et al., 2012). Residential photovoltaic systems are typically installed on the roofs of buildings. A basic configuration is shown in Fig. 1.6. It essentially consists of a PV array and a hybrid inverter integrated with a DC/DC converter. A meter measures the flow of electricity between consumption and electricity exported to the grid. Such a configuration reduces the amount of the bill and does not require specific maintenance.

Commercial photovoltaic systems

Commercial PV power plants are medium-sized solar power systems installed by businesses, industries, or institutions primarily for self-consumption purposes. These systems involve installing solar panels on commercial properties to generate electricity, providing a sustainable and cost-effective energy solution for businesses. While the initial investment can be significant, the long-term savings in energy costs, along with potential tax incentives, make commercial PV systems a financially viable option (Wang and Sueyoshi, 2017; De Wild-

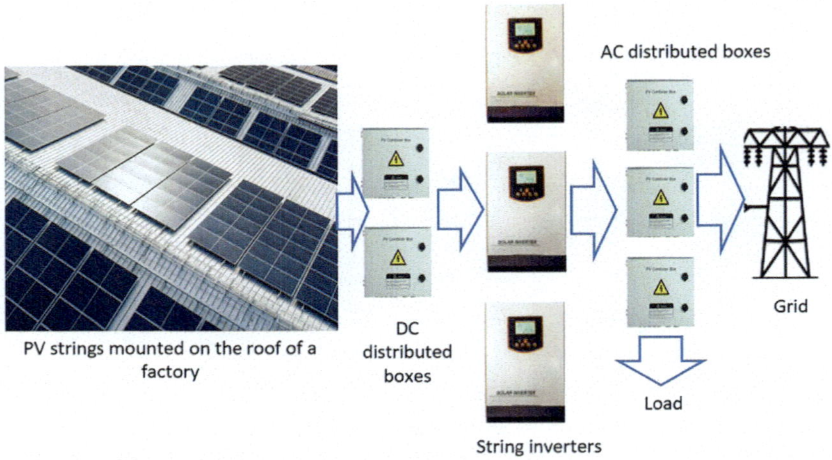

FIGURE 1.6 Residential PV system configuration. The figure shows the configuration of a residential photovoltaic system, including the PV array, inverter, and connection to the home's electrical loads. It illustrates how solar-generated power is integrated into a household, often with optional grid interaction or energy storage. *PV*, photovoltaic.

Scholten, 2013). Commercial PV systems are typically connected to the electrical grid, and central inverters are the most commonly used in large PV installations. These systems are generally larger than residential systems but smaller than utility-scale solar farms, typically ranging in capacity from 20 kW to 1 MW. While large factories or industrial areas can deploy systems ranging from 1 MW to 10 MW.

Industrial PV plants are typically configured as grid-connected systems, often without batteries, to directly offset a facility's electricity consumption and feed excess power into the grid. The configuration of an industrial PV system is shown in Fig. 1.7. It regroups various medium-sized PV strings and solar inverters with distributed generation.

Utility-grid photovoltaic plants

A utility-scale solar PV system refers to a large-scale solar power facility specifically designed to generate electricity for sale to the public electricity grid. Unlike residential or commercial PV systems, which are typically intended for on-site consumption, utility-scale PV systems deliver power directly to the transmission or distribution grid. These systems are often deployed under power purchase agreements (PPAs) or similar long-term contractual arrangements with utilities or large energy buyers, ensuring a steady stream of revenue. Utility-scale PV systems are characterized by high power generation capacities, typically starting from 1 MW and extending up to several hundred MW or even GW in some large solar parks (Hernandez

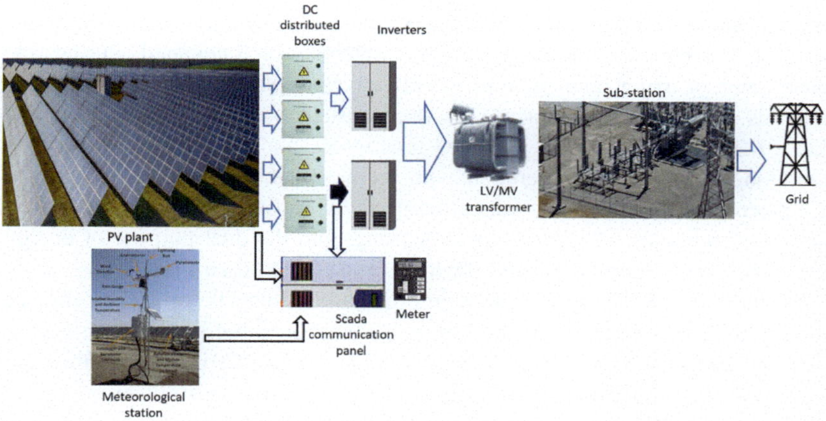

FIGURE 1.7 Utility grid PV plant configuration. The figure depicts the configuration of a utility-scale grid-connected photovoltaic plant, showing large PV arrays linked to inverters and transformers that feed electricity into the utility grid. It highlights the integration of solar generation at a utility level, enabling power delivery on a large scale. *PV*, photovoltaic.

et al., 2014; De Marco et al., 2014). They are operated under a PPA, which typically lasts 10–25 years, during which the system benefits from predictable energy prices.

These systems require large land areas, advanced grid integration strategies, and central inverter technologies. These systems play a key role in achieving national renewable energy targets, providing economies of scale and significantly contributing to grid decarbonization and energy diversification. Fig. 1.7 illustrates the main components of a utility-scale solar PV power plant. The plant consists of solar cell strings assembled in DC combiner boxes, inverters, AC distribution boxes, a supervisory control and data acquisition system, and low/medium voltage transformers. Utility-scale solar PV plants also include a weather station.

Utility-scale grid-connected PV systems can generate hundreds of megawatts, benefiting from economies of scale, low-cost per watt, and low carbon emissions, while providing stable, long-term renewable energy. These systems integrate directly with the grid, eliminating the need for local storage. However, they face challenges such as extensive land requirements, power intermittency, high initial investment costs, grid dependency, and potential transmission losses. Moreover, they can cause environmental impacts due to land use and ecosystem disruption.

Building integrated photovoltaic systems

Building-integrated PV (BIPV) systems refer to PV technologies that are seamlessly integrated into a building envelope, serving as both structural elements and electricity-generating components. Unlike conventional PV systems installed on existing buildings, BIPV systems are integrated during the construction or major renovation phase, effectively replacing traditional building materials such as roof tiles, facade cladding, curtain walls, skylights, or window glass. BIPV systems offer esthetic, functional, and economic benefits by combining solar energy generation with architectural design. They contribute to net-zero-energy buildings, reduce the need for separate mounting structures, and lower material and labor costs by serving both the dual purpose of energy generation and building envelope (Fuentes et al., 2018) (See Fig. 1.8).

Furthermore, BIPV systems may improve a building's thermal performance and daylighting, especially when semitransparent or tinted PV glass is used in windows and facades (Agathokleous and Kalogirou, 2020). Because they are integrated into the building structure, BIPV systems must meet electrical standards (e.g., IEC 61730 and IEC 61215) and building codes (related to fire safety, mechanical strength, and weather resistance) (Jaeckel et al., 2018). This technology is particularly promising in urban environments where roof space is limited, and pressure is increasing for sustainable architectural solutions (Compilation and Analysis of User Needs for BIPV and its Functions, 2019). BIPV offers the main advantage of seamlessly combining energy generation with building elements, reducing the need for separate mounting structures and enhancing esthetics. However, their main limitation is the typically higher cost and lower efficiency compared to conventional PV systems, due to design constraints and suboptimal orientation or shading.

FIGURE 1.8 BIPV system. The figure illustrates a BIPV system, where solar modules are integrated directly into the building's envelope, such as the roof or facade. *BIPV*, building-integrated photovoltaic.

Floating photovoltaic systems

Floating photovoltaic (FPV) systems consist of PV modules mounted on floating platforms above bodies of water such as lakes, reservoirs, and ponds (Floating Photovoltaic Power Plants: A Review of Energy Yield, Reliability, and Maintenance, 2025; Where Sun Meets Water: Floating Solar Market Report, 2025) (See Fig. 1.9). These systems are designed to generate electricity without occupying valuable land, making them a practical option in situations where land availability or cost is a constraint. The deployment of floating PV systems is expected to gain momentum in the coming years; however, several barriers to their widespread adoption remain. These include regulatory and legislative challenges, cost-competitiveness compared to traditional land-based PV systems, and ongoing uncertainty regarding environmental impacts, system reliability, and long-term performance (Floating PV plants., 2020).

Currently, most commercial floating PV systems are located on sheltered or calm bodies of water, but there is no standardized classification framework for their deployment across different types of aquatic environments. FPV systems benefit from improved efficiency due to water-based cooling and land space savings, but face greater installation and maintenance challenges due to complex structures and water conditions (Cazzaniga et al., 2018).

Agri-voltaics

Agri-voltaic systems refer to the innovative and dual-purpose use of land for both solar PV electricity generation and agricultural production. In these systems, solar modules are strategically installed above, between, or around

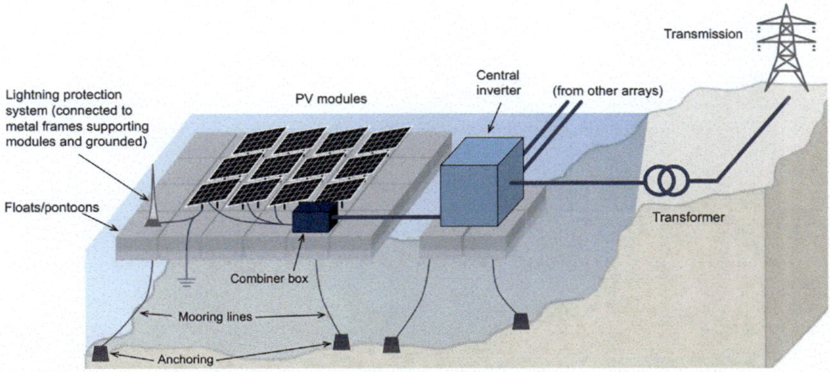

FIGURE 1.9 Typical large-scale FPV system. The figure shows a typical large-scale floating photovoltaic (FPV) system, with solar panels installed on buoyant structures over a water body. It highlights the use of water surfaces for solar energy generation, which can reduce land use and improve panel cooling for better efficiency.
Where Sun Meets Water: Floating Solar Market Report. (2025). http://documents.worldbank.org/curated/en/670101560451219695/Floating-Solar-Market-Report.

agricultural crops, allowing sunlight to be shared between plant growth and power generation (Dinesh and Pearce, 2016) (See Fig. 1.10). This approach optimizes land productivity by colocating food and energy systems, especially in regions facing land-use constraints or competition between agriculture and renewable energy expansion. Agri-voltaic systems offer a key advantage in improving land use by enabling both solar energy production and crop cultivation (Sirnik et al., 2023). This improves land productivity, supports farmers' income, and reduces heat stress on some crops through partial shading. However, their main drawback is the potential conflict between energy generation and agricultural production (Grid Parity unveils new glass-glass module for agrivoltaic systems, 2025). Shading from solar panels can reduce the amount of sunlight reaching crops, requiring careful system design, crop selection, and management to achieve an effective balance between the two functions (Dual Land Use for Agriculture and Solar Power Production: Overview and Performance of Agrivoltaic Systems, 2025).

Solar PV is expected to grow rapidly, driven by supportive local policies, land-use constraints, and favorable project economics. This growth will create new opportunities for developers and farmers alike. Standardized design frameworks are likely to emerge, reducing costs while adopting proven configurations. Meanwhile, demand for pilot and customized systems, incorporating modern PV technologies and farm support equipment to meet specific needs, will increase. Designs will remain highly regional, adapted to local agricultural practices and conditions. As costs decline and the sector matures, solar PV projects are expected to grow significantly, contributing to national solar deployment targets

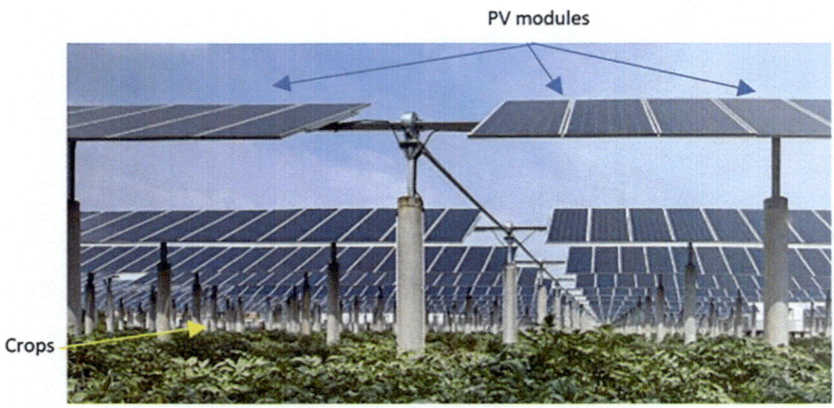

FIGURE 1.10 Agri-voltaics PV system. The figure illustrates an agrivoltaic PV system, where solar panels are installed above agricultural land. It demonstrates the dual use of land for both crop cultivation and solar energy generation, optimizing land productivity and providing shade benefits for certain crops. *PV*, photovoltaic.
Grid Parity unveils new glass-glass module for agrivoltaic systems. (2025). https://www.pv-magazine.com/2025/03/04/grid-parity-unveils-new-glass-glass-module-for-agrivoltaic-systems/.

(Dual Land Use for Agriculture and Solar Power Production: Overview and Performance of Agrivoltaic Systems, 2025).

Bifacial photovoltaic

Bifacial PV (BPV) systems use solar panels capable of capturing sunlight from both their front and rear sides, increasing overall energy yield. While the front side captures direct and diffuse sunlight, the rear side collects reflected light (albedo) from the ground, nearby surfaces, or even snow. Bifacial modules are commonly used in conjunction with ground-mounted, elevated, or tracking systems to maximize rear-side exposure (Mahim et al., 2024). Fig. 1.11 illustrates the basic working principle of a BPV system (Bifacial Solar Panels: What are They and Are They Worth It?, 2024).

BPV systems offer the advantage of increased energy production, typically 5%–30% more, by capturing sunlight on both their front and rear surfaces, making use of reflected and scattered light to improve overall efficiency. However, their performance depends heavily on-site conditions such as ground reflectivity, panel tilt, and spacing; in areas with low reflectivity or closely spaced panels, the rear-side energy gain can be minimal, limiting the system's benefits and complicating accurate energy yield predictions (Equipment, 2021; Guerrero-Lemus et al., 2016).

The global market forecast is still quite positive, with bifacial installations expected to account for more than half of new capacity additions by 2030, despite issues with soiling, shading, rear-side modeling, and the requirement for standardized testing techniques (Bifacial Photovoltaic Modules and Systems: Experience and Results from International Research and Pilot Applications, 2021).

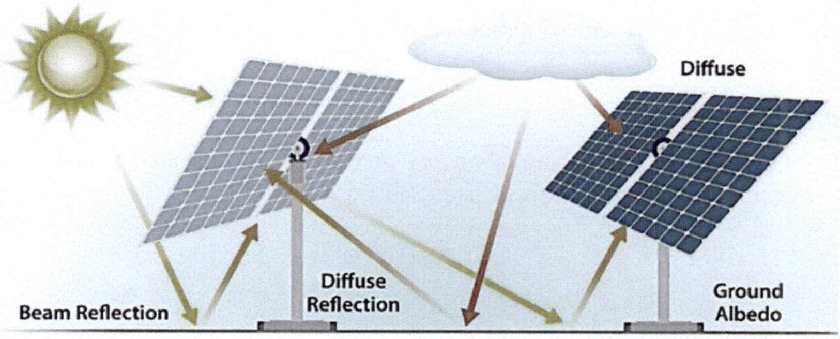

FIGURE 1.11 BPV system. The figure depicts a BPV system, where PV modules are integrated with building components while also serving additional functional purposes. It highlights the combination of energy generation with other building functionalities, such as structural support or shading. *BPV*, bifacial photovoltaic.
Bifacial Solar Panels: What are They and Are They Worth It? (2024). https://www.alternergy.co.uk/blog/post/bifacial-solar-panels-what-are-they.

Key performance indicators for photovoltaic systems

Reference yield

Reference yield (Y_r) represents the equivalent number of hours that the irradiance would need to be at 1 kW/m² to equal the solar irradiation received during a given period (See Eq. 1.7).

$$Y_r = \frac{H_t}{G_{stc}} \tag{1.7}$$

where, H_t is the total in-plane solar irradiance kWh/m^{2,} and G_{stc} is the standard irradiance 1 kW/m^2

Final yield

Final yield (Y_f) represents the total AC energy output delivered to the grid, normalized by the system rated power (See Eq. 1.8).

$$Y_f = \frac{E_{AC}}{P_{rated}} \tag{1.8}$$

where E_{AC} is the AC energy output of the system (kWh), and Prated is the PV system rated power (kWp)

Array yield

Array yield (Y_a) represents the DC energy produced by the PV array normalized by the system rated power (See Eq. 1.9).

$$Y_a = \frac{E_{DC}}{P_{rated}} \tag{1.9}$$

where E_{DC} is the DC energy produced by the PV array

Specific yield

Specific yield (SY) is the annual energy production per installed kWp, useful for benchmarking plant performance (See Eq. 1.10).

$$SY = \frac{E_{annual}}{P_{rated}} \tag{1.10}$$

Performance ratio

Performance ratio (PR) quantifies the overall efficiency of the photovoltaic system by comparing the actual energy output with the theoretical energy output under ideal conditions.

$$PR = \frac{Y_f}{Y_r} \times 100\% \tag{1.11}$$

Photovoltaic installation capacity and cost

According to the International Energy Agency, the cumulative installed PV capacity worldwide surpassed 2.2 terawatts (TW) in 2024 (See Fig. 1.12), with projections indicating continued exponential growth through 2030 due to falling costs and favorable policy frameworks (Snapshot, 2025). It took more than 40 years to reach a cumulative capacity of 1.18 TW (in 2022), but just 2 years to double this.

As shown in Fig. 1.13, in 2024, PV once again supplied more than three-quarters of new renewable generation capacity.

As shown in Fig. 1.14, the global weighted average, the levelized cost of electricity (LCoE) for 2023 for large PV systems is 0.036 €/kWh (= 36 €/MWh). LCoE has decreased by 16% year-on-year over the last 13 years (Nagaraja et al., 2025). The LCoE represents the average cost of generating 1 kWh of electricity over the system's lifespan. It can be expressed mathematically as follows:

$$LCoE = \frac{C_{tot}}{\sum_{n=1}^{N} E_n} \tag{1.11}$$

where C_{tot} is the total lifecycle cost of the PV system ($), and En is the energy produced per year (kWh)

Decarbonization, environment, and carbon footprint

The transition toward renewable energy sources, including PV systems, is at the forefront of global efforts to decarbonize the energy sector and

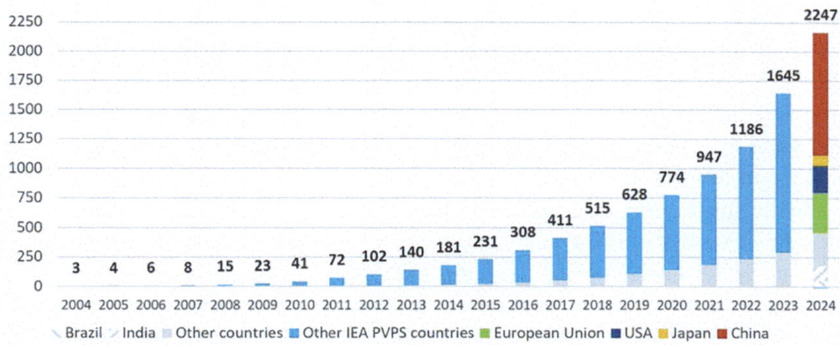

FIGURE 1.12 Evolution of cumulative PV installations. The figure shows the evolution of cumulative photovoltaic installations over time, illustrating the rapid growth of global solar capacity. It highlights trends in adoption, reflecting increasing deployment of PV systems worldwide. *PV*, photovoltaic.
Snapshot. (2025). https://iea-pvps.org/snapshot-reports/snapshot-2025/.

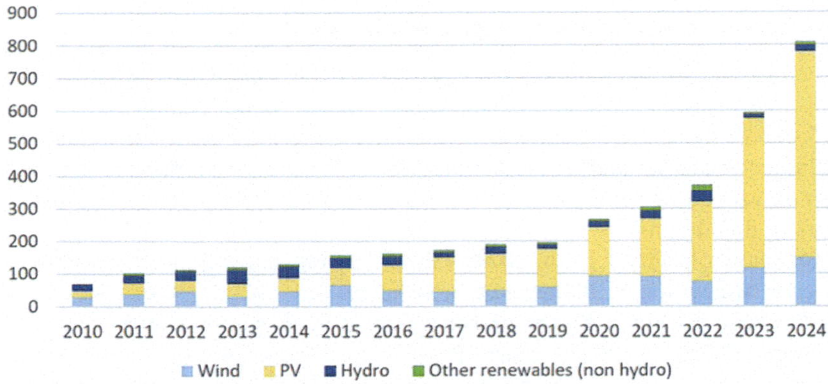

FIGURE 1.13 Evolution of annual renewable energy production. The figure illustrates the evolution of annual renewable energy production, showing contributions from various sources such as solar, wind, hydro, and biomass. It highlights the overall growth of renewable energy generation over time and the increasing share of solar and wind in the energy mix.
Snapshot. (2025). https://iea-pvps.org/snapshot-reports/snapshot-2025/.

mitigate climate change. Decarbonization in this context refers to the system-wide reduction of greenhouse gas (GHG) emissions, primarily carbon dioxide (CO_2), resulting from the combustion of fossil fuels in power generation, transportation, and industrial activities. PV systems offer a clean and sustainable electricity generation option, producing energy with near-zero GHG emissions during operation (Sameti and Syron, 2022). Beyond operational benefits, life cycle assessments demonstrate that solar PV technologies exhibit a significantly lower carbon footprint over their entire lifetime compared to conventional fossil-fuel-based energy systems (Li et al., 2025). While PV modules embody carbon emissions associated with raw material extraction, manufacturing, transportation, installation, and end-of-life processing, these emissions are typically offset within a short carbon payback period, resulting in low life cycle carbon intensity.

Furthermore, the large-scale deployment of PV technologies contributes to environmental preservation by reducing air pollutant emissions, lowering water consumption, and minimizing ecological disturbance when systems are appropriately designed and sited. As countries strive to meet the climate targets set by the Paris Agreement, solar energy has become a central pillar of national and regional decarbonization strategies. However, the rapid scaling of PV deployment also necessitates careful consideration of land use, material sustainability, and end-of-life management to ensure a truly low-carbon, environmentally responsible, and circular energy transition (Peters, 2025).

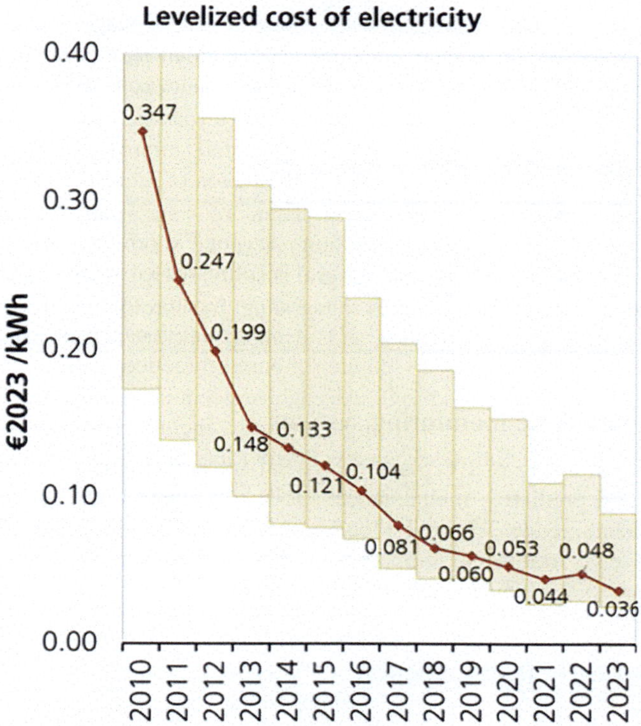

FIGURE 1.14 Global weighted average LCOE for large PV systems. The figure presents the global weighted average levelized cost of electricity (LCOE) for large-scale photovoltaic systems over time. It shows a significant decline in the cost of solar electricity, reflecting improvements in technology, manufacturing, and economies of scale.
Photovoltaics report. (2025). https://www.ise.fraunhofer.de/en/publications/studies/photovoltaics-report.html.

Challenges in photovoltaics

At the cell level, the primary limitation remains the energy conversion efficiency. Although extensive research has been conducted over the past decades, and significant advancements have been achieved through novel materials, improved manufacturing techniques, and advanced cell architectures (e.g., Passivated Emitter and Rear Cell, Tunnel Oxide Passivated Contact, Heterojunction with Intrinsic Thin layer and Perovskites), the efficiency of commercial PV cells is still below the theoretical limits. Factors such as recombination losses, optical losses, and thermalization of carriers contribute to this gap. Continuous efforts are being made globally by research institutions and industry players to push the boundaries of cell efficiency while maintaining cost-effectiveness and long-term stability. While at the system level, the deployment of PV technologies introduces a new set of challenges that go beyond the performance of individual modules. These challenges can be broadly categorized into the following domains (Mellit and Kalogirou, 2022):

Accurate photovoltaic power forecasting

The intermittent and weather-dependent nature of solar irradiance introduces significant variability in power generation. Accurate short-term and long-term forecasting of PV output is essential for grid stability, energy trading, and optimal energy management. This requires integrating meteorological data, satellite imagery, historical PV performance, and advanced machine learning models.

Smart photovoltaic monitoring systems

Real-time monitoring of PV plants is crucial for performance tracking, fault detection, and predictive maintenance. Modern systems leverage Internet of Things sensors, edge AI, and cloud platforms to monitor environmental conditions (e.g., irradiance, temperature), electrical parameters (e.g., voltage, current, power), and PV module health indicators.

Automated early fault detection

Identifying anomalies at an early stage, such as shading, delamination, soiling, or electrical faults, can significantly reduce energy losses and prevent catastrophic failures. Traditional inspection methods are labor-intensive and time-consuming. Today, automated diagnostic techniques based on thermography, electroluminescence, and AI-driven image analysis offer scalable alternatives.

Efficient fault diagnosis

Once faults are detected, diagnosing the root cause is critical for proper intervention. Advanced diagnostic frameworks incorporate AI-based classifiers, decision trees, data-driven modeling, and multimodal to distinguish between different fault types (e.g., potential induced degradation, microcracks, inverter failures, and more) and recommend corrective actions.

Optimized maintenance strategies

Preventive and predictive maintenance strategies are increasingly replacing reactive approaches (corrective) in modern PV systems. Condition-based maintenance, supported by predictive analytics and digital twins, enables operators to minimize downtime, reduce operational and maintenance costs, and extend system lifetime.

Cyber-physical security

With the increasing digitalization and connectivity of PV systems, cybersecurity and physical security are emerging as critical concerns. PV systems are vulnerable to cyberattacks targeting monitoring platforms, inverters, and communication networks. Additionally, physical tampering or theft of components poses risks, particularly in remote or unattended sites. Comprehensive security frameworks involving intrusion detection systems, anomaly detection algorithm-based AI, and secure communication protocols are essential to safeguard PV assets.

Fault prediction and defective component identification in photovoltaic systems

Predicting faults and identifying defective components in PV systems remains a significant challenge due to the nonlinear behavior of solar modules under constantly changing environmental conditions. Variations in solar irradiance, temperature, shading, and soiling lead to fluctuations in performance indicators, making it difficult to distinguish between normal variations and early signs of degradation or faults such as hotspots, bypass diode failures, or module mismatches. Traditional monitoring methods based on fixed thresholds often fail to detect early degradation trends. While machine learning and deep learning approaches offer improved fault prediction capabilities, their effectiveness is limited by the availability of clean, high-quality datasets and standardized fault classifications.

The following chapters will explore how emerging artificial intelligence techniques can further enhance the performance and sustainability of PV systems. In addition to established approaches such as machine learning, deep learning, reinforcement learning, and edge AI, recent advances in large language models (LLMs) and multimodal AI offer new opportunities for intelligent PV operation and management. By combining heterogeneous data sources, including sensor measurements, weather data, thermal and visual imagery, and system logs, multimodal AI can enable more accurate diagnostics, performance assessment, and decision support. LLMs can facilitate intelligent monitoring, automated reporting, and human–machine interaction, supporting operators and engineers in system analysis and maintenance.

Future development pathways

The PV sector continues to advance through the development of emerging technologies, each offering unique advantages while facing specific commercialization challenges (Jean et al., 2015). Unlocking their full potential will require coordinated progress in materials science, manufacturing techniques, and system-level integration (Ali et al., 2025; Green, 2019). The successful transition from laboratory prototypes to commercially viable solutions hinges on advancements in three critical areas: operational stability, scalability of production, and cost-effectiveness.

As these innovations mature, they are poised to reshape the renewable energy landscape, complementing established silicon-based technologies and accelerating the global shift toward sustainable power generation (Trends in photovoltaic applications, 2024). Future trends may include the emergence of more advanced AI algorithms, including Generative AI and embedded AI for making autonomous decisions in real-time, and continuous learning systems able to adapt to ever-evolving environmental and technological changes.

Conclusion

This chapter provided a comprehensive overview of PV technology, starting from the fundamental principles of the PV effect and the historical development of solar energy conversion. Various solar cell technologies were discussed, ranging from conventional monocrystalline and polycrystalline silicon cells to emerging thin-film and next-generation materials, illustrating the continuous pursuit of higher efficiency and material sustainability. Diverse PV system configurations, off-grid, residential, commercial, and utility-scale, as well as advanced concepts such as bifacial modules, agri-voltaics, floating PV, and BIPV, were presented to highlight the versatility of PV technology and its expanding applications. Essential key performance indicators for evaluating PV system productivity and operational efficiency were introduced, enabling standardized assessment across different installations.

Global trends showed an acceleration in PV deployment driven by decreasing system costs, ambitious decarbonization targets, and supportive energy transition policies. However, challenges remain, including limitations in efficiency, environmental sustainability, integration into existing infrastructure, and the need for robust performance monitoring and maintenance strategies. Looking ahead, innovation in materials, digital transformation through artificial intelligence and the Internet of Things, hybrid system integration, and circular/low-impact manufacturing pathways will play a crucial role in shaping the next generation of PV systems. In general, PV technology is a cornerstone of future energy systems, enabling the generation of clean, distributed, and cost-effective electricity and supporting global efforts toward climate neutrality.

Solar PV will continue to play a pivotal role in the energy transition as countries expand their renewable energy deployments. Advanced economies, such as China, the US, and the EU, are scaling up large-scale PV installations, while many developing countries are adopting solar power to improve energy access and reduce reliance on fossil fuels. With declining costs, advances in storage technologies, and the integration of smart grids, solar PV is becoming one of the most competitive and reliable energy sources. As climate policies strengthen globally, PV technology will remain essential for reducing emissions and enabling the development of sustainable energy.

AI disclosure

During the preparation of this work, the author(s) used ChatGPT in order to improve English writing. After using this tool/service, the author(s) reviewed and edited the content as needed and take(s) full responsibility for the content of the publication.

References

Agathokleous, R.A., Kalogirou, S.A., 2020. Status, barriers and perspectives of building integrated photovoltaic systems. Energy 191. Available at https://doi.org/10.1016/j.energy.2019.116471, https://www.journals.elsevier.com/energy.

Ali, A.O., Elgohr, A.T., El-Mahdy, M.H., Zohir, H.M., Emam, A.Z., Mostafa, M.G., et al., 2025. Advancements in photovoltaic technology: A comprehensive review of recent advances and future prospects. Energy Conversion and Management: X 26. Available at https://doi.org/10.1016/j.ecmx.2025.100952, https://www.sciencedirect.com/science/journal/25901745.

Anderson, W.W., Chai, Y.G., 1976). Becquerel effect solar cell. Energy Conversion 15 (3–4), 85–94. Available at https://doi.org/10.1016/0013–7480(76)90020-6.

Becquerel, E., 1839. On electric effects under the influence of solar radiation. Compt. Rend. 9, 561 Compt. Rend. **9**.

Bifacial Photovoltaic Modules and Systems: Experience and Results from International Research and Pilot Applications 2021. Available at https://iea-pvps.org/wp-content/uploads/2021/04/IEA-PVPS-T13–14_2021-Bifacial-Photovoltaic-Modules-and-Systems-report.pdf

Bifacial Solar Panels: What are They and Are They Worth It? 2024. Available at https://www.alternergy.co.uk/blog/post/bifacial-solar-panels-what-are-they

Cazzaniga, R., Cicu, M., Rosa-Clot, M., Rosa-Clot, P., Tina, G.M., Ventura, C., 2018. Floating photovoltaic plants: Performance analysis and design solutions. Renewable and Sustainable Energy Reviews 81, 1730–1741. Available at https://doi.org/10.1016/j.rser.2017.05.269.

Chopra, K.L., Paulson, P.D., Dutta, V., 2004. Thin-film solar cells: an overview. Progress in Photovoltaics: Research and Applications 12 (2–3), 69–92. Available at https://doi.org/10.1002/pip.541.

Compilation and Analysis of User Needs for BIPV and its Functions, 2019. Available at https://iea-pvps.org/wp-content/uploads/2020/02/IEA_PVPS_Task_15_STC_C1_Report_20190216.pdf

De Marco, A., Petrosillo, I., Semeraro, T., Pasimeni, M.R., Aretano, R., Zurlini, G., 2014. The contribution of Utility-Scale Solar Energy to the global climate regulation and its effects on local ecosystem services. Global Ecology and Conservation 2, 324–337. Available at https://doi.org/10.1016/j.gecco.2014.10.010, https://www.journals.elsevier.com/global-ecology-and-conservation.

De Wild-Scholten, M.J., 2013. Energy payback time and carbon footprint of commercial photovoltaic systems. Solar Energy Materials and Solar Cells 119, 296–305. Available at https://doi.org/10.1016/j.solmat.2013.08.037.

Dinesh, H., Pearce, J.M., 2016. The potential of agrivoltaic systems. Renewable and Sustainable Energy Reviews 54, 299–308. Available at https://doi.org/10.1016/j.rser.2015.10.024, https://www.journals.elsevier.com/renewable-and-sustainable-energy-reviews.

Dual Land Use for Agriculture and Solar Power Production: Overview and Performance of Agrivoltaic Systems, 2025. Available at https://iea-pvps.org/key-topics/dual-land-use-agriculture-solar-power-production/.

Equipment, V.P., 2021. International technology roadmap for photovoltaic (ITRPV). Results 12.

Floating Photovoltaic Power Plants: A Review of Energy Yield, Reliability, and Maintenance, 2025. Available at https://iea-pvps.org/key-topics/t13-floating-pv-plants-review-2025/.

Floating PV plants. Academic Press, 2020.

Fuentes, M., Vivar, M., de la Casa, J., Aguilera, J., 2018. An experimental comparison between commercial hybrid PV-T and simple PV systems intended for BIPV. Renewable and Sustainable Energy Reviews 93, 110–120. Available at https://doi.org/10.1016/j.rser.2018.05.021.

Ghafoor, A., Munir, A., 2015. Design and economics analysis of an off-grid PV system for household electrification. Renewable and Sustainable Energy Reviews 42, 496–502. Available at https://doi.org/10.1016/j.rser.2014.10.012, https://www.journals.elsevier.com/renewable-and-sustainable-energy-reviews.

Green, M.A., 2019. Photovoltaic technology and visions for the future. Progress in Energy 1 (1), 013001. Available at https://doi.org/10.1088/2516-1083/ab0fa8.

Green, M.A., Dunlop, E.D., Yoshita, M., Kopidakis, N., Bothe, K., Siefer, G., et al., 2025. Solar Cell Efficiency Tables (Version 66). Progress in Photovoltaics: Research and Applications 33 (7), 795–810. Available at https://doi.org/10.1002/pip.3919.

Grid Parity unveils new glass-glass module for agrivoltaic systems, 2025. Available at https://www.pv-magazine.com/2025/03/04/grid-parity-unveils-new-glass-glass-module-for-agrivoltaic-systems/.

Guerrero-Lemus, R., Vega, R., Kim, T., Kimm, A., Shephard, L.E., 2016. Bifacial solar photovoltaics - A technology review. Renewable and Sustainable Energy Reviews 60, 1533–1549. Available at https://doi.org/10.1016/j.rser.2016.03.041, https://www.journals.elsevier.com/renewable-and-sustainable-energy-reviews.

Hernandez, R.R., Easter, S.B., Murphy-Mariscal, M.L., Maestre, F.T., Tavassoli, M., Allen, E.B., et al., 2014. Environmental impacts of utility-scale solar energy. Renewable and Sustainable Energy Reviews 29, 766–779. Available at https://doi.org/10.1016/j.rser.2013.08.041.

Jaeckel, B., Volberg, G., Monokroussos, C., Muelhofer, G., Roth, A., 2018. Type approval and safety considerations for bifacial PV modules: requirements for IEC 61215 and IEC 61730. In: 35th European PV Solar Energy Conference and Exhibition, 1689–1699.

Jean, J., Brown, P.R., Jaffe, R.L., Buonassisi, T., Bulović, V., 2015. Pathways for solar photovoltaics. Energy and Environmental Science 8 (4), 1200–1219. Available at https://doi.org/10.1039/c4ee04073b, http://pubs.rsc.org/en/journals/journal/ee.

Kaygusuz, K., 2001. Renewable Energy: Power for a Sustainable Future. Energy Exploration & Exploitation 19 (6), 603–626. Available at https://doi.org/10.1260/0144598011492723.

Lee, T.D., Ebong, A.U., 2017. A review of thin film solar cell technologies and challenges. Renewable and Sustainable Energy Reviews 70, 1286–1297. Available at https://doi.org/10.1016/j.rser.2016.12.028, https://www.journals.elsevier.com/renewable-and-sustainable-energy-reviews.

Leloux, J., Narvarte, L., Trebosc, D., 2012. Review of the performance of residential PV systems in Belgium. Renewable and Sustainable Energy Reviews 16 (1), 178–184. Available at https://doi.org/10.1016/j.rser.2011.07.145.

Li, X., Yang, Q., Duan, H., Wu, C., 2025. Carbon footprint and environmental impact of China's PV production and reduction potential of PV power generation. Environmental Impact Assessment Review 115.

Lim, J., Park, N.G., Il Seok, S., Saliba, M., 2024. All-perovskite tandem solar cells: from fundamentals to technological progress. Energy and Environmental Science 17 (13), 4390–4425. Available at https://doi.org/10.1039/d3ee03638c, http://pubs.rsc.org/en/journals/journal/ee.

Liu, Z., Lin, R., Wei, M., Yin, M., Wu, P., Li, M., et al., 2025. All-perovskite tandem solar cells achieving >29% efficiency with improved (100) orientation in wide-bandgap perovskites. Nature Materials 24 (2), 252–259. Available at https://doi.org/10.1038/s41563-024-02073-x, https://www.nature.com/nmat/.

Luque, A., Hegedus, S., 2011. Handbook of Photovoltaic Science and Engineering. Wiley. Available at https://doi.org/10.1002/9780470974704.

Mahim, T.M., Rahim, A.H.M.A., Rahman, M.M., 2024. Review of Mono-and Bifacial Photovoltaic Technologies: A Comparative Study. IEEE Journal of Photovoltaics 14 (3), 375–396. Available at https://doi.org/10.1109/JPHOTOV.2024.3366698, https://ieeexplore.ieee.org/xpl/RecentIssue.jsp?punumber=5503869.

Mellit, A., Kalogirou, S., 2022. Handbook of Artificial Intelligence Techniques in Photovoltaic Systems: Modeling, Control, Optimization, Forecasting and Fault Diagnosis Handbook of Artificial Intelligence Techniques in Photovoltaic Systems: Modeling, Control, Optimization, Forecasting and Fault Diagnosis. Elsevier, 1–358, Available at https://www.sciencedirect.com/book/9780128206416. 10.1016/C2019-0-00960-0.

Nagaraja, M.R., Biswas, W.K., Selvan, C.P., 2025. Advancements and challenges in solar photovoltaic technologies: enhancing technical performance for sustainable clean energy – A review. Solar Energy Advances 5. Available at https://doi.org/10.1016/j.seja.2024.100084, https://www.sciencedirect.com/science/journal/26671131.

Nelson, J.A., 2003. 2003. The physics of solar cells. World Scientific Publishing Company.

NREL Best Research-Cell Efficiency Chart, 2024. Available at https://www.nrel.gov/pv/cell-efficiency

Peters, I.M., 2025. Strategic Global Deployment of Photovoltaic Technology: Balancing Economic Capacity and Decarbonization Potential. Advances in Atmospheric Sciences 42 (2), 261–268. Available at https://doi.org/10.1007/s00376-024-4176-9, https://www.springer.com/journal/376.

Photovoltaics report, 2025. Available at https://www.ise.fraunhofer.de/en/publications/studies/photovoltaics-report.html.

Priya, P., Stonier, A.A., 2025. Emerging innovations in solar photovoltaic (PV) technologies: The perovskite solar cells and more. Energy Reports 14, 216–242. Available at https://doi.org/10.1016/j.egyr.2025.06.003, https://www.sciencedirect.com/science/journal/23524847.

Raisa, A.T., Sakib, S.N., Hossain, M.J., Rocky, K.A., Kowsar, A., 2025. Advances in multijunction solar cells: An overview. Solar Energy Advances 5. Available at https://doi.org/10.1016/j.seja.2025.100105, https://www.sciencedirect.com/science/journal/26671131.

Rappaport, P., 1959. The photovoltaic effect and its utilization. Solar Energy 3 (4), 8–18. Available at https://doi.org/10.1016/0038–092x(59)90002-7.

Saga, T., 2010. Advances in crystalline silicon solar cell technology for industrial mass production. NPG Asia Materials 2 (3), 96–102. Available at https://doi.org/10.1038/asiamat.2010.82.

Sameti, M., Syron, E., 2022. 100% renewable industrial decarbonization: Optimal integration of solar heat and photovoltaics. Smart Energy 8, 100087. Available at https://doi.org/10.1016/j.segy.2022.100087.

Sirnik, I., Sluijsmans, J., Oudes, D., Stremke, S., 2023. Circularity and landscape experience of agrivoltaics: A systematic review of literature and built systems. Renewable and Sustainable Energy Reviews 178, 113250. Available at https://doi.org/10.1016/j.rser.2023.113250.

Snapshot, 2025. Available at https://iea-pvps.org/snapshot-reports/snapshot-2025/.

Snapshot of Global PV Markets, 2023. Available at https://iea-pvps.org/key-topics/snapshot-of-global-pv-markets-2023/.

Starowicz, A., Rusanowska, P., Zieliński, M., 2023. Photovoltaic cell – the history of invention – review. Polityka Energetyczna – Energy Policy Journal 26 (1), 169–180. Available at https://doi.org/10.33223/epj/161290.

Trends in photovoltaic applications, 2024. Available at https://iea-pvps.org/wp-content/uploads/2024/10/IEA-PVPS-Task-1-Trends-Report-2024.pdf

Wang, D.D., Sueyoshi, T., 2017. Assessment of large commercial rooftop photovoltaic system installations: Evidence from California. Applied Energy 188, 45–55. Available at https://doi.org/10.1016/j.apenergy.2016.11.076, http://www.elsevier.com/inca/publications/store/4/0/5/8/9/1/index.htt.

Where Sun Meets Water: Floating Solar Market Report, 2025. Available at http://documents.worldbank.org/curated/en/670101560451219695/Floating-Solar-Market-Report

Yamaguchi, M., Takamoto, T., Araki, K., Ekins-Daukes, N., 2005. Multi-junction III-V solar cells: Current status and future potential. Solar Energy 79 (1), 78–85. Available at https://doi.org/10.1016/j.solener.2004.09.018.

Chapter 2

Fundamentals of embedded artificial intelligence

Adel Mellit[1] **and Maamar Bettayeb**[2]
[1]*Faculty of Sciences and Technology, University of Jijel, Jijel, Algeria,* [2]*Department of Electrical Engineering, University of Sharjah, Sharjah, United Arab Emirates*

Introduction

The rapid advancements in artificial intelligence (AI) have brought about transformative changes across various industries, with intelligent systems now central to automation, decision-making, and data analysis. This chapter provides a comprehensive overview of key AI technologies and their applications, focusing on the convergence of machine learning (ML), deep learning (DL), generative AI (GenAI), and embedded systems (ESs). We begin by introducing fundamental AI methodologies (supervised learning (SL), unsupervised learning (USL), semisupervised learning (SSL), and reinforcement learning (REL)), each offering distinct strategies for solving data-driven problems. The discussion then delves into DL architectures such as convolutional neural networks (CNNs), hybrid CNN models, and advanced frameworks like You Only Look Once (YOLO) and vision transformers (ViT), which have revolutionized computer vision and pattern recognition. The chapter also highlights the rise of GenAI, focusing on the capabilities of large language models (LLMs) and generative adversarial networks (GANs) in producing synthetic content and enhancing human–computer interaction. With the increasing integration of AI systems into physical devices, we transition to embedded AI, explaining the architecture, types, and operational constraints of ESs, including limitations related to memory, storage, and processing power.

Significant emphasis is placed on lightweight models (tiny machine learning [TinyML]) such as MobileNets, EfficientNets-lite, and ShuffleNet, which enable intelligent processing on resource-constrained edge devices. Optimization techniques such as quantization and pruning are presented as essential tools for the efficient deployment of embedded AI. The comparative tables will illustrate the performance and trade-offs of these models. The chapter also provides a brief overview of explainable AI (XAI), highlighting

Embedded Artificial Intelligence and the Internet of Things for Photovoltaic Systems.
DOI: https://doi.org/10.1016/B978-0-443-44835-5.00006-0

its role in enhancing transparency and interpretability in ESs. The chapter assesses the advantages and limitations of embedded AI and identifies future trends toward scalable, autonomous, and energy-efficient AI solutions. By combining these topics, readers will gain a solid understanding of the technologies shaping modern AI and its potential to redefine embedded computing and GenAI.

Artificial intelligence techniques

AI is an area of computer science concerned with building computers and machines that can reason, learn, and act in a way resembling human intelligence, or systems that involve data whose scale exceeds what humans can analyze. The field includes many different disciplines, including data analytics, statistics, hardware and software engineering, neuroscience, and even philosophy (Russell and Norvig, 2021). AI focuses on developing systems that act and think like humans and make rational decisions. Fig. 2.1 shows the relationship between the key branches of AI. At the broadest level is AI, which encompasses all techniques for creating intelligent behavior in machines. Within AI, ML focuses on enabling systems to learn from data (Kissinger et al., 2021). DL is a subset of ML that uses multilayered neural networks (NNs) to model complex patterns. GenAI, the most specific subset, refers to DL models that generate new content such as text, images, or audio. Thus GenAI is part of DL, which is a part of ML, all within the broader field of AI.

Machine learning

ML is a subset of AI that uses algorithms trained on datasets to create models capable of performing tasks that would otherwise be possible only for humans

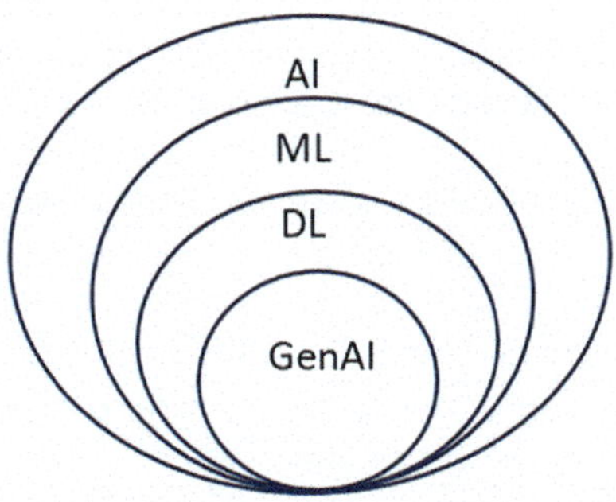

FIGURE 2.1 **Venn diagram of AI.** The figure illustrates that AI is the broadest field, within which machine learning is a subset; deep learning is a subset of machine learning; and generative AI is a subset of deep learning (*AI*, Artificial intelligence).

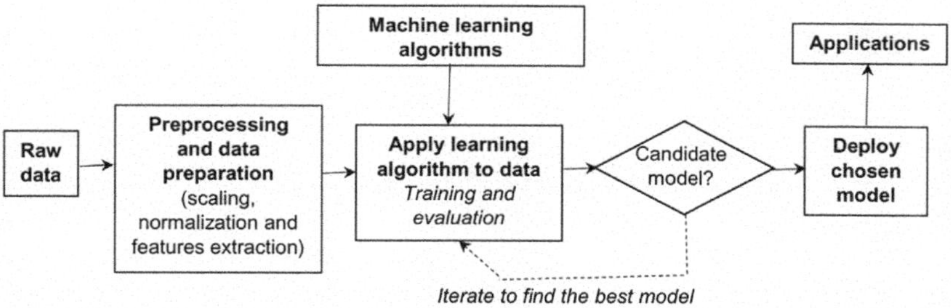

FIGURE 2.2 **Configuration of ML workflow process.** This block diagram illustrates the typical steps in a machine learning workflow, starting from data collection, through data preprocessing, model training, evaluation, and deployment (*ML*, Machine learning).

(Shalev-Shwartz and Ben-David, 2014). ML is one of the most widely used forms of AI today, supporting many of the digital products and services. Fig. 2.2 illustrates the main stages of developing an ML system, starting with raw data collection, followed by feature extraction and preprocessing, model training and evaluation, and finally deployment (Alpaydin, 2020; Mellit and Kalogirou, 2022).

ML can be globally classified into four key learning algorithms: (1) SL, (2) USL, (3) SSL and (4) REL. Most of these algorithms have been used to address various tasks in the photovoltaic area, ranging from the nanoscale (materials) to the large scale (photovoltaic power plants) (Mellit and Kalogirou, 2022).

Supervised learning

SL is a type of ML where the model is trained on labeled data (input–output pairs) to predict outputs based on unseen inputs. Fig. 2.3 shows the basic workflow of SL, which is mainly used to address tasks related to regression and classification problems, such as predicting continuous numerical values or identifying the type of defects in images. SL operates in two key paradigms: classification, where inputs map to discrete classes, and regression, where inputs map to continuous values. The most widely used supervised algorithms include linear regression, logistic regression, decision trees, random forests, support vector machines, k-nearest neighbors, gradient boosting, and more (Mueller and Guido, 2016).

SL offers significant advantages; it learns patterns from labeled examples to make accurate predictions on unseen data. With sufficient training data, its performance improves, making it versatile for tasks. However, it requires high-quality labeled datasets and is prone to labeling errors. It also struggles with complex or abstract tasks beyond its training scope and risks overfitting to noisy data. To maintain accuracy, models often need continuous updates with new labeled data. SL algorithms have been widely applied to solve major

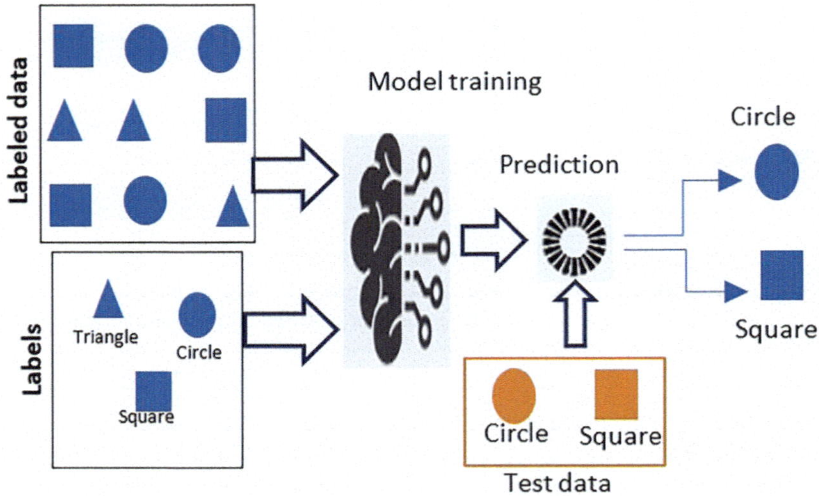

FIGURE 2.3 **Basic supervised learning workflow.** The figure illustrates the basic steps of supervised learning, where labeled data is used to train the model. It also shows how the model learns from input and output examples, and is then tested to make predictions on new data.

challenges in the photovoltaic field, including solar power prediction, anomaly detection, energy management optimization, and photovoltaic system fault diagnosis based on image classification (Mellit and Kalogirou, 2022).

Unsupervised learning

USL discovers hidden patterns or structures in unlabeled data without predefined output labels. Fig. 2.4 outlines the basic workflow of USL. Essentially used to address tasks related to dimensionality reduction, clustering, and anomaly detection. Most widely used unsupervised algorithms include K-means, hierarchical clustering, principal component analysis, singular value decomposition, etc.

USL offers key advantages; it eliminates the need for labeled data, enabling rapid analysis of large datasets; efficiently reduces data complexity while preserving critical patterns; and uncovers hidden relationships, revealing actionable insights and trends. However, it faces significant drawbacks: the absence of labels makes validation difficult, often resulting in lower precision for complex tasks; postprocessing (e.g., labeling clusters) can be labor-intensive; and results are highly sensitive to data quality issues like noise, outliers, or missing values. USL has become instrumental in photovoltaics for automating fault diagnostics, identifying performance degradation trends, denoising sensor data, and compressing high-dimensional PV system measurements (Mellit and Kalogirou, 2022).

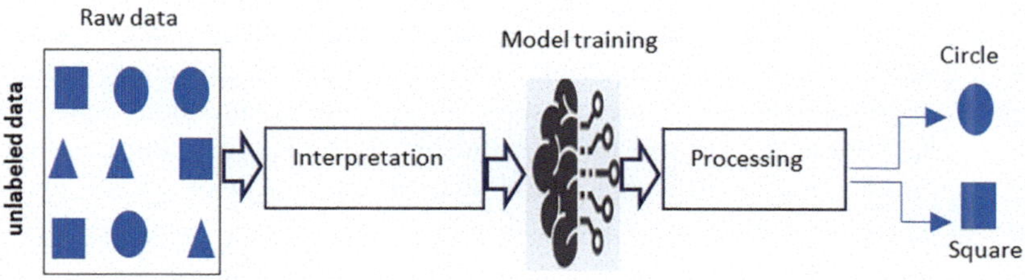

FIGURE 2.4 Basic supervised learning workflow. The figure outlines the basic supervised learning workflow, starting with labeled data that is split into training and testing sets. The model learns from the training data and is later evaluated on the test data to assess its ability to make accurate predictions.

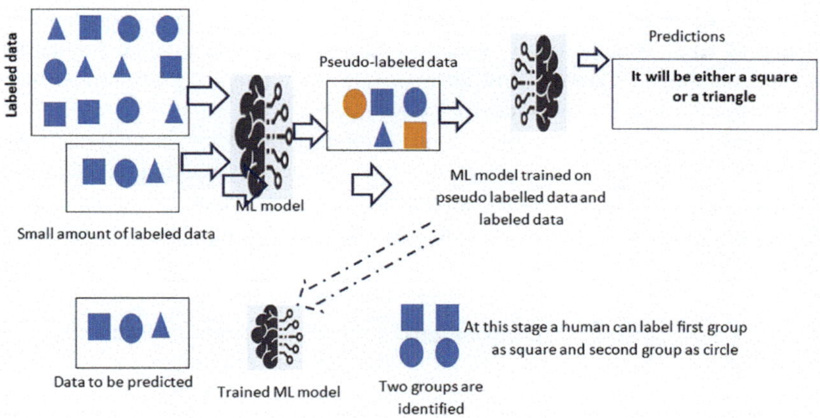

FIGURE 2.5 Basic semisupervised learning workflow. The figure shows the basic semisupervised learning workflow, where a small portion of labeled data is combined with a larger amount of unlabeled data. The model uses both sources to improve learning and generate predictions more effectively than with labeled data alone.

Semisupervised learning

An ML technique called SSL uses both labeled and unlabeled data to train models (Alpaydin, 2020; Shalev-Shwartz and Ben-David, 2014). It falls between SL and USL (See Fig. 2.5). When labeled data is hard to come by or prohibitively expensive, it makes use of the advantages of both SL and USL to create more accurate and effective models. However, unlike SL, the algorithm is trained on a dataset that contains both labeled and unlabeled data. When a lot of unlabeled data is available, but labeling it all would be too costly or challenging, SSL might be especially helpful.

By mixing a small quantity of labeled data with a large collection of unlabeled data, SSL provides the benefit of lowering the requirement for huge labeled datasets while increasing accuracy and generalization and lowering explanation costs. This is especially helpful when obtaining labeled data is difficult or costly. However, the quality of the unlabeled data and specific presumptions on the data distribution, like smoothness or clustering, have a significant impact on how effective it is. Model performance may suffer if these presumptions are not met or if the unlabeled data is noisy. Furthermore, SSL algorithms can be difficult to build and run the danger of overfitting or error propagation if improperly handled.

Reinforcement learning

REL is an ML paradigm where an agent learns to make sequential decisions by interacting with an environment to maximize cumulative rewards through trial and error (Alpaydin, 2020; Mueller and Guido, 2016). Fig. 2.6 shows the basic concept of REL. ReL is mainly used to address issues that involve sequential decision-making in dynamic and uncertain environments (e.g., robotics, self-driving, game playing, and industrial control).

REL excels at solving complex, nondeterministic problems through real-time environment interaction and adaptive error correction, making it ideal for dynamic real-world applications. However, it is inefficient for simple tasks where traditional methods suffice, demands substantial computational resources,

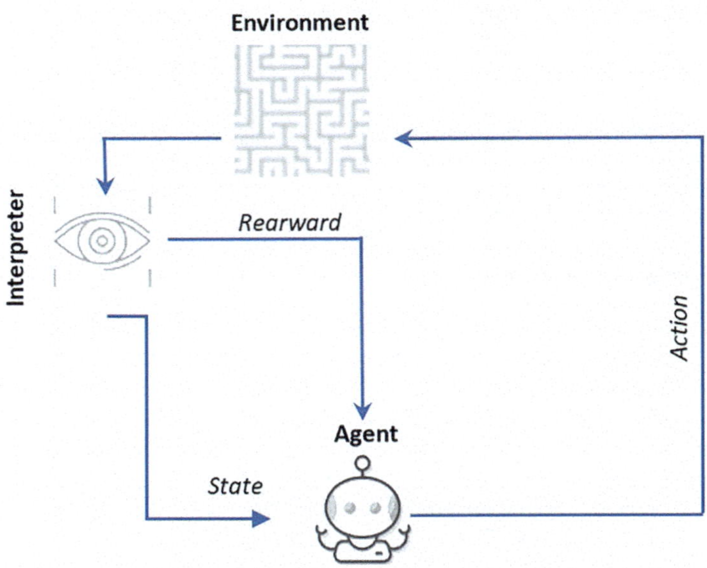

FIGURE 2.6 **Basic reinforcement learning workflow.** This figure illustrates the basic reinforcement learning workflow, where an agent interacts with an environment by taking actions and receiving feedback in the form of rewards or penalties. Through repeated interactions, the agent learns to choose actions that maximize its cumulative reward.

relies critically on precise reward function design (risking suboptimal behavior if poorly defined), and suffers from opaque decision-making that complicates debugging. REL optimizes photovoltaic systems by enabling adaptive control for maximum power pPoint tracking under shading/varying weather, dynamic energy dispatch in grid-connected PV-battery hybrids, and automated anomaly identification in large-scale solar farms.

Deep learning

DL is a form of ML that enables computers to learn from experience and understand the world in terms of a hierarchy of concepts (Goodfellow et al., 2016). DL is a specialized subset of ML that utilizes NNs with multiple layers to analyze data and learn complex patterns, much like the human brain. DL algorithms are typically trained on large datasets of labeled data. The algorithms learn to associate features in the data with the correct labels (Lecun et al., 2015). Fig. 2.7 illustrates a simplified process diagram of DL. As can be seen, the feature extraction step is removed as it is not needed in this kind of learning; this is because DL algorithms are able to learn and automatically extract features from raw (original) input data.

Convolutional neural network

An example CNN is shown in Fig. 2.8. This is the most commonly used configuration for image classification tasks. Its key components are convolutional layers, pooling layers, and fully connected layers (Mellit and Kalogirou, 2025).

Other versions of CNN (1989), like LeNet-5 (1998), AlexNet (2012), VGGNet (2014), ResNet (2015), DenseNet (2016), and lightweight CNN versions (e.g., Xception, MobileNet, and EfficientNet). These configurations are the most commonly used in photovoltaic applications, especially for anomaly detection and fault diagnosis of photovoltaic modules based on image classification.

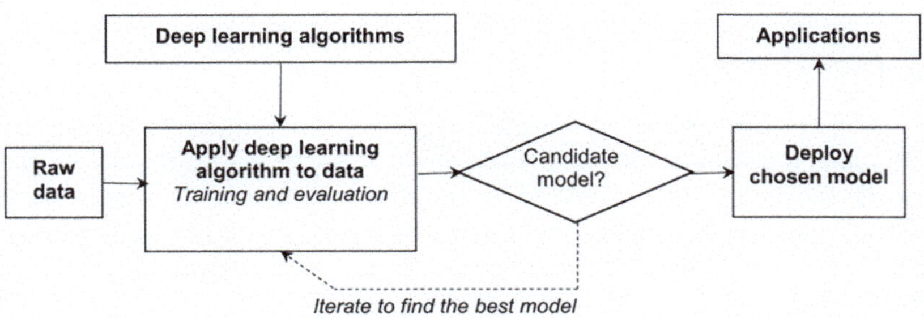

FIGURE 2.7 **Deep learning workflow.** The figure illustrates the deep learning workflow, where raw data is fed into a neural network that automatically extracts features through multiple layers. The model is trained, evaluated, and adjusted until it achieves the desired performance for prediction or classification.

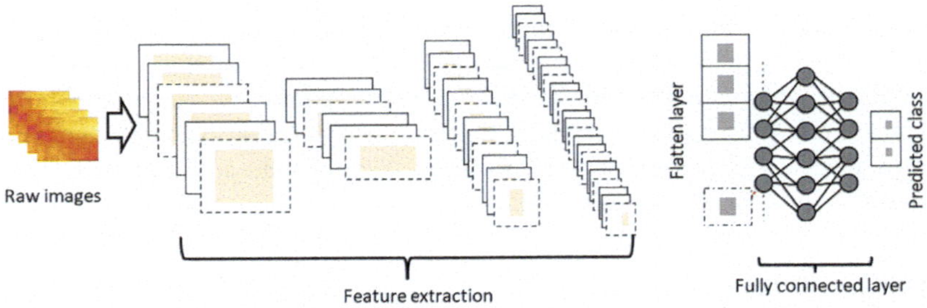

FIGURE 2.8 **Basic CNN configuration.** This figure shows a basic CNN architecture, where input images pass through convolution and pooling layers to extract important features. These features are then fed into fully connected layers to produce the final prediction or classification result (*CNN*, Convolutional neural network).

Hybrid convolutional neural network-machine learning

Hybrid CNN-ML models combine the feature extraction capabilities of CNNs with ML algorithms for classification or other tasks. These hybrid models aim to leverage the strengths of both CNNs and ML techniques (e.g., decision tree, random forest, k-nearest neighbors, support vector machine, and other algorithms) to improve performance, especially when dealing with smaller datasets or specific application domains (Drir et al., 2025a), such as the diagnosis and classification of anomalies based on images (Jena et al., 2021). The idea is to replace the fully connected layers with an ML algorithm (See Fig. 2.9). The key advantages of the structure are: (1) Leverages CNNs for feature extraction and ML models for accurate, robust classification, (2) helps reduce overfitting on smaller datasets, and (3) faster to train ML classifiers compared to fine-tuning large CNNs.

Hybrid CNN-ML-based classifiers were employed to improve the accuracy in various applications in photovoltaic systems, such as fault diagnosis of PV plants.

You Only Look Once

The YOLO architecture is designed for real-time object detection by performing localization and classification in a single forward pass of the network (Redmon, 2016). YOLO treats object detection as a regression problem, making it significantly faster compared to earlier region-based methods (e.g., R-CNN). The architecture of the first introduced YOLO is shown in Fig. 2.10.

Key stages in YOLO architecture are: (1) Extract features from the image, capturing spatial information, (2) predicts bounding boxes and confidence scores, (3) determines class probabilities for each detected object, and (4) filters overlapping boxes, keeping the most confident predictions.

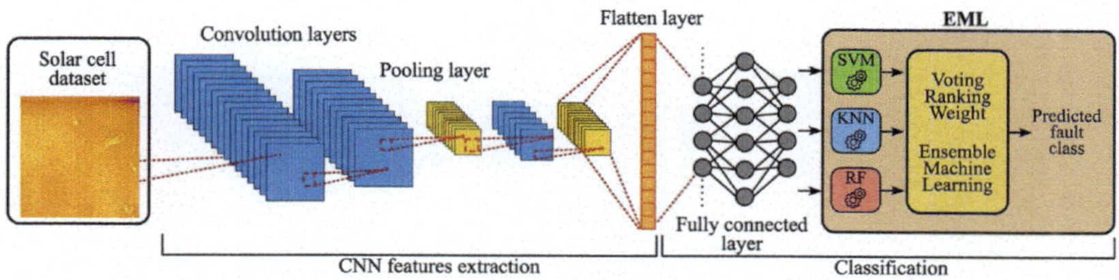

FIGURE 2.9 General architecture hybrid CNN-ML (Case of CNN with *SVM*, Support vector machine, *KNN*, K-nearest neighbor, and *RF*, random forest). The figure presents a hybrid architecture where a CNN is used to extract features from images, and instead of the usual fully connected classifier, the extracted features are passed to traditional machine learning models. These models, such as SVM, KNN, or Random Forest, perform the final classification based on the CNN-derived features (*CNN,* Convolutional neural network; *ML*, machine learning).

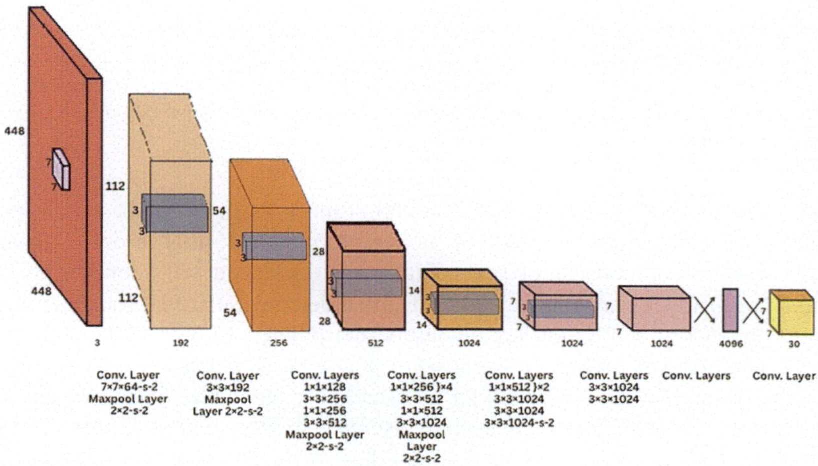

FIGURE 2.10 YOLO architecture. The figure illustrates the YOLO architecture, where an input image is processed by a convolutional neural network to detect multiple objects at once. The model divides the image into grid cells and predicts bounding boxes and class probabilities in a single, unified step (*YOLO*, You Only Look Once).

For instance, in the inspection of photovoltaic plants, various versions of YOLO were employed to detect photovoltaic panels using visual images. In the case of IR thermographic images, YOLO was utilized to detect and localize hotspots in photovoltaic plants.

Vision transformer

Self-attention is a mechanism that allows a model to weigh the importance of different parts of the input sequence when processing it. Originally popularized by the transformer model in natural language processing (NLP), self-attention has been adapted to computer vision tasks as well (Vaswani et al., 2017).

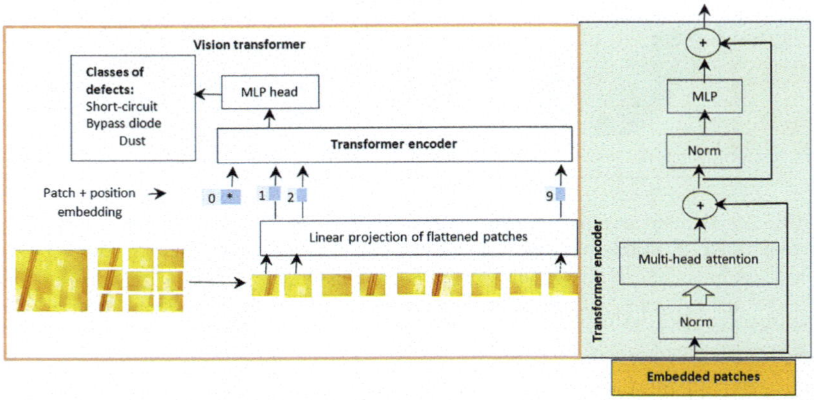

FIGURE 2.11 ViT architecture for IR image classification. The figure shows the ViT architecture, where an infrared image is divided into small patches that are converted into embeddings and passed through a transformer encoder. The model uses attention mechanisms to learn relationships between patches and produces a final classification output (*IR*, Infrared; *ViT*, vision transformer).

ViT is an architecture that uses self-attention mechanisms to process images. The ViT architecture consists of a series of transformer blocks. Each transformer block consists of two sublayers: a multihead self-attention layer and a feed-forward layer (See Fig. 2.11). ViT demonstrated that transformers could outperform CNNs on large datasets (Dosovitskiy et al., 2020).

Indeed, for real-time fault diagnosis of photovoltaic plants, it is essential to develop a lightweight ViT that can be effectively embedded into low-power microcontrollers (MCUs). This design ensures computational efficiency while achieving high accuracy in detecting and diagnosing faults in real time using infrared (IR) thermographic images.

Generative AI

(GenAI refers to a class of AI that creates original, high-dimensional content (such as, text, images, or audio) by learning patterns from training data. Unlike discriminative models that predict labels, generative models synthesize new data instances resembling their training distribution (Foster, 2022). GenAI boasts diverse applications. In computer vision, it generates realistic images, creates variations, or completes missing sections. In NLP, it powers language translation, text synthesis, and conversational agents that produce human-like responses (Vajjala et al., 2023). GenAI models use ANNs to find patterns in data and create new content. Trained with unsupervised/semisupervised methods on large unlabeled datasets, they form foundation models. Examples of foundation models are LLM, GANs, variational autoencoders, and multimodal models.

Large language models

An LLM was designed to handle diverse NLP tasks. Built on transformer architectures and trained on vast datasets, LLMs excel in recognizing, translating, predicting, and generating text or other forms of content (Brown et al., 2020). The main advantages of LLMs are:

- They enable automation of various language-related tasks, such as code generation, content creation, and complex aspects of a project.
- They can perform zero-shot learning, enabling generalization to tasks for which they were not explicitly trained.
- They can be fine-tuned on specific datasets or domains, facilitating continuous learning and adaptation to particular use cases.
- They efficiently handle vast amounts of data, making them ideal for tasks that require a deep comprehension of extensive text quantities.

Several popular general-purpose LLMs dominate the current AI landscape, including OpenAI's GPT-4 and GPT-5, Google DeepMind's Gemini, Anthropic's Cloud, and Meta's Llama 2 and Llama 3. Other influential models include the highly efficient DeepSeek-V2 and DeepSeek-R1, as well as Cohere's Command-R and Command-R+, which are widely used in enterprise applications. Collectively, these models represent leading advancements in natural language understanding and generation. Beyond mainstream general-purpose models, recent advances in LLMs include the development of multimodal models capable of understanding text, images, audio, and video; efficient lightweight models optimized for on-device and edge deployment. Significant progress has also been made in reasoning-optimized models, open-source architectures, and alignment techniques that improve safety, reliability, and factual accuracy. Furthermore, innovations such as long-context models, memory-augmented systems, and REL-based training continue to expand the capabilities and practical applications of LLMs across research, industry, and everyday use.

Generative adversarial network

A GAN is a DL architecture (Goodfellow et al., 2014); it trains two NNs to compete against each other to generate more authentic new data from a given training dataset. It comprises two main components: (1) The generator takes noise, referred to as the latent vector, as input and attempts to generate data (e.g., IR thermography images) that mimic the original dataset and (2) the discriminator receives input that could either be original data or synthetic data produced by the generator. Its task is to determine whether the input is real or fake. For instance, Fig. 2.12 illustrates a synthetic IR image generated using a deep convolutional generative adversarial network (DCGAN) (Radford et al., 2016). This model combines two NNs (a generator and a discriminator) trained

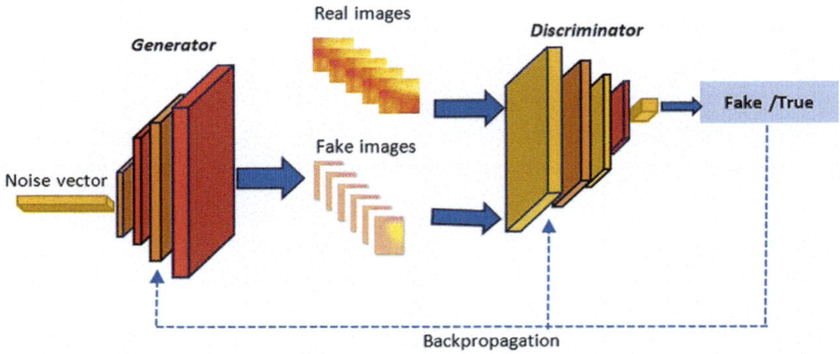

FIGURE 2.12 **Basic structure of DCGAN.** The figure depicts the basic structure of a deep convolutional generative adversarial network, consisting of two competing models: a generator that creates synthetic data and a discriminator that evaluates whether the data is real or generated. Through this adversarial process, the generator learns to produce increasingly realistic outputs (*DCGAN*, Deep convolutional generative adversarial network).

adversarially to create high-quality outputs. DCGANs incorporate convolutional layers in both networks, making them particularly effective for image-related applications. The generator output is connected directly to the discriminator input. Through backpropagation, the discriminator's classification provides a signal that the generator uses to update its weights. Other versions of GAN are: CycleGAN, StyleGAN, Progressive GAN, and Pix2-PixGAN.

Applications that employ GenAI approaches in solar photovoltaic power plants have not yet been put into service. Even though these methods are widely used in many different sectors, there is still unrealized potential for them to enhance photovoltaic power plant monitoring, problem detection, and maintenance procedures. Nonetheless, there have been some recent initiatives to use LLM and GANs in photovoltaics for fault detection (Guo and Weng, 2025) and the creation of synthetic data (Drir et al., 2025b) to increase the dataset and deal with the unbalanced datasets.

Embedded systems

ESs encompass a variety of fundamental concepts for the design, development, and deployment of specialized computing devices. These systems are characterized by their integration with larger systems or products, often performing specific functions under immediate and resource-limited constraints (Marwedel, 2021). An ES is a dedicated computer system designed to perform specific tasks within a larger system, usually with real-time constraints. They are not traditional computer systems or general-purpose computers, but rather information storage units that operate independently or are connected to a larger system to perform specific functions. These units operate with little or no human intervention. Some of these ESs are equipped with a real-time operating system (RTOS). Some small ESs may not have this system. Therefore, an ES can be defined as a real-time, software-driven, reliable control system that relies on a microprocessor (CPU) or MCU. Lightweight operating systems made for low-power devices are called embedded operating systems. These include Linux Embedded, Zephyr OS, and FreeRTOS and are crucial for resource management and effectively executing AI tasks. ESs are used in a wide range of applications, from consumer electronics to industrial automation and health care. Some common applications include automotive, consumer electronics, health care, industrial automation, Internet of Things (IoT), defense and aerospace, and telecommunications (Noergaard, 2012).

Basic structure of embedded systems

An ES consists of three key components (Liu, 2003): (1) Hardware, (2) software, and (3) firmware. A block structure of an ES is illustrated in Fig. 2.13. The hardware part includes the MCU or CPU, memory, input/output interfaces, sensors, actuators, and other supporting electronic components. The software or hardware is the firmware that controls the hardware, processes inputs, and produces the desired outputs, often optimized for a specific task. A RTOS or control logic is used to manage task scheduling, timing, and resource allocation, ensuring the system meets real-time performance requirements.

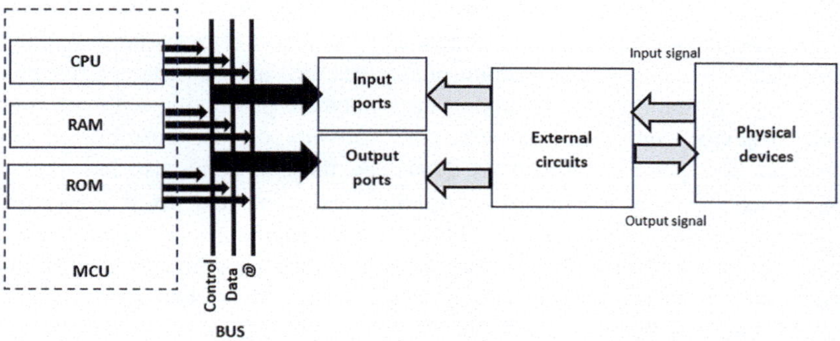

FIGURE 2.13 **Basic structure of an ES.** The figure shows the basic structure of an expert system, where a user interacts with the system through a user interface. The system relies on a knowledge base and an inference engine to analyze the input and provide decisions or recommendations, similar to human expert reasoning (*ES*, Expert system).

An ES operates using an MCU or CPU to run custom software that processes input from sensors or other devices (output signal), performs specific programmed tasks, and sends output signals to motors or displays. The hardware and software are tightly integrated and optimized for a single purpose, enabling efficient, reliable, and often real-time operation. Its benefits include ease of use, improved performance in terms of execution times, power consumption, and memory capacity, and reduced size and increased flexibility (Zurawski, 2018).

Type of embedded systems

ESs can be classified into four key classes:

- Real-time ESs: These systems respond to inputs and produce outputs within strict time constraints, which is critical for applications such as automotive control systems and medical devices.
- Networked ESs: These systems connect to other devices via wired or wireless networks, enabling communication and data exchange. This is common in IoT devices and smart home systems.
- Mobile ESs: Optimized for mobility and portability, the power devices such as smartphones, tablets, and wearable technology, they provide user interfaces and processing capabilities on the go.
- Application-specific ESs: Specifically designed for specialized tasks or industries such as aerospace, health care, and industrial automation, they provide customized solutions for specific functions and requirements.

Advantages and limits of embedded systems

ESs offer numerous advantages that make them indispensable in modern technology. They deliver high performance for specific tasks while maintaining low-power consumption, small size, and cost-effectiveness, making them ideal for mobile or mass-produced devices. ESs are also characterized by reliability and stability, often supporting immediate processing with deterministic behavior, and their hardware and software can be customized and optimized for specific applications. However, these systems also have some limitations. They have limited computing resources and fixed hardware, making upgrades or expansion difficult. Development can be complex, requiring expertise in both hardware and low-level programming, and it is often inflexible and designed for a single task. Other challenges include security vulnerabilities, limited user interfaces, compatibility issues, and, in some cases, shorter lifespans due to hardware limitations. Despite these limitations, ESs remain crucial for applications where efficiency, reliability, and immediate operation are paramount.

Embedded AI

Despite its strength, AI is limited in real-time, offline, and power-constrained applications. Herein lies the role of EAI. The integration of AI capabilities into specialized devices with constrained resources is known as EAI (Boruah et al., 2025). It consists of three main modules: (1) Data, (2) model, and (3) inference. The integration of AI is revolutionary as it combines the power of decision-making and learning with the efficiency and speed of embedded hardware. When ESs are combined with AI, these systems become faster, smarter, and more effective, providing new possibilities for industries (Huang et al., 2025). AI enables ESs to make decisions in real time, adapt to changing conditions, and improve their performance over time. Furthermore, AI improves the entire lifecycle of ESs, from design to operation and maintenance. It enables faster development, smarter operations, and longer system life.

Tiny machine learning

TinyML is an example of EAI in small edge devices (e.g., MCU) (Warden and Situnayake, 2019). It is a branch of edge AI, where sensors generate data with very low-power consumption, ultimately enabling ML to be deployed on devices. The concept of edge AI derives from edge computing, which is based on the same principle: storing, processing, and managing data directly at IoT endpoints. All TinyML solutions are a form of EAI, but not all EAI solutions are TinyML. It is a rapidly growing field, with applications in a wide range of industries, including health care, consumer electronics, and industrial automation. This allows for applications like always-on sensor analytics, predictive maintenance, and other low-power, intelligent devices. NNs have historically required a lot of resources and strong hardware, but with the development of TinyML, inferences may now be performed directly on devices with lower processing power. TinyML is an intersection between ML and ES (See Fig. 2.14).

TinyML enables ultralow-power, persistent AI for use in practical applications by bringing the intelligence of ML models to a confined, embedded environment. From an ES's perspective, TinyML integrates intelligent decision-making directly into the device, leveraging the real-time sensing and control capabilities of ES platforms. This reduces latency, bandwidth consumption, and privacy concerns by eliminating the need to send raw data to the cloud (Banbury et al., 2022). From the broadest at the top to the most detailed at the bottom, Fig. 2.15 illustrates some of the key ideas in edge AI in relation to one another (Situnayake and Plunkett, 2023).

The workflow of TinyML is shown in Fig. 2.16.

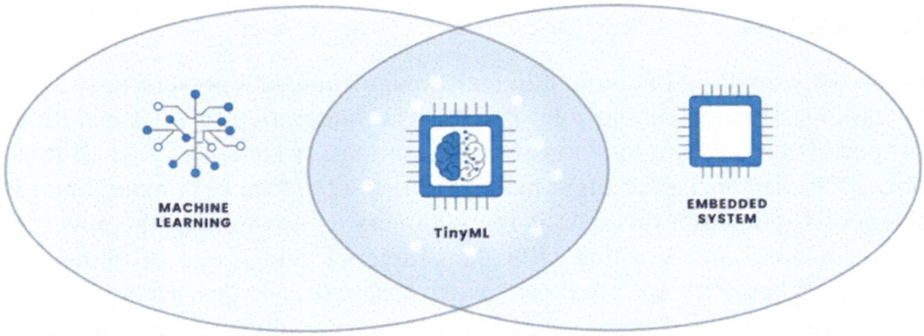

FIGURE 2.14 **The intersection of ML and embedded systems.** The figure shows how machine learning intersects with embedded systems, illustrating that ML models can be deployed directly on small, resource-constrained hardware to enable intelligent decision-making on devices (*ML*, Machine learning).

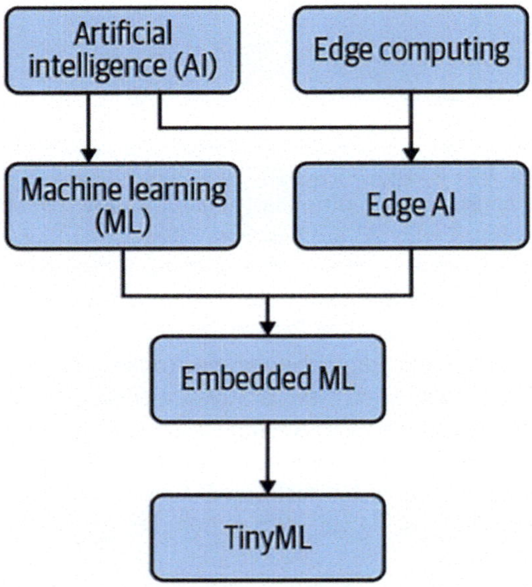

FIGURE 2.15 **Concept of edge AI in context with each other (edge computing, AI, ML, embedded ML, and TinyML).** The figure illustrates how the concepts of edge computing, artificial intelligence, machine learning, embedded machine learning, and TinyML relate to each other. It shows that TinyML is the most specialized form, enabling ML models to run directly on very small, low-power edge devices (*AI*, Artificial Intelligence; *ML*, Machine Learning).

The key steps in TinyML are:

- Data collection: It consists of collecting data from sensors, such as sound, video, or images, to be used to feed the ML model.
- Preprocessing & feature extraction: This block aims to convert raw data (structural or unstructured) into usable features.

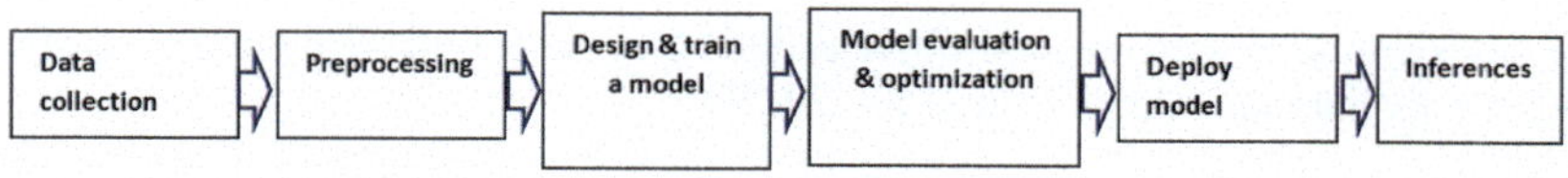

FIGURE 2.16 **TinyML pipeline.** The figure illustrates the TinyML pipeline, starting from data collection and model training on powerful computers, followed by model optimization and deployment onto small embedded devices. It shows how the trained model runs locally on the device to make real-time predictions with very low power consumption.

- Model selection and training: This step aims to use frameworks, such as TensorFlow, to train a NN or classical ML model.
- Model evaluation: Test the trained model on new unseen data to measure its predictive accuracy.
- Model optimization: It aims to reduce the size of the ML model to a few KB in order to be uploaded into an edge device (e.g., a low-power and low-cost MCU) with limited resources.
- Conversion to deployment format: In this step, the mode needs to be converted to an appropriate format, such as C arrays (for ultralow-level deployment).
- Deployment on MCU: This is the last step, aiming at uploading the optimized model onto a CPU for on-device inference. Then the MCU will perform predictions instantly using live sensor data.

ML workflows are a comprehensive guide applicable across various platforms, including cloud solutions, edge computing, and TinyML. However, EAI workflows present unique complexities and challenges, making them an attractive field and paving the way for significant innovations (Somvanshi, 2025).

Among TinyML's many advantages are its compact model sizes that scale to large storage units, on-device processing that reduces latency and improves privacy, and its ultralow-power consumption that enables it to run on MCUs and battery-powered devices. It delivers real-time performance for applications such as voice activation and anomaly detection and enables cost-effective deployment on low-cost devices. Its drawbacks include the need for offline training because devices typically only handle inference, the potential for loss of accuracy due to model compression, and limited model complexity due to memory and computing capacity constraints. Furthermore, TinyML development requires specialized knowledge of ESs and ML optimization, and its capabilities are better suited to specific, well-defined tasks than complex, multifunctional AI applications.

Lightweight CNN models

Lightweight CNN models are CNNs designed to reduce parameters and computational requirements while maintaining high accuracy. They are optimized for deployment on resource-constrained devices, such as ESs, mobile phones, and

IoTs devices, enabling fast and efficient inference with limited memory, power, and processing capabilities (Liu et al., 2024).

SqueezeNet

SqueezeNet is a CNN for image classification released in 2016. It was developed by researchers at DeepScale, the University of California, Berkeley, and Stanford University (Iandola et al., 2016). It was designed to achieve AlexNet-level accuracy on ImageNet while reducing parameter usage by a factor of 50. This architecture works by replacing most of the large convolution filters with smaller, more efficient "Fire" units, consisting of: (1) a compression layer (1x1 convolutions) that reduces the number of input channels and (2) a dilation layer (a combination of 1x1 and 3x3 convolutions) that increases the size of the feature map to improve representation. SqueezeNet reduces model size while maintaining accuracy.

MobileNet

MobileNet is a family of lightweight CNN architectures designed specifically for mobile and embedded devices, where computational resources are limited. Introduced by Google, it uses efficient techniques to achieve good performance with fewer parameters and lower computational cost, making it suitable for real-time applications (Howard et al., 2017). The main difference between MobileNet and traditional CNNs is that MobileNet replaces a single 3x3 convolution layer with a 3x3 depthwise convolution followed by a 1x1 pointwise convolution, improving efficiency. The main improvements over the v2 architecture include: (1) the use of HardSwish instead of ReLU6, (2) the addition of the Squeeze-and-Excitation module, (3) the incorporation of MnasNet and NetAdapt, and (4) the redesign of expensive layers.

CondenseNet

CondenseNet (Huang et al., 2018) is a condensed variant of DenseNet (Huang et al., 2017) that reduces computation and parameters by using learnt group convolutions. It reorganizes the network into effective group convolutions for inference after identifying and eliminating less significant connections (condensing stage) during training. This makes the model faster and smaller while maintaining excellent accuracy, making it perfect for embedded and low-power devices.

EfficientNet

EfficientNet was introduced in 2019 (Tan and Le, 2019). For maximum accuracy and efficiency, EfficientNet uses a compound scaling technique to scale a CNN's depth, breadth, and input resolution all at once. Using MBConv blocks with depthwise separable convolutions and squeeze-and-excitation layers, it starts from a modest baseline (B0) and achieves excellent accuracy with fewer parameters, making it perfect for embedded and mobile AI.

ShifNetNet

ShiftNet was introduced in 2018 (Yan et al., 2018). With zero FLOPs and no additional parameters, ShiftNet substitutes a straightforward shift operation for spatial convolutions, moving feature map channels in various directions without the need for multiplications. After that, it mixes channels using thin 1x1 convolutions. ShiftNet is perfect for embedded and mobile devices because of its design, which significantly decreases processing and model size while preserving acceptable accuracy.

Table 2.1 lists a comparison between Shiftnet, MobileNet, EfficientNet, SqueezeNet, and CondenseNet in terms of key idea, efficiency, accuracy, computational cost, size of the model, common use cases, and special strength.

ShiftNet uses channel shifts instead of spatial convolutions to achieve ~69% ImageNet accuracy with only ~1 M parameters and minimum FLOPs. MobileNet V1, which uses depthwise separable convolutions, provides a decent compromise between speed and accuracy (~70.6% top-1). EfficientNet-B0 scales depth, width, and resolution proportionately to achieve the greatest accuracy (~77.1%) with ~5.3 M parameters. SqueezeNet 1.1 uses Fire modules to retain ~58% accuracy while being incredibly small (~1.2 M parameters). CondenseNet is effective for embedded and mobile AI applications because it combines learnt group convolutions with dense connectivity to achieve approximately 74% accuracy with over 1 million parameters.

Model compression

Deploying DL models directly to mobile and embedded devices is challenging because of the high computational cost of these models and their constrained resources. We can speed up and compress models without sacrificing performance, as some DL model parameters may duplicate (Nagel et al., 2021). The most commonly used optimization techniques for NNs are pruning and quantization. In DL, the three most commonly used metrics for model compression are memory access cost, floating-point operations, and multiply-accumulate operations (MACs).

Pruning

Pruning is a widely used method for model compression that reduces the network's size by eliminating unnecessary and ineffective parameters. The pruning technique eliminates redundant parameters by assessing their influence on model performance. Three steps make up a standard pruning algorithm: determining which parameters are significant, eliminating those that are not, and fine-tuning to restore accuracy (See Fig. 2.17) (Nagel et al., 2021).

TABLE 2.1 Comparison between existing lightweight CNN-based models.

Feature	SqueezeNet	MobileNet	CondenseNet	ShiftNet	EfficientNet
Ref.	Iandola et al. (2016)	Howard et al. (2017)	Huang et al. (2018)	Yan et al. (2018)	Tan and Le (2019)
Key idea	Fire modules (squeeze & expand layers)	Depthwise separable convolutions	Learned group convolutions with pruning	Channel shifting (zero-FLOP spatial ops)	Compound scaling (depth, width, resolution)
Efficiency	Very high	High	High	Very high (zero FLOPs on spatial conv)	Very high
Computational cost	Low	Low	Low	Extremely low	Moderate to low (scalable)
Accuracy	Good	Good	High	Good (close to standard CNNs)	State-of-the-art (top accuracy)
Size of the model	Very small	Small	Small	Very small	Varies (B0-B7)
Common usage cases	Embedded, TinyML	Mobile & embedded AI	Embedded & mobile	Mobile & embedded devices	Mobile, embedded & cloud AI
Special strength	Ultrasmall size with AlexNet-level accuracy	Efficient depthwise convolutions	Dynamic pruning for efficiency	Zero spatial convolution FLOPs	Balanced scaling for efficiency and accuracy

AI, Artificial intelligence; *CNN*, convolutional neural network; *FLOP*, floating-point operation.

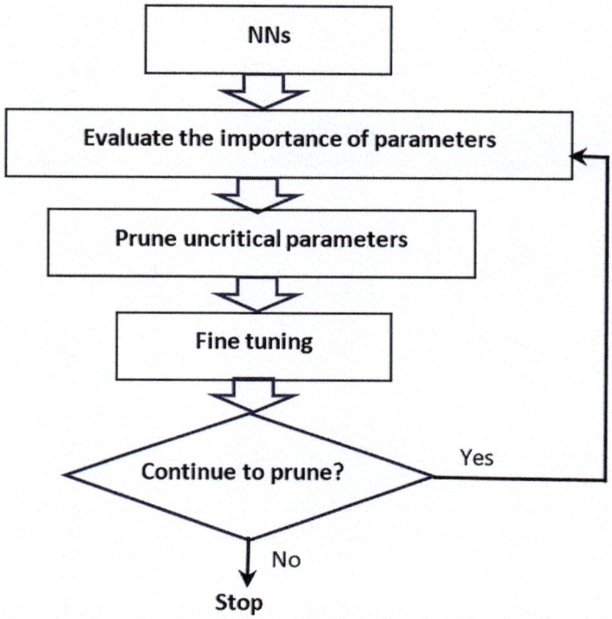

FIGURE 2.17 **Pruning steps.** The figure shows the pruning steps, where a trained neural network has less important weights removed to reduce model size. After pruning, the model is fine-tuned to maintain accuracy while becoming lighter and more efficient.

Pruning is the process of setting a node or weight to zero to turn it off. Pruning aims to decrease the number of parameters, which minimizes the size of a model and increases computational performance. A sparse model, where sparsity is defined as the number of pruned parameters, is the outcome of pruning as opposed to an unpruned dense model. An NN both before and after trimming is depicted in Fig. 2.18.

Quantization

Quantization is the process of compressing floating-point data bits in NN parameters to minimize the number of bits utilized by floating-point numbers while preserving the highest level of model accuracy (Nagel et al., 2021). This reduces the size and complexity of the model. Computations in NNs use floating-point numbers (32 bits). Three components that can be quantized in a NN: (1) weights, (2) activations, and (3) gradients. Weights and activations can be quantized to reduce the size of the model, and gradients can be quantized in situations where the training environment is dispersed to reduce communication costs. However, because high-precision gradients are required for backpropagation, quantizing gradients is typically more challenging than quantizing weights and activations. Post-

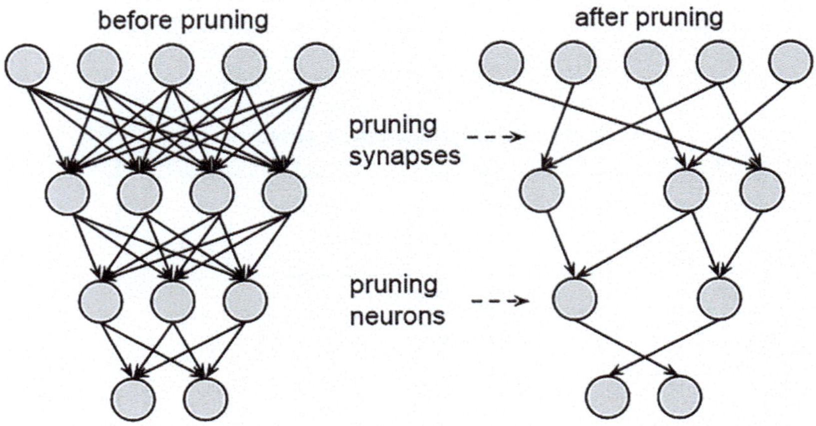

FIGURE 2.18 **NN model before and after pruning process.** The figure compares a neural network before and after pruning. In the pruned version, less important connections are removed, resulting in a smaller and more efficient model while keeping the main structure intact (*NN*, Neural network).

training quantization (PTQ) and quantization-aware training (QAT) are the two primary categories into which quantization techniques can be divided.

PTQ is an optimization technique applied to NNs after they have been trained. It consists to train the NN model using float32, then quantizing it. The fastest and simplest way to obtain a quantitative NN model, although it may lead to large differences in accuracy in some cases. Fig. 2.19 shows the process of PTQ (Ho and Chang, 2023). Various PTQ options are: (1) dynamic range quantization (DRQ), (2) full integer quantization (FIQ), and (3) float16 quantization (F16Q).

Dynamic range quantization

It provides lower memory usage and faster computation without the need for a representative dataset for calibration. This is achieved by statically quantizing weights from floating-point to 8-bit precision and dynamically quantizing activations during inference. While activations are always stored in float32, they are converted to 8-bit integers for processing and then converted back to floating-point after processing. This approach achieves latency close to that of pure fixed-point inferences. It is designed to reduce the model's memory footprint and accelerate inference speed while attempting to maintain the model's accuracy.

Full integer quantization

Further latency improvements, reduced peak memory usage, and compatibility with integer-only devices or accelerators can be achieved by ensuring that all

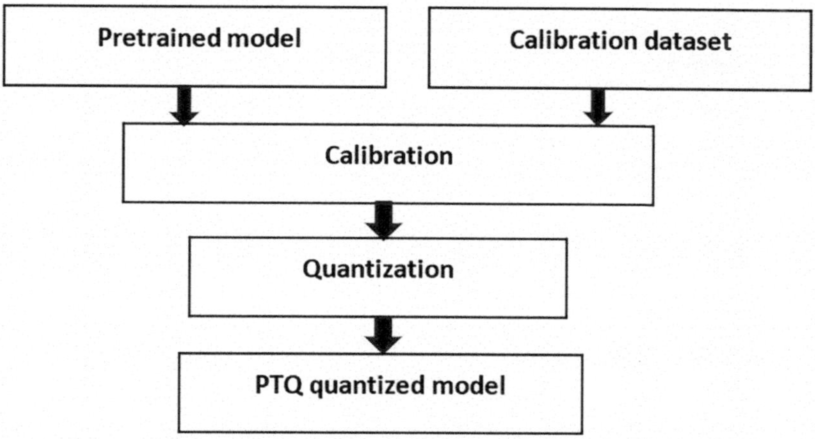

FIGURE 2.19 **PTQ workflow.** The figure shows the post-training quantization workflow, where a trained neural network model is converted from high-precision weights to lower-precision values to reduce size and computation. The workflow demonstrates that quantization happens after training and results in a more efficient model suitable for deployment on resource-limited devices (*PTQ*, Post-training quantization).

model computations are integer quantized. This involves statically quantizing all model weights and activations to 8-bit integers. To achieve this, it is necessary to normalize or estimate the range (minimum and maximum values) of all floating-point tensors in the model, including fixed tensors such as weights and biases, as well as variable tensors such as model inputs, activations from intermediate layers, and model outputs.

Float16 quantization

The size of a floating-point model can be reduced by quantizing its weights to float16, which halves the model size while minimizing the loss of accuracy. Some devices operate directly on float16 data, enabling faster execution compared to float32 calculations. However, this approach has drawbacks: it does not reduce latency as much as fixed-point quantization, and by default, a float16 quantizer model running on a CPU resets the weights to float32 before processing, as CPUs typically optimize float16 values to float32 during execution.

QAT is a technique that incorporates quantization into the training phase of NN models. Unlike PTQ, which applies quantization after the model is trained, QAT enables the model to learn and adapt to the effects of quantization during training. This is accomplished by adding dummy quantization nodes that mimic the low-resolution behavior in the forward pass while keeping the backward pass in float mode. The quantizer determines how to transform high-resolution inputs into quantized outputs, ensuring that all weight adjustments are made during training with the knowledge that the model will eventually be quantized. Fig. 2.20 illustrates the workflow of QAT (Hasan, 2024).

Table 2.2 summarizes a comparison between QAT and PTQ in terms of accuracy, process, complexity, computation, and best use case (Hasan, 2024).

QAT incorporates quantization effects directly into the training process by simulating low-precision computations in the forward pass while maintaining full accuracy in the backward pass. This enables the model to adapt to decreasing accuracy during learning, generally resulting in higher accuracy compared to PTQ. In contrast, quantization is applied after the model has been

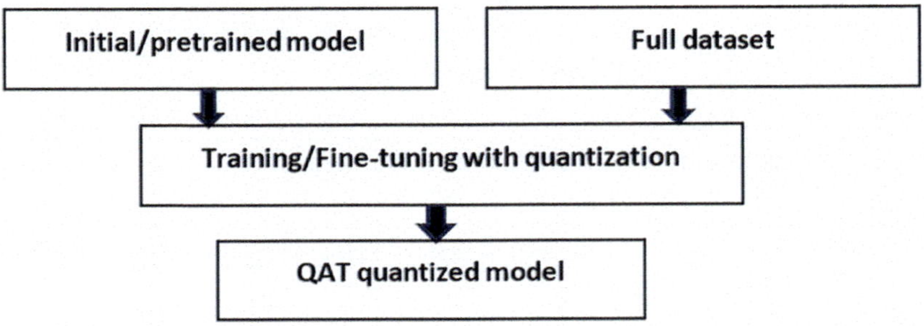

FIGURE 2.20 **QAT workflow.** The figure illustrates the quantization-aware training workflow, where the model is trained while simulating lower-precision operations so it can learn to adapt to quantization effects. After training, the model is converted to a quantized version that maintains higher accuracy when deployed on resource-constrained devices (*QAT*, Quantization-aware training).

TABLE 2.2 Comparison between QAT and PTQ techniques.

	QAT	PTQ
Accuracy	Typically higher, close to perfect accuracy	Slightly lower, and may deteriorate with aggressive quantization.
Complexity	More complex, requires training setup	Simpler, requires no retraining
Process	Simulates quantization effects in forward pass	Converts weights/activations to lower precision without retraining
Computation during training	Higher (due to simulated quantization)	No additional training computation is required.
Best use case	When accuracy is critical and retraining is possible	When rapid deployment is needed and a small loss in accuracy is acceptable.

PTQ, Post-training quantization; *QAT*, quantization-aware training.

fully trained, making it easier and faster to implement, but it can lead to greater accuracy loss, especially in models sensitive to accuracy changes (Hasan, 2024).

Explainable AI for embedded systems

XAI refers to a branch of AI that focuses on making the decisions, predictions, and actions of AI systems understandable and interpretable by humans. XAI aims to provide clear insights into how an AI model arrives at a particular outcome, enabling users to trust, verify, and interact effectively with the system (Minh et al., 2022). Unlike traditional AI models, which often operate as black boxes, XAI emphasizes transparency, accountability, and the ability to justify decisions, crucial for critical applications. In ESs, XAI is essential for ensuring reliability, safety, and trustworthiness. Techniques such as model-agnostic explainers, saliency maps, and interpretable surrogate models allow EAI to provide human-understandable insights without compromising real-time performance. By integrating explainability directly into the embedded framework, XAI facilitates better decision-making, simplifies debugging, and enhances confidence in AI-driven control and diagnostics, paving the way for responsible deployment in critical energy applications.

Key advantages and limits of embedded AI

EAI offers significant advantages by being directly integrated into devices. Among its key benefits is real-time decision-making with low latency, crucial for applications such as autonomous vehicles, robotics, industrial automation, and renewable energy systems like solar photovoltaic monitoring. By processing data locally, EAI also ensures enhanced privacy and security, as sensitive information doesn't need to be transmitted to external servers. This local processing also reduces network bandwidth and power consumption, making it ideal for battery-powered devices or remote devices in IoT networks. EAI can operate offline, providing autonomy in environments with intermittent or no internet connectivity, and can lower operating costs by reducing reliance on cloud computing. Its compact and energy-efficient design allows for deployment across a wide range of devices, from mobile phones and drones to peripheral sensors and MCU-based data loggers (Huang et al., 2025).

Despite these advantages, EAI also faces some inherent limitations. Hardware limitations, including limited processing power, memory, and storage, restrict the size and complexity of AI models that can be deployed. Optimizing AI models for ESs often requires techniques such as quantization, pruning, or compression, which can reduce accuracy or flexibility. Updating or upgrading EAI models is more challenging than cloud-based AI, and deployment often requires specialized expertise in both AI algorithms and embedded hardware design. Furthermore, because EAI systems are typically optimized for specific tasks, they may be less agile in adapting to new problems or changing operational requirements. Security also remains a concern, as on-premises processing alone does not provide complete protection against cyber threats. However, the combination of low latency, autonomy, privacy, and energy efficiency makes EAI a key enabler of modern intelligent systems in multiple fields (Seng and Ang, 2022).

Future trends in embedded and generative AI

The future of EAI and GenAI at the edge promises to transform how intelligent systems operate across industries. Edge devices will increasingly host powerful AI and GenAI models, enabling real-time decision-making, predictive analytics, and autonomous operations without relying on cloud connectivity (Vermesan et al., 2023). Advances in TinyML, low-power AI accelerators, and optimized hardware will allow sophisticated models to run on mobile devices, the IoT, and embedded platforms with minimal power consumption (Giorgetti and Pau, 2025). Privacy and security will be enhanced through on-device collaborative and unified learning, reducing the need to transfer sensitive data. Hybrid architectures combining native AI and cloud learning will support continuous updates, adaptive optimization, and efficient communication, while generative models at the edge will enable capabilities such as automated reporting, anomaly detection, and scenario simulation.

These developments will have a particular impact in sectors such as renewable energy, smart cities, health care, and industrial automation, where real-time, autonomous, and intelligent decision-making is essential for improving efficiency, reliability, and sustainability. Specialized AI accelerators will improve speed and power consumption. Privacy learning will generally be enhanced through collaborative learning devices without sharing raw data. Contributing to communication and compression will facilitate deployment, while hybrid electronics will combine local AI and interact with cloud computing to learn and update communications (Navardi et al., 2025). The ability of GenAI to analyze complex information, predict environmental consequences, and enhance energy generation and distribution underscores its pivotal role in the future (Mousavi et al., 2025).

Conclusion

This chapter explored the fundamental and emerging technologies driving AI techniques, with a particular focus on AI, ML, DL, GenAI, and EAI. By examining a range of learning paradigms, from SL and USL to REL, and advanced architectures such as CNNs, YOLO, and ViTs, we highlighted the increasing sophistication of AI in tackling complex tasks like image recognition, anomaly detection, and autonomous decision-making. The integration of GenAI, including LLMs, multimodal AI and GANs, represents a significant leap forward in content creation and human–computer interaction, while ESs and edge AI technologies like TinyML and lightweight CNNs enable intelligent processing at the device level. These innovations not only improve performance and efficiency but also make AI more accessible and scalable across diverse applications.

Looking ahead, the convergence of EAI and GenAI promises to redefine the boundaries of automation, enabling more intelligent and adaptive systems that operate with minimal human intervention. Furthermore, XAI enhances EAI systems by providing transparency and interpretability, supporting safer, more reliable, and efficient deployment. By facilitating debugging and informed decision-making, XAI accelerates the adoption of intelligent edge devices and ensures responsible AI implementation. This evolution will be pivotal in shaping the future of sustainable technologies, intelligent diagnostics, and real-time decision-making across a wide range of sectors.

AI Disclosure

During the preparation of this work, the author(s) used ChatGPT in order to do proofreading. After using this tool/service, the author(s) reviewed and edited the content as needed and take(s) full responsibility for the content of the publication.

References

Alpaydin, E., 2020. Introduction to Machine Learning, 4th. MIT Press, Cambridge, MA, USA.

Banbury, C., et al., 2022. TinyML Cookbook: Combine Artificial Intelligence and Ultra-Low-Power Embedded Devices to Make the World Smarter. Packt Publishing Ltd., Birmingham, UK.

Boruah, A.N., Goswami, M., Kumar, M., Loyola-González, O., 2025. Embedded Artificial Intelligence: Real-Life Applications and Case Studies Embedded Artificial Intelligence: Real-Life Applications and Case Studies. CRC Press, India, pp. 1–310. Available at https://www.taylorfrancis.com/books/edit/10.1201/9781003481089, http://doi.org/10.1201/9781003481089.

Brown T.B., Mann B., Ryder N., Subbiah M., Kaplan J., Dhariwal P., et al., 2020. 1 2020/01 Advances in Neural Information Processing Systems 10495258 Neural information processing systems foundation United States Language models are few-shot learners. https://papers.nips.cc/2020-.

Dosovitskiy, A., Beyer, L., Kolesnikov, A., Weissenborn, D., Zhai, X., Unterthiner, T., et al., 2020. An image is worth 16X16 words: Transformers for image recognition at scale. arXiv. Available at https://doi.org/10.48550/arxiv.2010.11929, https://arxiv.org.

Drir, N., Mellit, A., Bettayeb, M., Dhimish, M., 2025a. Enhanced Photovoltaic Defect Detection Using Perceptual Loss in DCGAN and VGG16-Integrated Models on Electroluminescence Images. IEEE Journal of Photovoltaics 15 (6), 759–769. Available at https://doi.org/10.1109/JPHOTOV.2025.3542829, https://ieeexplore.ieee.org/xpl/RecentIssue.jsp?punumber=5503869.

Drir, N., Chekired, F., Mellit, A., Blasuttigh, N., 2025b. Hybrid CNN-EML model for fault diagnosis in Electroluminescence images of photovoltaic cells. Renewable Energy.

Foster D., 2022. Generative Deep Learning, 2nd. O'Reilly Media, Inc., Sebastopol, CA, USA.

Giorgetti, G., Pau, D.P., 2025. Transitioning from TinyML to edge GenAI: a review. Big Data and Cognitive Computing 9 (3).

Goodfellow, I., Bengio, Y., Courville, A., 2016. Deep Learning. MIT Press, Cambridge, MA, USA.

Goodfellow I.J., Pouget-Abadie J., Mirza M., Xu B., Warde-Farley D., Ozair S., et al. 2014. 1 2014/01 Advances in Neural Information Processing Systems 10495258 January 2672–2680. Neural information processing systems foundation Canada Generative adversarial nets 3.

Guo, M., Weng, Y., 2025. Solar Photovoltaic Assessment with Large Language Model. arXiv preprint. 2507.19144

Hasan, J., 2024. Optimizing Large Language Models through Quantization: A Comparative Analysis of PTQ and QAT Techniques. arXiv. Available at https://doi.org/10.48550/arXiv.2411.06084, https://arxiv.org.

Ho, N.D., Chang, I.J., 2023. O-2A: Outlier-Aware Compression for 8-bit Post-Training Quantization Model. IEEE Access 11, 95467–95480. Available at https://doi.org/10.1109/ACCESS.2023.3311027, http://ieeexplore.ieee.org/xpl/RecentIssue.jsp?punumber=6287639.

Howard, A.G., Zhu, M., Chen, B., Kalenichenko, D., Wang, W., Weyand, T., et al., 2017. MobileNets: Efficient convolutional neural networks for mobile vision applications. arXiv. Available at https://doi.org/10.48550/arxiv.1704.04861, https://arxiv.org.

Huang G., Liu Z., Van Der Maaten L., Weinberger K.Q., 2017. 11 6 2017/11/06 Proceedings - 30th IEEE Conference on Computer Vision and Pattern Recognition, CVPR 2017 9781538604571 2261–2269. Institute of Electrical and Electronics Engineers Inc. United States Densely connected convolutional networks 2017. https://doi.org/10.1109/CVPR.2017.243.

Huang, G., Liu, S., der Maaten, L.V., Weinberger, K.Q., 2018. Condensenet: An efficient densenet using learned group convolutions. In: Proceedings of the IEEE conference on computer vision and pattern recognition, pp. 2752–2761.

Huang, X., Wang, H., Shiyin, Q., Su-Kit, T., T, 2025. Embedded Artificial Intelligence: A Comprehensive Literature Review. Electronics 14 (17).

Iandola, F.N., Han, S., Moskewicz, M.W., Ashraf, K., Dally, W.J., Keutzer, K., 2016. SqueezeNet: AlexNet-level accuracy with 50x fewer parameters. arXiv preprint. 1602.07360.

Jena, B., Saxena, S., Nayak, G.K., Saba, L., Sharma, N., Suri, J.S., 2021. Artificial intelligence-based hybrid deep learning models for image classification: The first narrative review. Computers in Biology and Medicine 137. Available at https://doi.org/10.1016/j.compbiomed.2021.104803, www.elsevier.com/locate/compbiomed.

Kissinger, H. A., Schmidt, E., & Huttenlocher, D., 2021. The Age of AI. Hachette UK., London, UK.

Lecun, Y., Bengio, Y., Hinton, G., 2015. Deep learning. Nature 521 (7553), 436–444. Available at https://doi.org/10.1038/nature14539, http://www.nature.com/nature/index.html.

Liu, H.-I., Galindo, M., Xie, H., Wong, L.-K., Shuai, H.-H., Li, Y.-H., et al., 2024. Lightweight Deep Learning for Resource-Constrained Environments: A Survey. ACM Computing Surveys 56 (10), 1–42. Available at https://doi.org/10.1145/3657282.

Liu, J., 2003. Real-Time Concepts for Embedded Systems. CRC Press., Boca Raton, FL, USA.

Marwedel P., 2021. Embedded System Design: Embedded Systems Foundations of Cyber-Physical Systems, and the Internet of Things, 4th. Springer Nature, Cham, Switzerland.

Mellit, A., Kalogirou, S., 2022. Handbook of Artificial Intelligence Techniques in Photovoltaic Systems: Modeling, Control, Optimization. Forecasting and Fault Diagnosis Handbook of Artificial Intelligence Techniques in Photovoltaic Systems: Modeling, Control, Optimization, Forecasting and Fault Diagnosis. Elsevier, undefined, pp. 1–358. Available at https://www.sciencedirect.com/book/9780128206416, http://doi.org/10.1016/C2019-0-00960-0.

Mellit, A., Kalogirou, S., 2025. Recent advances in the application of infrared thermographic imaging and embedded artificial intelligence for fault diagnosis and predictive maintenance of photovoltaic plants: Challenges and future directions. Renewable and Sustainable Energy Reviews 223, 116057. Available at https://doi.org/10.1016/j.rser.2025.116057.

Minh, D., Wang, H.X., Li, Y.F., Nguyen, T.N., 2022. Explainable artificial intelligence: a comprehensive review. Artificial Intelligence Review 55 (5), 3503–3568.

Mousavi, R., Mousavi, A., Mousavi, Y., Tavasoli, M., Arab, A., Kucukdemiral, I.B., et al., 2025. Revolutionizing solar energy resources: The central role of generative AI in elevating system sustainability and efficiency. Applied Energy 382, 125296. Available at https://doi.org/10.1016/j.apenergy.2025.125296.

Mueller, A.C., & Guido, S., 2016. Introduction to Machine Learning with Python: A Guide for Data Scientists. O'Reilly Media, Inc., Sebastopol, CA, USA.

Nagel, M., Fournarakis, M., Amjad, R.A., Bondarenko, Y., van Baalen, M., Blankevoort, T., 2021. A white paper on neural network quantization. arXiv preprint. 2106.08295, 4.

Navardi, M., Aalishah, R., Fu, Y., Li, H., Chen, Y., Mohsenin, T., 2025. GenAI at the edge: Comprehensive survey on empowering edge devices. In: Proceedings of the AAAI Symposium Series. 5 (1), 1180–1187.

Noergaard T., 2013. Embedded Systems Architecture: A Comprehensive Guide for Engineers and Programmers. Newnes, Waltham, MA, USA.

Radford, A., Metz, L., Chintala, S., 2016. 1 2016/01 4th International Conference on Learning Representations, ICLR 2016 - Conference Track Proceedings International Conference on Learning Representations, ICLR United States Unsupervised representation learning with deep convolutional generative adversarial networks. https://dblp.org/db/conf/iclr/iclr2016.html.

Redmon, J., 2016. You only look once: Unified, real-time object detection. In: Proceedings of the IEEE conference on computer vision and pattern recognition.

Russell S., & Norvig P., 2021. Artificial Intelligence: A Modern Approach, 4th. Pearson, Hoboken, NJ, USA.

Seng, K.P., Ang, L.M., 2022. Embedded Intelligence: State-of-the-Art and Research Challenges. IEEE Access 10, 59236–59258. Available at https://doi.org/10.1109/ACCESS.2022.3175574, http://ieeexplore.ieee.org/xpl/RecentIssue.jsp?punumber=6287639.

Situnayake, D., & Plunkett, J, 2023. AI at the Edge: Solving real-world problems with embedded machine learning. O'Reilly Media.

Shalev-Shwartz, S., Ben-David, S., 2014. Understanding Machine Learning: From Theory to Algorithms. Cambridge University Press. Available at http://doi.org/10.1017/CBO9781107298019.

Somvanshi, S., Islam, M.M., Chhetri, G., Chakraborty, R., Mimi, M.S., Shuvo, S.A., Islam, K.S., Javed, S., Rafat, S.A., Dutta, A., Das, S., 2025. From tiny machine learning to tiny deep learning: A survey. ACM Computing Surveys 58 (7), 1–33.

Tan, M., Le, Q.V., 2019. 1 2019/01 Proceedings of Machine Learning Research 26403498 6105–6114 ML Research Press United States EfficientNet: Rethinking Model Scaling for Convolutional Neural Networks. https://proceedings.mlr.press/97.

Vajjala, S., Majumder, B., Gupta, A., Surana, H., 2023. Applied Generative AI for Beginners. Springer Nature, Cham, Switzerland.

Vaswani, A., Shazeer, N., Parmar, N., Uszkoreit, J., Jones, L., Gomez, A.N., et al., 2017. 1 2017/01 Advances in Neural Information Processing Systems 10495258 5999–6009 Neural information processing systems foundation United States Attention is all you need 2017.

Vermesan, O., Nava, M.D., Debaillie, B., 2023. Embedded Artificial Intelligence: Devices, embedded Systems, and Industrial Applications. River Publishers. Available at http://doi.org/10.1201/9781003394440.

Warden, P., & Situnayake, D., 2019. Tinyml: Machine learning with tensorflow lite on arduino and ultra-low-power microcontrollers. O'Reilly Media.

Yan, Z., Li, X., Li, M., Zuo, W., Shan, S., 2018. Shift-net: Image inpainting via deep feature rearrangement. In: Proceedings of the European conference on computer vision (ECCV), pp. 1–17.

Zurawski R., 2018. Embedded Systems Handbook, 2nd. CRC Press, Boca Raton, FL, USA.

Chapter 3

Edge devices and platforms for embedded artificial intelligence

Adel Mellit[1], **Bekkay Hajji**[2], **and Lotfi Baghli**[3]
[1]*Faculty of Sciences and Technology, University of Jijel, Jijel, Algeria,* [2]*Department of Electronics, Computer Science and Telecommunications, National School of Applied Sciences, Mohammed First University, Oujda, Morocco,* [3]*Génie électrique, Université de Lorraine - GREEN – ENSEM, Vandoeuvre-lès-Nancy Cedex, France*

Introduction

The convergence of artificial intelligence (AI) and distributed computing has given rise to a new paradigm known as Edge AI, where intelligent data processing occurs directly on edge devices rather than on centralized cloud computing servers. This shift of intelligence from the cloud to the Edge is not accidental. It is driven by unavoidable technical imperatives: the need to reduce latency for real-time decisions, such as in an obstacle avoidance system for a drone; the desire to reduce bandwidth consumption by avoiding the transfer of continuous video streams to the cloud; the requirement to enhance data confidentiality by processing it locally; and the need to ensure that devices can operate autonomously even when connectivity is unavailable. However, this remote computing power faces severe hardware constraints. Unlike data centers, embedded devices must balance high computing performance with extremely low-power consumption, small footprint, and controlled costs. Meeting this challenge requires an innovative hardware approach, using a diverse range of specialized accelerators, from graphics processing units (GPUs) to neural processing units (NPUs) to field-programmable gate arrays (FPGAs), each optimized for specific tasks and trade-offs.

The objective of this chapter is to provide a comprehensive and structured overview of the embedded AI hardware ecosystem. We will begin by introducing the fundamental concepts of cloud, fog, and edge computing, highlighting their architectural differences and roles in modern data systems. We will then detail a comparative analysis illustrating the trade-offs between latency, bandwidth, scalability, and computational efficiency in these models. Building on this foundation, the chapter explores the concept of Edge AI,

Embedded Artificial Intelligence and the Internet of Things for Photovoltaic Systems.
DOI: https://doi.org/10.1016/B978-0-443-44835-5.00003-5

emphasizing its importance in enabling real-time, low-power intelligence at the device level of embedded devices. It provides a concrete overview of the main solutions offered by market players, including microcontroller units (MCUs) (e.g., Nano BLE Sense), single-board computers (e.g., Raspberry Pi), GPUs (e.g., Nvidia Jetson Nano), tensor processing units (e.g., Google Coral), FPGAs, and application-specific integrated circuits (ASICs), each offering unique capabilities for deploying AI applications.

To support the development of Edge AI, the chapter reviews key platforms and frameworks such as TensorFlow Lite, PyTorch Mobile, ONNX, OpenVINO, Edge Impulse, X-CUBE-AI, uTensor, and the embedded learning library (ELL). These tools facilitate model optimization, deployment, and inference on resource-constrained devices. Comparative tables are included to illustrate the performance, compatibility, and use cases of various edge devices and software platforms. The chapter also discusses the benefits, challenges, and future trends of edge AI, including its potential to transform industries through decentralized intelligence, while addressing concerns related to energy efficiency, security, and scalability.

Cloud, fog, and edge computing

The data processing ecosystem has radically evolved from a centralized model to a distributed architecture. Faced with the explosion of data and the need for responsiveness, three complementary approaches are emerging: cloud, fog computing, and edge computing (See Fig. 3.1). Cloud computing offers

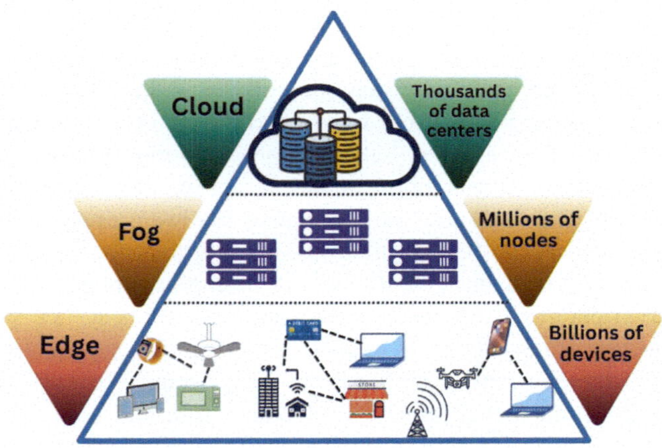

FIGURE 3.1 Cloud, fog, and edge. The figure illustrates the relationship between cloud, fog, and edge computing, showing how data processing can occur at different levels of proximity to the source. Edge devices process data closest to where it is generated, fog computing acts as an intermediate layer, and the cloud provides centralized high-power processing and storage.

scalability but increases latency because data is processed in centralized data centers. Fog computing reduces latency and bandwidth consumption by converging processing across intermediate nodes. For immediate response, edge computing processes data at or near the source. These elements work together to balance scalability, latency, and performance.

This computing hierarchy addresses modern challenges of latency, bandwidth, and data sovereignty, forming a coherent ecosystem where each layer has its specific role.

Cloud computing

Cloud computing is a paradigm that aims to provide reliable, dedicated, and quality-assured dynamic computing environments to the end user (Wang et al., 2008). It allows users and organizations to access scalable computing resources without having to own or manage the underlying physical infrastructure. This model thus offers increased flexibility, improved cost-efficiency, on-demand scalability, and universal remote access. Cloud architecture is based on large-scale distributed data center networks, which can span dozens or even hundreds of sites around the world. These infrastructures form the physical foundation of cloud computing, see Fig. 3.2 (What Is Cloud Computing? Definition, Benefits, Types, and Trends, 2021).

Cloud-hosted services can be broadly divided into three types of infrastructure:

- Infrastructure as a Service (IaaS): Access to virtualized resources (servers, storage, network);
- Platform as a Service (PaaS): Application development and deployment environment;
- Software as a Service (SaaS): Software accessible via a browser without local installation.

Based on the deployment model, clouds can also be classified as public (shared services open to all), private (infrastructure dedicated to a single organization), and hybrid clouds (a combination of the two models). Today, cloud computing is mainly offered by major players such as Microsoft Azure, Google Cloud Platform, and Amazon Web Services (AWS), as well as by other providers such as IBM Cloud (Gill et al., 2025). These dominant platforms offer a full range of services covering all deployment and service models.

Fog computing

Fog computing is positioned as a strategic extension of cloud computing, deploying computing, storage, and networking capabilities closer to data sources, between the central cloud and peripheral Internet of Things (IoT) devices (Yi et al., 2015). Its origins can be traced to a conference presentation

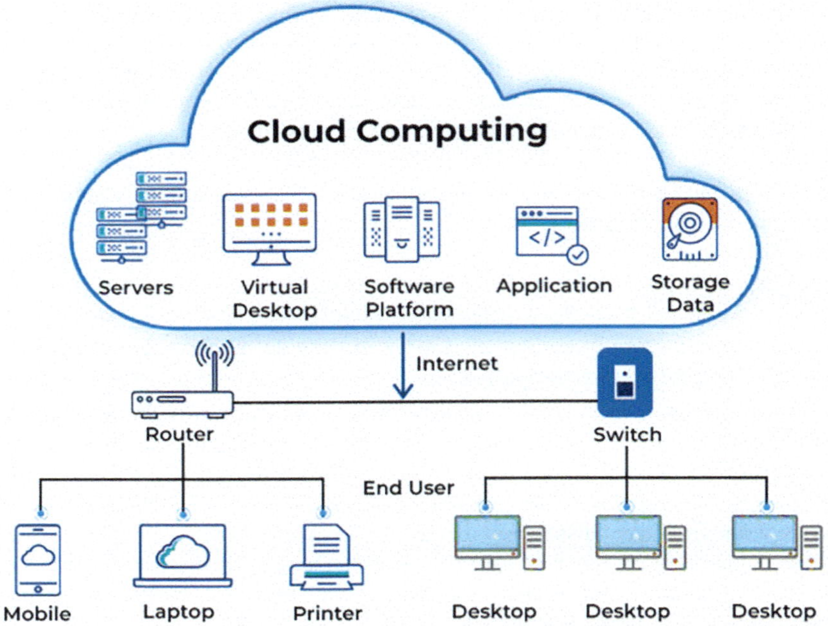

FIGURE 3.2 Cloud computing architecture. The figure illustrates the concept of cloud computing, where data storage, processing, and applications are hosted on remote servers accessible through the Internet. Users and devices send data to the cloud for computation, allowing high-performance processing without relying on local hardware resources.

by Cisco Systems, Inc., in 2012 (Bonomi et al., 2022). As shown in Fig. 3.3, the fog commuting is situated between the cloud and IoT devices. Fog computing establishes a processing hierarchy where data is analyzed at different levels of the network. Distinctive characteristics of fog are:

a) Low latency and location awareness: Enables real-time processing with precise location knowledge;
b) Wide geographic distribution: Nodes spread across the entire territory;
c) Mobility support: Support for devices and users in motion;
d) Massively distributed architecture: Thousands of interconnected nodes;
e) Autonomy and resilience: Ability to operate in degraded mode;
f) Strong presence of broadcast and real-time applications: Fog computing favors applications requiring massive data dissemination and strictly real-time processing, such as emergency systems;
g) Variability: The fog environment must constantly adapt to changing conditions in terms of connectivity, node load, and quality of service.

The fog ecosystem is built around four fundamental pillars: intelligent edge devices (sensors, cameras, and actuators), heterogeneous connectivity

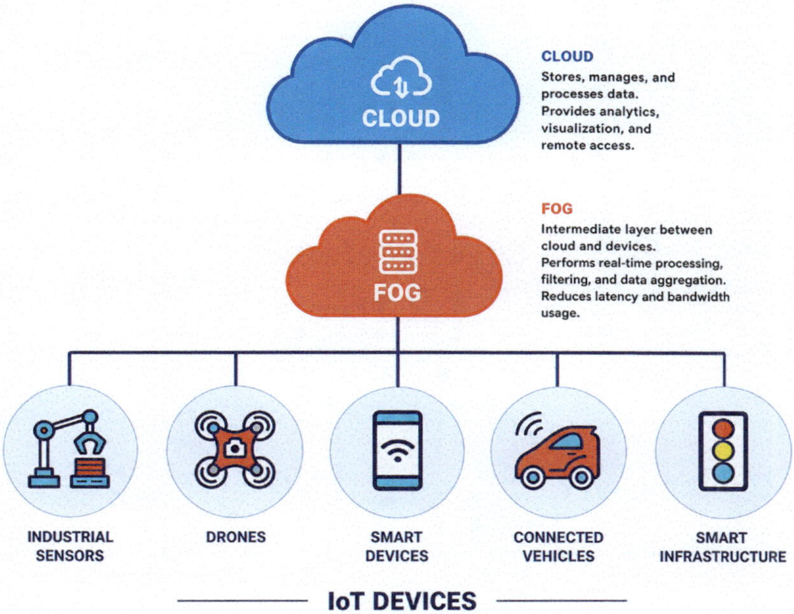

FIGURE 3.3 Fog computing architecture. The figure shows fog computing as an intermediate layer between edge devices and the cloud, enabling data processing closer to where data is generated. This reduces latency and bandwidth usage by handling part of the computation locally before sending only necessary data to the cloud.

(multiprotocol wired and wireless networks), distributed and tiered storage, and virtualized processing capabilities. Real-world examples of fog computing are IoT devices equipped with sensors and cameras Industrial Internet of Things (IIoT).

Edge computing

Edge computing refers to the enabling technologies that allow computing to take place at the edge of the network, on downstream data on behalf of cloud services and upstream data on behalf of IoT services, without transit through the cloud (Shi et al., 2016). As the name suggests, edge computing takes place right at the edge of the application network. Topologically, this means that the edge computer is located next to or even above endpoints (such as controllers and sensors) connected to the network. Data is then partially or fully processed and sent to the cloud for further processing or storage (See Fig. 3.4). For example, a smartphone is the edge between things in the body and the cloud, a gateway in a smart home is the edge between things in the home and the cloud, and a micro data center and micro cloud are the edge between a mobile device and the cloud (Satyanarayanan et al., 2009).

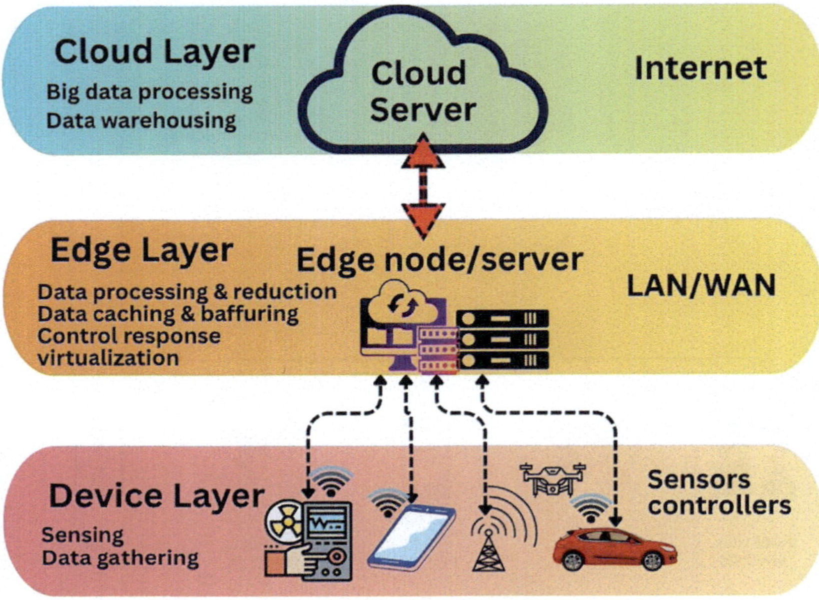

FIGURE 3.4 Edge computing architecture. The figure presents an edge computing architecture, where data is processed directly on or near the devices that generate it rather than being sent to the cloud. This structure reduces latency and improves real-time decision-making by keeping computation close to the data source.

This approach reduces latency, conserves bandwidth, improves real-time response, and enhances data privacy, making it suitable for applications that require rapid or continuous decision-making. Edge computing is ideal for IoT and real-time applications, enabling low-latency processing, lower bandwidth consumption, enhanced privacy, and offline operation. However, large-scale analytics can lead to data fragmentation due to resource limitations, complex management, security issues, and high deployment costs. Edge computing eliminates the need for a single central data processing center and increases computing efficiency by bringing data centers closer to where they are actually needed.

Comparison

Table 3.1 summarizes a comparison between the three paradigms in terms of latency, scalability, bandwidth, connectivity, privacy, and security.

Cloud computing processes data in centralized data centers, offering high scalability and processing power, but with higher latency, bandwidth utilization, and a reliance on stable internet connections. This makes it ideal for nontime-critical tasks such as big data analytics, training AI models, long-term storage,

TABLE 3.1 Key difference between cloud, fog, and edge computing.

Feature	Cloud computing	Fog computing	Edge computing
Processing location	Centralized data centers, often distant from the data source	An intermediate layer between the cloud and edge, closer to devices	Directly on or near the data source (IoT device, gateway)
Latency	High (due to long transmission distance)	Medium (shorter distance than cloud)	Very low (minimal transmission)
Bandwidth	High, as raw data is sent to the cloud	Moderate, as some preprocessing reduces the data sent	Low, as only the necessary data is sent
Scalability	Very high, virtually unlimited	High, but depends on the fog node distribution	Limited by local hardware capacity
Power	Very high (powerful servers)	Medium to high (local servers/gateways)	Low to medium (embedded devices)
Real-time capability	Poor for time-critical tasks	Good for near-real time	Excellent for real time
Applications	Big data analytics, AI training	Smart cities, industrial IoT, video analytics	Autonomous vehicles, robotics, and on-device AI
Connectivity	Requires a stable internet	Works with intermittent cloud connection	Can operate fully offline
Privacy and security	Data passes through public networks	Sensitive data can be processed locally	Maximum local data control
Architecture	Centralized	Distributed	Distributed

AI, Artificial intelligence; *IoT*, Internet of Things.

and traditional enterprise applications. Its centralized nature enables unified resource management and reduced infrastructure costs for organizations.

Fog computing acts as an intermediate layer between the cloud and the edge, reducing latency and bandwidth utilization by preprocessing data close to the source, making it suitable for near-real-time applications such as smart cities and the Industrial IoT. This architecture enables distributed data analysis while maintaining coordination with the central cloud for broader analytics. Fog computing is useful in situations where low latency is critical, but data still requires preprocessing or intermediate aggregation before being sent to the cloud, such as in the IIoT or smart cities.

Edge computing takes this logic even further by performing processing directly on the devices themselves or in their immediate vicinity, thus almost completely eliminating latency and making systems fully autonomous in the face of connection interruptions. It is ideal for real-time tasks such as autonomous vehicles and embedded AI, industrial robotics, augmented reality, and connected medical devices (Tank and Gandhi, 2023). When massive processing power, centralized storage, and comprehensive analytics are needed without strict real-time constraints, for example, in big data processing or AI model training, cloud computing is used (Dolui and Datta, 2017). It can be concluded that fog computing is preferred for use by organizations that deal with data processing, while it is recommended for middleware organizations, which act as the backend of the network.

Edge artificial intelligence

In contrast to traditional AI technologies, which depend on centralized cloud servers for data processing, edge AI leverages the computing power of local devices to run AI algorithms directly at the end of the network. This approach involves deploying AI algorithms and models directly to edge devices, physical computing equipment located at the network edge. This includes smartphones, IoT devices, and embedded systems (See Fig. 3.5) (Shankar, 2024). Essentially, edge AI represents the practical application of AI to devices operating in the physical world, granting them immediate decision-making autonomy.

This paradigm shift directly addresses the inherent limitations of server-based, centralized AI architectures, such as reliance on a permanent network connection, data security vulnerabilities, and bandwidth limitations (Singh and Gill, 2023). Driven by advancements in algorithm efficiency, the proliferation of connected devices, and the maturation of edge computing, the potential of edge AI is now an operational reality. Its advantages are numerous: extremely low latency for real-time decision-making, significant bandwidth reduction thanks to selective transmission of processed data, enhanced confidentiality through local information processing, high resilience in the event of degraded connectivity, and, in many cases, better energy efficiency compared to cloud processing. The key performance metrics to evaluate an edge AI are: Latency, Security and privacy, Energy consumption, Accuracy, Scalability, Bandwidth, Cost, and Resources utilization.

Despite the aforementioned high advantages and potential mentioned in the earlier section, edge AI faces challenges such as: (1) limited processing power of devices in the edge network, (2) management difficulties related to network heterogeneity, and (3) energy constraints resulting from limited resources. If future researchers succeed in solving these challenges, the

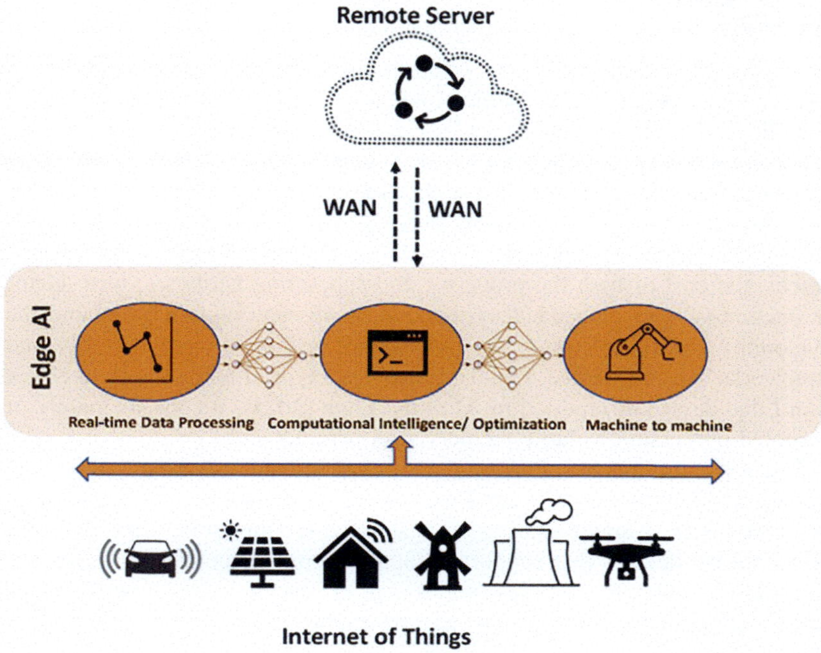

FIGURE 3.5 Edge AI paradigm. The figure illustrates the Edge AI paradigm, where artificial intelligence models are deployed directly on edge devices instead of relying on cloud processing. This enables faster real-time decisions, reduced latency, and improved privacy since data remains on the device.

application areas of edge AI should expand (Gill et al., 2025). Popular applications of Edge AI include: health care and medical devices (e.g., diagnosis), transportation (autonomous driving), agriculture (e.g., precision farming), smart cities (e.g., traffic control), energy (e.g., smart grid), and security (e.g., threat detection).

Among the major players in the Edge AI sector are: AWS, Dell, Google, IBM, Huawei, and Microsoft (Edge AI market report, 2025). Edge AI hardware is evolving rapidly, with new processors, AI accelerators, and neural processing units (NPUs) designed to efficiently handle ML workloads. Companies are developing specialized hardware that delivers high-performance AI inference at the network edge while maintaining energy efficiency. These advancements enable real-time processing for applications. For example, NXP unveiled the i.MX RT700 MCU, a multimodule MCU designed to enhance AI capabilities at the network edge. This MCU integrates two Arm Cortex-M33 cores, Cadence Tensilica HiFi digital signal processors (DSPs) and an eIQ Neutron NPU, offering up to 172x acceleration for AI tasks while reducing power consumption per inference by up to 119x (Edge AI market report, 2025).

Edge devices

Integrating AI with edge devices offers multiple benefits, such as enhanced privacy, reduced bandwidth, reduced latency, and real-time data processing capabilities (Dolui and Datta, 2017). Edge AI computing uses a variety of central processing units (CPUs), including specialized and standard processors with AI acceleration features. When choosing a device for edge AI, consider performance to meet the processing needs of tasks such as image or speech recognition, power consumption to ensure energy efficiency in constrained devices, the cost of both the processor and supporting hardware, form factor for integration into embedded systems, and software support to enable the development and deployment of AI models using compatible tools and frameworks. As an example, Fig. 3.6 illustrates a typical hardware architecture of an Edge device as reported in AI at the Edge (2023). It consists mainly of integrated processors, volatile memory, and peripherals.

The main limitations of Edge devices are related to power processing, memory availability, storage capacity, power consumption, processing support, and device resource sharing. This subsection aims to describe the most common edge devices that support AI applications, including CPUs, MCUs, GPUs, TPUs, and NPUs. These processors offer varying levels of performance and power efficiency for different edge AI applications.

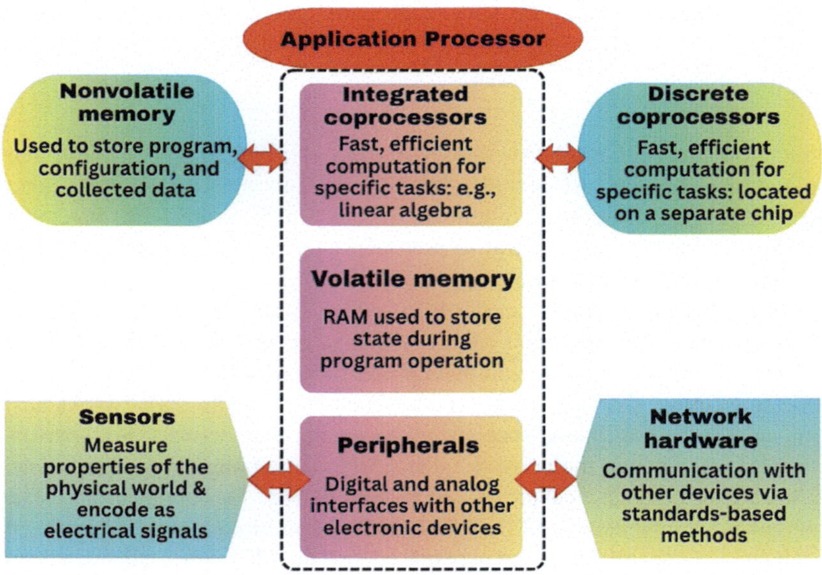

FIGURE 3.6 Basic architecture of an Edge device. The figure shows the basic architecture of an edge device, including sensors for data acquisition, a processing unit for local computation, and communication modules to connect with other systems. It illustrates how the device can analyze data on-site without depending on cloud processing.

Central processing and microprocessor units

The CPU is a hardware component responsible for executing instructions and performing calculations. It interprets and executes instructions from the computer's memory and handles arithmetic, logical, control, and input/output (I/O) operations (See Fig. 3.7) (Computer Organization and Design ARM Edition: The Hardware Software Interface, 2016). It contains a control unit, an arithmetic and control unit, a memory, and I/O units. The main types of CPUs are: single-core, dual-core, and quad-core.

A microprocessor unit (MPU) is an integrated circuit (IC) that contains one or more CPUs and supporting electronics. Often designed for general-purpose computing. MPU applications include acting as the central processing unit in computers, acting as the main processing unit in embedded systems (such as washing machines and microwaves), controlling industrial machinery in automation, operating electronic control units in automobiles, and working in communications devices such as routers, modems, and switches. Fig. 3.8 shows the generations of MPU evolutions (Raghunathan, 2021).

As an example of an MPU that supports Edge AI applications, we can find Raspberry Pi 4 or 5 (Raspberry Pi 5, 2023), which is a single-board computer (See Fig. 3.9). Its processor is a 64-bit system on chip (SoC), powerful enough to run a Linux Operating System. However, the following points should be considered (Steadman et al., 2024): (1) It lacks a built-in AI accelerator, making inference on large models relatively slow (2) it is not suitable for training DNNs and is mainly limited to inference, and (3) it consumes more power (around 3–7 W) than ultralow-power MCU-based TinyML solutions. The main specifications of the Raspberry Pi 4 are shown in Table 3.2.

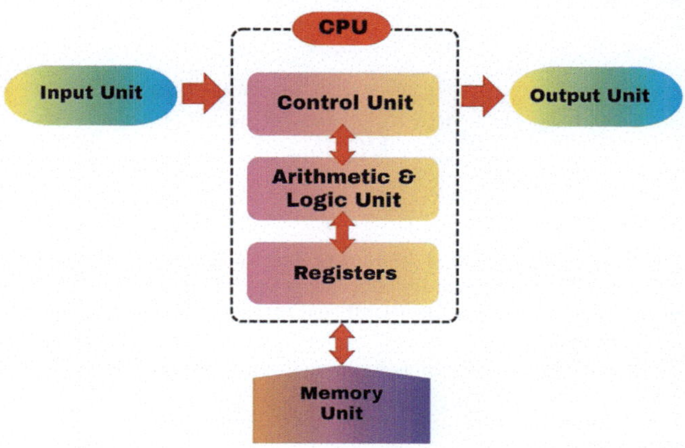

FIGURE 3.7 Basic structure of a CPU. The figure shows the basic structure of a CPU, with components such as the control unit, Arithmetic and Logic Unit (ALU), and registers working together to process instructions. Data flows through these units to execute operations and manage tasks within the computer system.

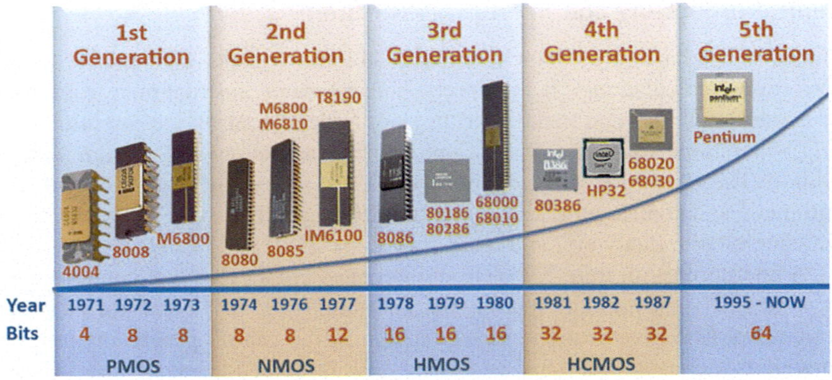

FIGURE 3.8 Timeline of MPU evolution. The figure displays a timeline showing the evolution of microprocessor units (*MPUs*) over the years. It highlights key milestones and advancements in processing power and architecture as technology progresses.

FIGURE 3.9 Photo of a Raspberry Pi 4. The figure shows a photo of a Raspberry Pi 4, a small single-board computer commonly used for edge computing and embedded AI applications. The board includes visible ports such as USB, HDMI, and Ethernet, along with components mounted on the compact circuit board.
Raspberry Pi 5 (2023). https://www.raspberrypi.com/products/raspberry-pi-5/.

Microcontroller units

An MCU is a small computer on a single (IC, often described as a SoC. While it does contain an MPU or CPU core, a large amount of memory, and peripherals, the key defining feature is the integration of all these essential

TABLE 3.2 Raspberry Pi 4 specifications.

CPU	**Broadcom BCM2711, Quad-core ARM Cortex-A72 @ 1.5 GHz**
GPU	VideoCore VI @ 500 MHz, OpenGL ES 3.0
RAM	LPDDR4–3200, 8 GB option
Storage	microSD card slot (UHS-I support)
Connectivity	Wi-Fi 802.11ac, Bluetooth 5.0, Gigabit Ethernet
Camera/Display	2× MIPI (1× DSI, 1× CSI)
Price	$75

components onto a single chip, making it a self-contained system. MCUs are designed for embedded applications, meaning they are built to perform specific, dedicated tasks within a larger electronic device, and are optimized for efficiency, low-power consumption, and real-time responsiveness, rather than general-purpose computing. Main peripherals are (I/O) Ports (GPIO), timers, communication interfaces (e.g., UART, SPI, I2C, USB): Allow the MCU to send and receive data from other chips or devices, analog-to-digital converter (ADC), and pulse width modulation to control power electronic devices. Some MCUs are called digital signal controller when they integrate a DSP core, allowing them to perform somme of product operation on a single clock period. They often have floating point unit that can do multiplication and division of floating point variables on a single clock period. A basic schematic of an MCU is shown in Fig. 3.10. It consists basically of a CPU, RAM, ROM, EPROM, IO ports, timing circuit, interrupt control, ADC/DAC, and a serial port (e.g., I2C, SPI, and UART) (Gridling, 2007). Often designed for one or a few specific tasks.

Common MCU Kits that support TinyML (Soro, 2021) applications are: Nano BLE Sense (Arduino nano family, 2020), ESP32 (ESP32, 2016), Portenta H7 (Portenta family, 2020), Nicla Vision (Nicla vision, 2022), Seeed Studio XIAO (Seeed XIAO, 2019), STM32 Nucleo-L476RG (nucleo-l476rg, 2016), and Raspberry Pi Pico (raspberry-pi-pico, 2021) (see Fig. 3.11). Table 3.3 shows a benchmark between various Edge AI devices.

The most popular TinyML MCU boards include the Arduino Nano 33 BLE Sense for all-in-one sensor applications, the ESP32 DevKitC for connected IoT ML nodes. The Arduino Portenta and the Arduino Nicla Vision for high-performance vision-based machine learning. The Seeed Studio XIAO nRF52840 Sense is designed for integrated wearable nodes, and the STM32 Nucleo-L476RG board is suitable for battery-powered IoT ML applications.

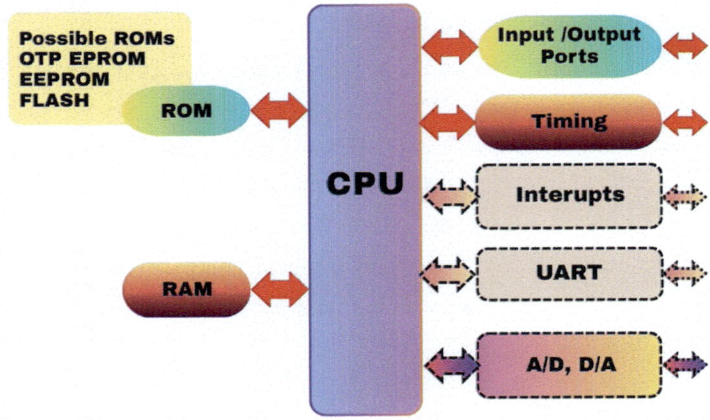

FIGURE 3.10 Basic layout of an MCU. The figure shows the basic layout of a microcontroller unit (*MCU*), including a processor core, memory, and input/output interfaces on a single chip. It illustrates how an MCU integrates computation, storage, and communication to control embedded systems.

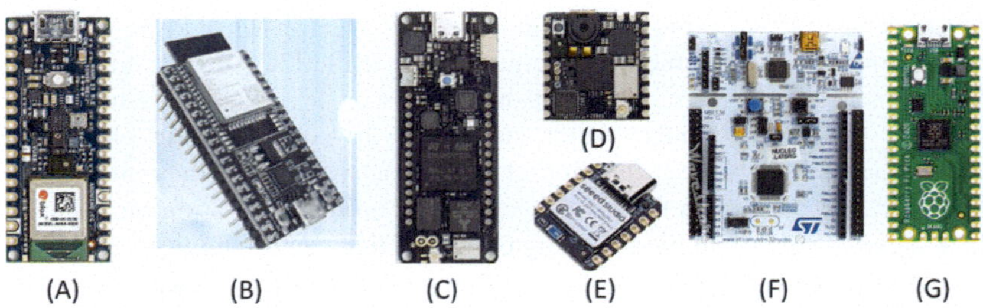

FIGURE 3.11 The figure displays various microcontroller and embedded development boards used in IoT and embedded applications: (A) Nano BLE Sense, (B) ESP32, (C) Portenta H7, (D) Nicla Vision, (E) Seeed Studio XIAO, (F) STM32 Nucleo-L476RG, and (G) Raspberry Pi Pico. It highlights the diversity of platforms available for prototyping, sensing, and edge computing tasks.

Raspberry Pi Pico is the lowest-cost MCU, good for small sensor ML applications (Jayanth et al., 2024). Most of these MCUs are based on Cortex-M cores, which are a family of ARM CPU cores designed for use in MCUs, ASICs, FPGAs, and SoCs (Reay, 2015).

To choose an MCU for TinyML applications, you have to consider computing, memory, sensor, connectivity, power, and cost requirements. For wearables or sensor-intensive devices with BLE technology, boards like the Arduino Nano 33 BLE Sense, ESP32, or the Seeed XIAO nRF52840 Sense are

TABLE 3.3 Most commonly used MCUs for TinyML applications.

Device	MCU	Clock Freq.	RAM	Flash	Onboard sensors	Connectivity	Price (€)
Arduino Nano 33 BLE Sense	Nordic nRF52840	Cortex-M4F @ 64 MHz	256 KB	1 MB	9-axis IMU, mic, temperature, humidity, pressure, color/gesture	BLE	38
ESP32	Espressif ESP32	Dual-core Xtensa LX6 @ 240 MHz	~520 KB SRAM + ext. PSRAM	4–16 MB	None (depends on dev board)	Wi-Fi, BLE	10
Arduino Portenta H7	STM32H747	Cortex-M7 @ 480 MHz + Cortex-M4 @ 240 MHz	1 MB + 8 MB ext. SDRAM	16 MB	None onboard (camera support via shield)	Wi-Fi, BLE	100
Arduino Nicla Vision	STM32H747AII6	Cortex-M7 @ 480 MHz + Cortex-M4 @ 240 MHz	1.5 MB SRAM	2 MB + 16 MB QSPI Flash	Camera, IMU, mic	Wi-Fi, BLE	80
Seeed Studio XIAO nRF52840 Sense	Nordic nRF52840	Cortex-M4F @ 64 MHz	256 KB	1 MB	Mic, 6-axis IMU	BLE	15
STM32 Nucleo-L476RG	STM32L476RG	Cortex-M4F @ 80 MHz	128 KB	1 MB	None onboard	USB (ST-LINK), optional shields for wireless	30
Raspberry Pi Pico	RP2040	Dual-core Cortex-M0+ @ 133 MHz	264 KB	External 2 MB (up to 16 MB)	None onboard	USB, optional external modules	5

MCU, Microcontroller unit.

ideal. In the case of IoT technologies that require wireless or cloud connectivity, the ESP32 or Portenta H7 Connect are suitable. For optical or audio technologies, the Nicla Vision or Portenta H7 provides the necessary CPU and memory. The case of ultralow-power sensor integration, the STM32 Nucleo-L476RG, or the XIAO nRF52840 Sense Excel is suitable. Some versions of ESP32, like the Lolin32 lite and Lolin32 C3 Pico, are equipped with an onboard BMS with a PH2 port to connect a 1-cell Lipo battery, and for ML tasks on a limited budget or simpler, the Raspberry Pi Pico is suitable, although external sensors may be required. In fact, the choice depends on balancing model size, real-time constraints, built-in sensors, connectivity, battery requirements, and overall cost.

Graphical processing units

A graphical processing unit (GPU) is an electronic circuit capable of performing high-speed mathematical calculations. Computing tasks such as graphics rendering, ML, and video editing require applying similar mathematical operations to large datasets. It is a specialized circuit, or single-chip processor, designed to speed up the output of the image into the frame buffer for output to the display (Dally et al., 2021). The GPU specializes in highly parallel computations, so it is designed with more transistors dedicated to processing data rather than buffering and controlling its flow (Cuda-c-programming-guide, 2007). Fig. 3.12 shows an example of the on-chip resource allocation for the CPU compared to the GPU. GPUs from NVIDIA

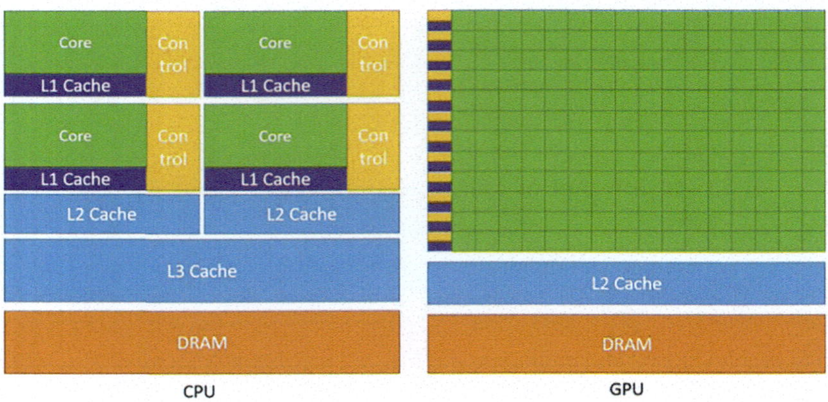

FIGURE 3.12 Distribution of chip resources for a CPU versus a GPU. This figure shows that a CPU dedicates most of its chip area to control logic and large caches with few cores, optimizing for low-latency, sequential tasks. In contrast, a GPU allocates most of its area to many simple cores with smaller caches, maximizing throughput for highly parallel workloads. *Cuda-c-programming-guide (2007). https://docs.nvidia.com/cuda/cuda-c-programming-guide/.*

have many processors, called streaming processors (SP). Each SP is capable of executing a sequential thread.

The main difference between a GPU and a CPU is that a GPU has a highly parallel architecture, making it more efficient than a CPU when used on partitionable data and processing it in parallel. More specifically, a GPU has high optimizations for performing advanced calculations, such as floating-point and matrix calculations, among others (Buber, 2018). In general, an application consists of a combination of parallel and serial components, so systems are designed with a mix of GPUs and CPUs to maximize overall performance. Highly parallel applications can exploit this massive parallelism of the GPU to achieve higher performance than the CPU (Dally et al., 2021). As an example of GPUs employed in Edge AI, we can find Nvidia Jetson developed kits (e.g., Nano, Orin, and Xavier), see Fig. 3.13. The key specifications of this board are listed in Table 3.4.

NVIDIA's parallel computing platform and programming approach, known as computing unified device architecture (CUDA), enables programmers to leverage the massive parallel processing capacity of NVIDIA GPUs for general computing activities beyond graphics (GPU parallel program development using CUDA, 2018). It enables programmers to run highly parallel algorithms (such as deep learning, image processing, and simulation) directly on thousands of GPU cores using languages such as C, C++, Python, and

FIGURE 3.13 Photo of an Nvidia Jetson Nano development kit. This figure shows a photo of the NVIDIA Jetson Nano development kit, a compact single-board computer designed for AI and edge computing, featuring a GPU, CPU, memory, and I/O ports for rapid prototyping of AI applications (*AI*, Artificial intelligence; *GPU*, graphical processing unit; *I/O*, input/output).

TABLE 3.4 NVIDIA Jetson Nano specifications.

GPU	128-core NVIDIA Maxwell GPU
CPU	Quad-core ARM Cortex-A57 @ 1.43 GHz
Memory (RAM)	4 GB LPDDR4 (25.6 GB/s bandwidth)
Storage	microSD card slot (boot & storage)
Connectivity	GB Ethernet (no built-in Wi-Fi/Bluetooth on base kit)
Camera	2-lane or 4-lane MIPI CSI-2 camera interface
Performance	~472 GFLOPS (FP16), suitable for lightweight AI inference
Price	~$99

CPU, Central processing unit; *GPU*, graphical processing unit.

Fortran, resulting in significant CPU performance acceleration. CUDA forms the foundation of edge AI, accelerating training and inference on devices like the NVIDIA Jetson series.

Hardware accelerating edge artificial intelligence

Neural processing units

An NPU is a specialized hardware accelerator designed to efficiently handle the computational demands of AI tasks, particularly those involving artificial neural networks (Lee, 2021). Unlike general-purpose processors such as CPUs and GPUs, NPUs are optimized for the parallel and matrix operations common in AI algorithms, resulting in faster processing speeds and improved energy efficiency for AI workloads. In short, an NPU contains several built-in processors called multiplication and summation units, designed to maximize efficiency in the most common calculations in machine learning. NPUs also have dedicated internal memory that allows them to quickly access model weights and other key AI data points, bypassing the need to retrieve them from RAM. Fig. 3.14 displays the architecture diagram of an NPU (NPU, 2017).

NPUs offer significant advantages for edge AI, providing high-performance, energy-efficient acceleration for deep learning tasks such as image, speech, and natural language processing. They also reduce latency by enabling real-time inference on the device, offloading workloads from CPUs/GPUs, improving overall system performance, and enhancing security by keeping data local. However, NPUs are primarily optimized for inference rather than training, have limited flexibility compared to general-purpose GPUs, and often face compatibility issues and vendor lock-in due to proprietary architectures and software systems. Additionally, their resource constraints, such as limited

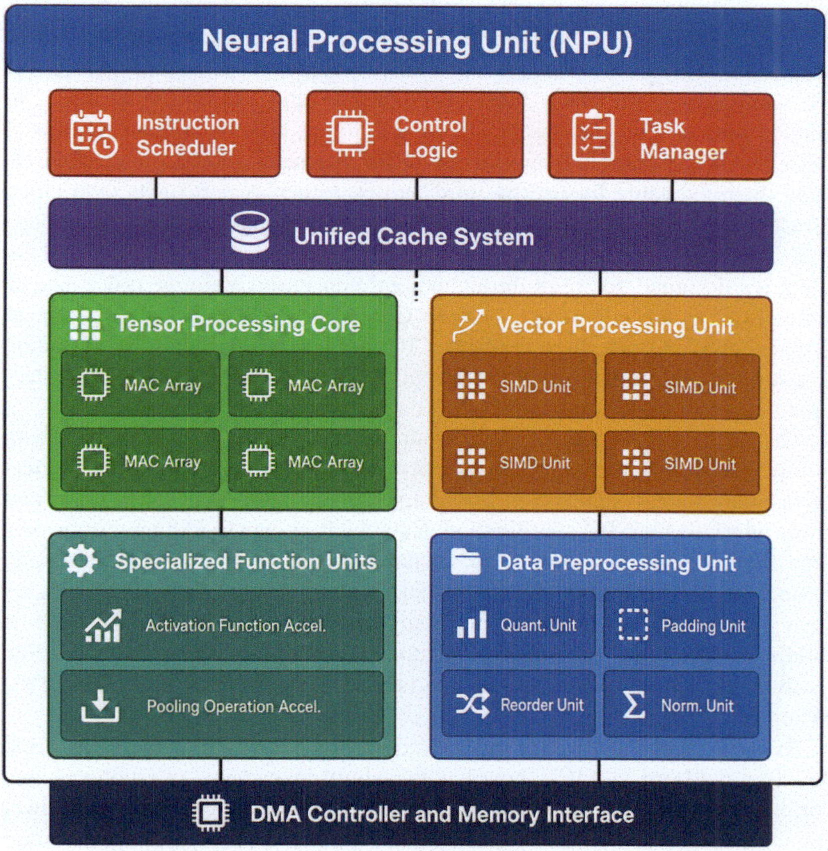

FIGURE 3.14 Architecture diagram of an NPU. This figure illustrates the architecture of a neural processing unit (*NPU*), showing specialized hardware blocks for matrix multiplication, convolution, and activation functions, along with memory buffers and control logic, optimized to accelerate deep learning inference efficiently.
NPU (2017). https://www.utmel.com/blog/categories/integrated%20circuit/neural-processing-unit-npu-explained.

memory bandwidth and reliance on low precision (e.g., INT8/FP16), can limit the deployment of large or complex models.

As an example, the Maixduino dev board is an Arduino-compatible board that supports AI and the IoT. It is powered by the Kendryte K210 chipset, which combines a dual-core RISC-V processor and a hardware NPU for edge AI tasks such as image and speech recognition. The board features Wi-Fi, cameras, and displays, and can be programmed using MaixPy (MicroPython), the Arduino development environment, or the Kendryte software development kit. Designed for TinyML and AIoT projects, this board provides a compact, low-cost platform for real-time edge AI (Sipeed-Maix, 2018).

Tensor processing units

Tensor processing unit (TPU) is an ASIC developed by Google to accelerate ML workloads, particularly those using TensorFlow. TPUs are designed to handle complex calculations in DL models more efficiently than traditional CPUs or GPUs. Google began using TPUs internally in 2015 and, in 2018, made them available for use by third parties, either as part of its own cloud infrastructure or by offering a smaller version of the chip for sale (Carrión and Prohaska, 2023). TPUs are optimized for frameworks like TensorFlow and are primarily designed for high-throughput inference and training tasks in data centers and at the edge (e.g., Google Coral Edge TPU). Taking tensors as virtual inputs and outputs makes the CPU architecture and its corresponding execution model fundamentally different from traditional CPU/GPU architectures that compute on standard vector data pairs (Jouppi et al., 2017).

The CPU architecture is built around a matrix multiplication unit with systolic matrices for parallel tensor operations, supported by on-chip buffers for fast data access, standard vector and vector units for non-matrix tasks, and a host interface for CPU coordination. External high-bandwidth memory stores large weights and activations for models, enabling efficient inference and training for deep learning. Unlike NVIDIA and Intel Edge platforms, Google's Edge TPU cannot run models other than TensorFlow at the edge. Google is building a seamless pipeline that automates and simplifies the workflow involved in training models in the cloud and deploying them to the Edge TPU (Wang et al., 2020).

As an example of an edge AI TPU development board, depicts the Coral Dev Board Micro, a MCU board with a built-in camera, microphone, and Coral Edge TPU, allowing you to quickly prototype and deploy low-power ESs with on-device ML inference. Table 3.5 shows the technical specification of this board (Coral dev board datasheet, 2020).

Field-programmable gate arrays

FPGAs are used in edge AI to accelerate AI computations in resource-constrained environments, offering a balance between performance and power efficiency. FPGAs are ideally suited for edge AI, combining high performance, low latency, power efficiency, and flexible hardware reconfiguration. Unlike fixed-function accelerators such as TPUs and NPUs, FPGAs can be customized to fit specific neural network architectures, enabling optimized data flows and enhanced parallelism for tasks such as image recognition, sensor fusion, and real-time analytics (Sipola et al., 2022). This adaptability makes them valuable in diverse edge applications, including robotics, autonomous vehicles, smart cameras, and the IIoT, where workloads and requirements often evolve over time (Pham-Quoc, 2023).

TABLE 3.5 Coral Dev board micro specification.

MCU	NXP i.MX RT1176 (Cortex-M7 and Cortex-M4)
ML accelerator	Coral Edge TPU coprocessor: 4 TOPS (int8); 2 TOPS per watt
RAM	64 MB
Flash memory	128 MiB
Sensors	Himax color camera (324 x 324 px); PDM mono microphone
Input/output	(2x) 12-pin GPIO header; (2x) 100-pin board-to-board connectors; USB Type-C (USB 2.0)
User interface	(4x) LEDs; (2x) switch buttons
Accessories	Coral Wi-Fi/Bluetooth Module Board; Coral Ethernet/POE Board
MCU	NXP i.MX RT1176 (Cortex-M7 and Cortex-M4)

A good example is Lattice Semiconductor's iCE40 UltraPlus FPGA (Lattice Intelligent Edge AI and FPGA-Solutions, 2018), which is well-suited for low-power, always-on edge AI applications such as pattern matching and AI inference. Another example is AMD's Versal AI Edge series (Adaptive socs and FPGAs, 2024), which combines FPGA logic, ARM cores, and AI cores. The flexibility of the Versal architecture lies in its programmable logic, which enables the integration of any sensor, connectivity to any interface, and the flexibility to handle any workload. Thanks to its parallelism and determinism, the programmable logic can implement sensor fusion and adaptation algorithms, accelerate pre- and post-pipeline data processing, implement deterministic networks and motor control for immediate response, isolate safety-critical functions to ensure safe operation, and enable hardware redundancy and failure tolerance.

FPGAs are revolutionizing the deployment of AI at the edge, offering unparalleled flexibility, low latency, and high-power efficiency. As AI applications at the edge continue to expand, the ability to leverage FPGA accelerators will become a critical skill for embedded systems engineers. While challenges remain, advances in FPGA technology and development tools are making it easier than ever to leverage their potential (Chi et al., 2024; Li et al., 2020; Li and Li, 2025).

Platforms and frameworks

Several software platforms are available for developing and deploying AI solutions at the edge. These platforms simplify the process of building, training, and deploying machine learning models on edge devices, from MCUs to powerful edge servers. Key frameworks and tools facilitating Edge AI development include TensorFlow Lite, PyTorch Mobile, OpenVINO, and more. These frameworks are critical for creating and deploying lightweight models that can operate efficiently with limited resources. They also ensure efficient inference and learning processes in resource-constrained environments by optimizing model size, complexity, and computational requirements (Jouini et al., 2024).

TensorFlow Lite

TensorFlow is an open-source framework for ML and DL applications, originally developed by Google Brain (Tensorflow, 2017). It provides a flexible architecture for numerical computation based on dataflow graphs, where nodes represent operations and edges represent tensors. TensorFlow supports both low-level search operations and high-level APIs, such as Keras, for rapid prototyping. Thanks to its scalability across CPUs, GPUs, and GPUs, as well as its ability to be deployed on servers, mobile devices, and embedded systems, TensorFlow is one of the most widely used platforms for AI applications, including computer vision, natural language processing, and predictive analytics.

While TensorFlow Lite (Tinyml: Machine learning with tensorflow lite on arduino and ultra-low-power microcontrollers, 2019) is an open-source, lightweight version of TensorFlow designed also for deploying pre-trained ML models on mobile, embedded, and edge devices, offering low-latency and low-power inference through model optimization and hardware acceleration. TensorFlow Lite supports various optimizations, including model quantization, which reduces model size and inference time without significant loss of accuracy, making it well-suited for resource-constrained environments. Fig. 3.15 illustrates TensorFlow Lite workflow (Top 8 TinyML Frameworks and Compatible Hardware Platforms, 2024). The key advantages of TensorFlow Lite are: fast inference, flexibility, and ease of implementation.

PyTorch Mobile

PyTorch Mobile is a lightweight runtime for the PyTorch framework that enables machine learning models to be deployed on mobile and edge devices, supporting efficient inference through model optimization, quantization, and hardware acceleration on Android and iOS platforms (Welcome to the ExecuTorch Documentation, 2025). Thanks to its flexibility, ease of use, and compatibility with the broader PyTorch ecosystem, PyTorch Mobile enables developers to build and deploy a wide range of AI-powered applications directly on edge devices, from pattern recognition to robotics and augmented reality. Fig. 3.16 illustrates the PyTorch pipeline (Pytorch, 2016).

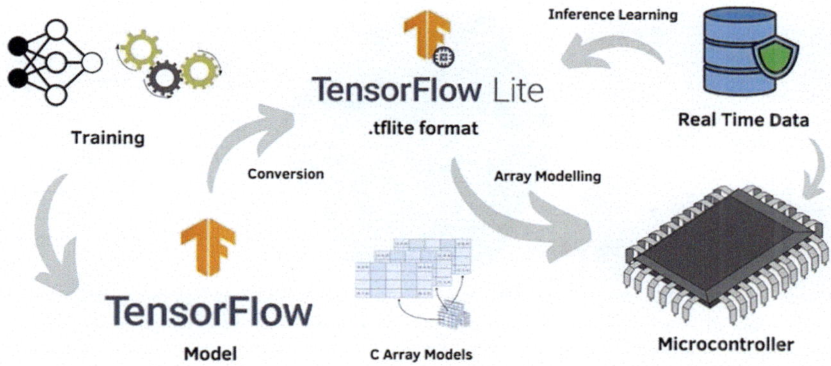

FIGURE 3.15 TensorFlow Lite workflow. This figure illustrates the TensorFlow Lite workflow, showing how a trained TensorFlow model is converted to a lightweight TFLite format, then deployed on edge devices where it can perform inference efficiently using the TensorFlow Lite interpreter or delegate accelerators like NPUs or Edge TPUs (*NPU*, Natural processing unit; *TPU*, tensor processing unit).
Tensorflow (2017). https://www.tensorflow.org/api_docs/python/tf/lite.

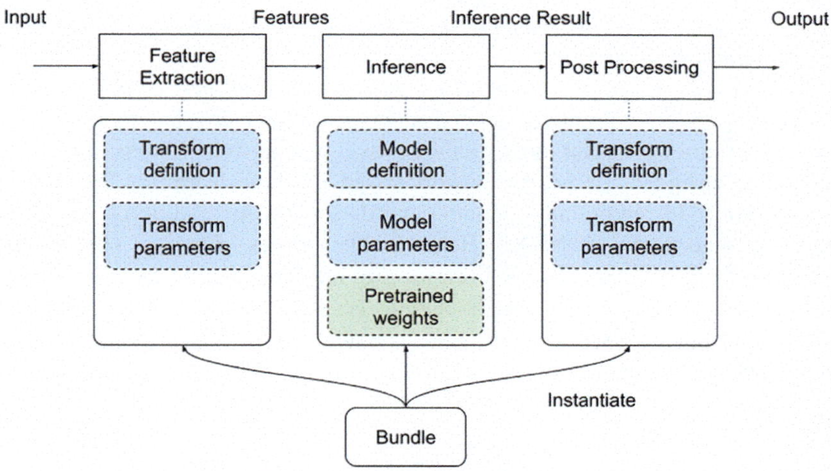

FIGURE 3.16 Pytorch pipeline. This figure shows the PyTorch pipeline, illustrating how data is loaded and preprocessed, fed into a neural network model for training or inference, and then optimized or deployed for edge or server-based applications using PyTorch's runtime and tools.

Open neural network exchange

The ONNX (ONNX, 2017) is an open-source standard format developed in collaboration between Microsoft and Meta, and supported by Huawei, IBM, AMD, ARM, Intel, and Qualcomm (See Fig. 3.17) (ONNX: Helping

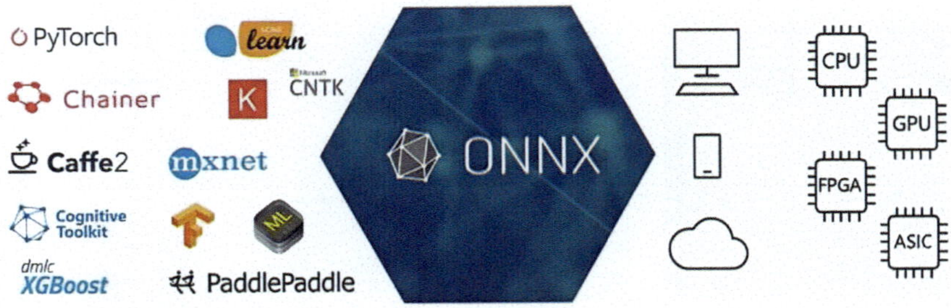

FIGURE 3.17 ONNX configuration. This figure illustrates the ONNX configuration, showing how models from different frameworks (like PyTorch or TensorFlow) are exported to the ONNX format for interoperability, then loaded and executed on various runtimes or hardware accelerators for inference.
ONNX: Helping Developers Choose the Right Framework (2019). https://www.opensourceforu.com/2019/06/onnx-helping-developers-choose-the-right-framework/.

Developers Choose the Right Framework, 2019). It allows developers to seamlessly exchange models between frameworks such as TensorFlow, PyTorch, and MXNet without the need for extensive model transformation or retraining (Shankar, 2024).

With a wide range of frameworks, transformers, runtimes, compilers, and visualizers, ONNX eliminates dependencies and allows developers to choose their own frameworks and tools. ONNX is available in two versions: the basic version is designed for neural network-based machine learning. The other, ONNX-ML, adds additional operators and data types for traditional machine learning algorithms (ONNX: Helping Developers Choose the Right Framework, 2019). The ONNX libraries provide functionality for converting models from various frameworks, including TensorFlow and PyTorch, into the ONNX format. These libraries are available in multiple programming languages, such as Python, C++, and C#.

Open visual inference and neural network optimization

OpenVINO is an open-source toolkit developed by Intel that enables the optimization and deployment of DL models for high-performance inference on Intel hardware, including CPUs, GPUs, and FPGAs (See Fig. 3.18). This toolkit supports models from various frameworks, such as TensorFlow, PyTorch, and ONNX, providing tools for model transformation, acceleration, and deployment in CV and AI applications. It aims to improve performance, reduce model footprint, and streamline the development and deployment of AI solutions. OpenVINO enables developers to take full advantage of Intel technology to develop and implement edge AI applications across a variety of

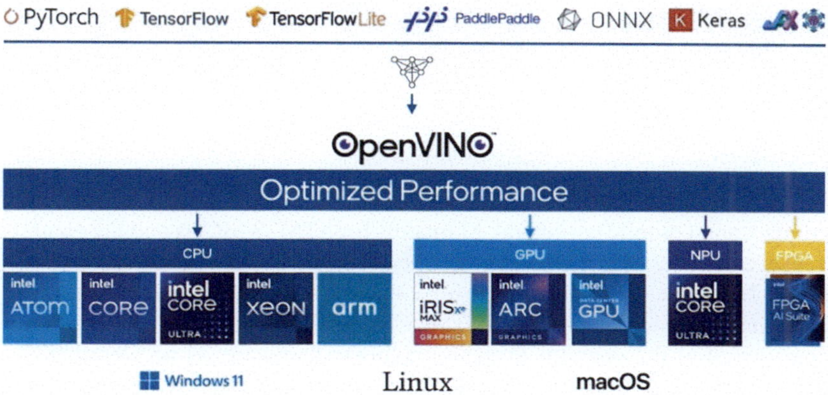

FIGURE 3.18 OpenVINO architecture. This figure shows the OpenVINO architecture, depicting how a trained deep learning model is optimized and converted using the OpenVINO toolkit, then deployed on Intel hardware (CPU, GPU, VPU: vision processing unit, or FPGA) through the Inference Engine for efficient edge or server-side AI inference. (*AI*, Artificial intelligence; *GPU*, graphical processing unit; *FPGA*, field-programmable gate arrays; *OpenVINO*, Open visual inference and neural network optimization).
OpenVino 2025.4. (2025). https://docs.openvino.ai/2025/index.html.

industries by focusing on optimization, flexibility, and cross-platform compatibility (Shankar, 2024).

Edge impulse

Edge impulse is a development platform for building, training, and deploying machine learning models, specifically designed for edge devices and embedded systems. It provides tools for data collection, model creation, optimization, and deployment, enabling real-time AI inference on MCUs, sensors, and other resource-constrained devices (Edge Impulse, 2019; AI at the Edge, 2023). The main parts are to (1) collect a dataset, (2) train the model and create an impulse, (3) validate and test the model, and (4) model deployment on an edge device. Thus the edge impulse platform bridges the gap between data collection, model training, and deployment on resource-constrained devices, making AI practical at the edge.

Edge Impulse supports a wide range of devices for TinyML and edge AI applications, including MCUs such as the Arduino Nano 33 BLE Sense, STM32 Nucleo boards, the Nordic nRF52 series, and the Kendryte K210; single-board computers such as the Raspberry Pi, BeagleBone, and Google Coral Dev Board; and AI accelerators such as the Coral USB Accelerator and Intel Neural Compute Stick 2. It also integrates with numerous sensors and edge devices, cameras, microphones, and IMUs, enabling real-time, low-power AI inference across diverse edge and embedded platforms (Hymel et al., 2022).

X-CUBE-AI

X-CUBE-AI is an STM32Cube expansion package designed to evaluate, optimize, and compile peripheral AI models for STM32 MCUs and the Neural-ART accelerator. When optimizing NN models for the Neural-ART accelerator's NPU, the tool generates precise code that maps AI operations to the NPU whenever possible and back to the CPU when necessary. This scheduling is performed at the operator level to maximize hardware AI acceleration. X-CUBE-AI is part of the STM32Cube.AI ecosystem and extends the capabilities of STM32CubeMX by automatically converting pretrained AI algorithms to C code. A generated optimizer library is also integrated into the user's project (AI expansion pack for STM32CubeMX, 2018) (See Fig. 3.19).

X-CUBE-AI supports models trained using TensorFlow, Keras, PyTorch, Caffe, and others. Regardless of the file type used to train the model, the model file must be in Keras (.h5), TensorFlow Lite (.tflite), or ONNX (.onnx) format (TinyML: Getting Started with STM32 X-CUBE-AI, 2020).

The ST Edge AI Suite supports a variety of STMicroelectronics products: STM32 MCUs and microprocessors, Neural-ART accelerators, Stellar MCUs, and smart sensors. It represents a strategic step toward popularizing edge AI technology, making it an essential resource for developers looking to harness the power of AI in embedded systems efficiently and effectively.

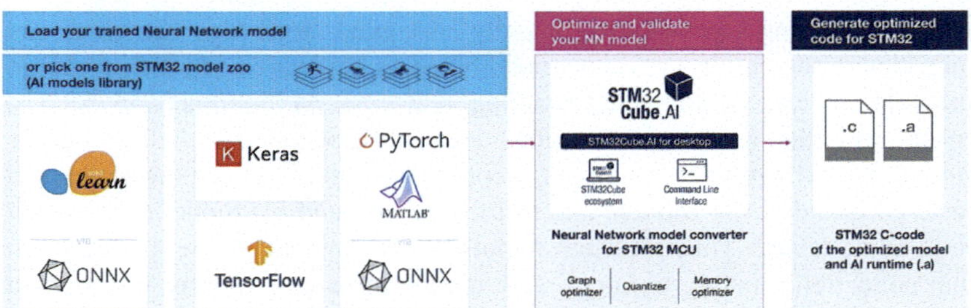

FIGURE 3.19 Steps to design and deploy TinyML into STM32 devices. This figure illustrates the steps to design and deploy a TinyML application on STM32 devices, showing the workflow from data collection and preprocessing to model training and conversion, and finally deployment onto the STM32 microcontroller for on-device inference.

STM32Cube.AI Developer Cloud generates AI workloads for STM32 microcontrollers. (2025). https://www.cnx-software.com/2023/01/31/stm32cube-ai-developer-cloud-generates-ai-workloads-for-stm32-microcontrollers/.

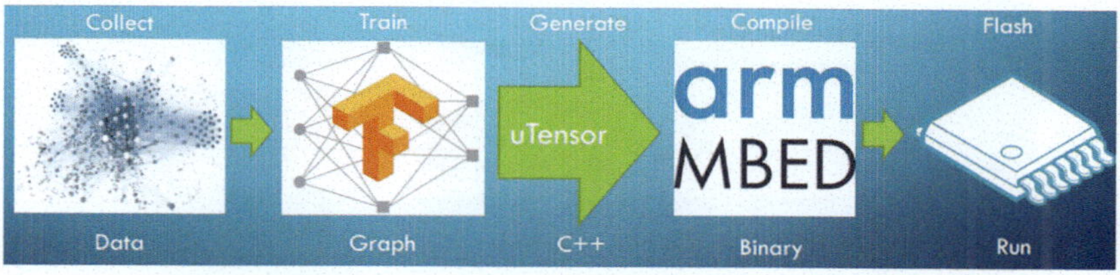

FIGURE 3.20 uTensor workflow. The figure illustrates the uTensor workflow, showing how a trained neural network model is converted into a lightweight format, integrated into an embedded application, and deployed onto microcontrollers to perform on-device inference with minimal memory and compute requirements.
uTensor (2018). https://github.com/uTensor/uTensor.

uTensor

uTensor is an open-source, lightweight framework for machine learning inference, optimized for MCUs, particularly those based on the ARM Cortex-M architecture. This framework enables the deployment of ML models on resource-constrained ESs by converting trained models into efficient C++ code (uTensor, 2018). The basic workflow of uTensor is shown in Fig. 3.20. It consists of four key steps: (1) collect data, (2) train the model using TensorFlow, (3) generate the C++, and (4) upload the code into ARM Mebd for real-time applications.

It runs on Mbed-compatible boards, such as the STM32 Nucleo/Discovery, Nordic nRF52, and NXP FRDM, making it suitable for TinyML applications on low-power devices. uTensor helps convert the model for proper deployment on Mbed, ST, and K64 boards. uTensor is a small module that requires only 2 KB of disk space (Top 8 TinyML Frameworks and Compatible Hardware Platforms, 2024).

Embedded learning library

Microsoft developed the ELL library to support the TinyML-blended learning environment. The library provides support for the Raspberry Pi, Arduino, and micro_bit platforms. The models used on these devices are not Internet-based and therefore do not require cloud access. The library currently supports image and audio classification. It also provides a set of programming tools and an optional Python interface, written in modern C++ (ELL, 2017). Fig. 3.21 displays the ELL workflow.

Compatible hardware platforms

The described edge AI frameworks can be classified into three groups: (1) MCU-focused TinyML, (2) mobile general edge AI, and (3) accelerator-

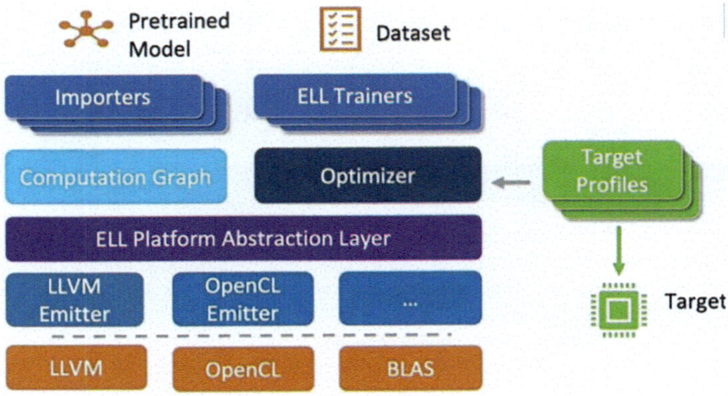

FIGURE 3.21 ELL workflow. This figure illustrates the embedded learning kibrary (*ELL*) workflow, showing how a trained machine learning model is imported, compiled, and optimized for deployment, then executed on embedded devices for efficient on-device inference.
ELL (2017). https://microsoft.github.io/ELL/.

industrial edge AI. Table 3.6 lists different edge AI frameworks, the supported hardware platforms, potential applications, and complexity level. The complexity was classified into three levels (low, medium, and high), based on ease of use of graphical tools, coding requirements, complex toolchains, and specialized devices.

MCU-specific frameworks are specifically designed for resource-constrained MCUs. These frameworks boast low power consumption, making them ideal for real-time sensor data processing. Their advantages include ease of deployment, ease of integration with embedded devices, and broad support for common MCUs. Limitations include less flexibility for complex models, limited support for large neural networks, and sometimes steeper learning curves for programming and optimization. Mobile general-purpose edge AI frameworks are specifically designed for smartphones, Raspberry Pi, and ARM Cortex-A devices, supporting more complex models and a wider range of applications, such as NLP, audio, and video tasks.

Advantages include flexibility, support for multiple frameworks, and relatively high model capacity. Limitations include higher computational requirements, power consumption, and the need for additional deployment effort on embedded systems compared to MCU-specific frameworks. Industrial accelerators-frameworks leverage dedicated hardware (such as Intel CPUs, GPUs, and FPGAs) to enable high-performance AI applications at the edge, such as industrial vision, robotics, and IoT analytics. Benefits include improved inference, hardware acceleration, and support for complex models. Limitations also include greater configuration complexity, hardware dependency, and reduced portability on very low-end devices.

TABLE 3.6 Edge AI frameworks with the corresponding hardware.

TinyML frameworks	Hardware platforms	Applications	Complexity
MCU-focused TinyML			
Edge impulse	Espressif ESP32, Arduino Nano 33 BLE Sense, Nicla Vision, Arduino Arduino Nicla Sense ME, H7 + Vision Shield Arduino Portenta, Nordic Semi Nrf52840 DK, Himax WE-I Plus, Nordic Semi Nrf5340 DK.	End-to-end pipeline for audio for vision, and sensor fusion (Classification, object detection, anomaly, pattern recognition, and speech)	Low
TensorFlow Lite	STM32F746, Arduino Nano 33 BLE Sense, Discovery Kit, Adafruit TensorFlow Lite for Microcontrollers Kit, Sparkfun Edge, DFRobot firebeetle ESP32, Espressif ESP-EYE, ARM Cortex-M MCUs, Raspberry Pi Pico.	Gesture recognition, image and audio classification, speech and gesture recognition, keyword spotting, optical character recognition, object detection, anomaly detection, video and text classification, segmentation, and pattern recognition.	Medium
uTensor	ARM Cortex-M (Mbed OS boards: STM32 Nucleo, Nordic nRF52, NXP FRDM)	Anomaly detection, IoT sensing, image classification. Image, gesture recognition, acoustic detection, and Motion Analysis.	Medium
X-CUBE-AI	STMicroelectronics STM32 MCUs (Cortex-M series)	Converting pretrained networks into optimized STM32 code for TinyML applications (Condition monitoring, anomaly detection, predictive maintenance, tracking people, counting, asset, and activity recognition.	Medium

(*Continued*)

TABLE 3.6 (Continued)

TinyML frameworks	Hardware platforms	Applications	Complexity
Mobile General Edge AI			
Pytorch mobile	Supports ARM64 processors, Android, and iOS devices.	Speech recognition, mobile vision, custom ML apps and NLP.	High
ONNX	Android, iOS, Linux-based edge devices, ARM64, x86.	Cross-framework deployment (TF, PyTorch, and Scikit-learn models in ONNX)	Medium
ELL	Raspberry Pi, Arduino, micro-bit.	Audio & vision classification, predictive IoT, and keyword spotting.	High
Accelerator-Industrial Edge AI			
OpenVINO	Intel CPUs, FPGAs, VPUs (Myriad X), Neural Compute Stick (NCS2), GPUs, and Raspberry Pi.	Edge AI acceleration, computer vision, and industrial IoT.	Medium

AI, Artificial intelligence; *ELL*, embedded learning library; *MCU*, microcontroller unit.

Benefits, challenges, and future trends of edge artificial intelligence

The popularity of edge AI has increased with the increase in computing power and energy efficiency of embedded processors. Edge AI has many advantages, mainly due to its ability to process data locally on devices at the edge of the network (i.e., near where the data is generated). This approach reduces latency, lowers costs, improves security, and enhances scalability. By processing data locally, edge AI reduces the need to transfer large amounts of data to a central server (cloud), resulting in faster response times, lower bandwidth consumption, and reduced dependence on cloud infrastructure (Meuser et al., 2024). Reducing dependence on network connectivity. Edge AI can be used in applications where cloud access is not possible, reducing dependence on the internet and potential downtime caused by network outages. Edge AI improves energy efficiency in battery-powered or resource-constrained devices and ensures reliable operation even without a network connection. In addition, it supports context-aware intelligence, allowing devices to adapt to local conditions and user contexts, while enabling scalable AI solutions that can be distributed across device fleets (Meuser et al., 2024).

Edge AI performs calculations directly on local devices rather than relying on cloud servers, and faces many technical challenges. Edge devices typically have limited computing resources and memory, requiring efficient, lightweight algorithms, and optimized models. Energy efficiency is critical, especially for battery-powered devices, to ensure continuous operation without excessive power consumption. Bandwidth limitations and intermittent connectivity make local processing necessary to minimize data transfer. Edge AI must also address data privacy and security, as sensitive information is processed on the device. Applications such as autonomous vehicles and industrial automation require real-time, low-latency inferences, while devices must remain adaptable to dynamic environments with changing conditions. Techniques such as model compression, quantization, and pruning help tailor AI models to constrained devices. In addition, distributed device management requires effective lifecycle management and over-the-air updates to maintain model performance and reliability (Gill et al., 2025).

Current trends in edge AI focus on delivering efficient, low-latency intelligence that preserves privacy directly on devices. This includes the emergence of TinyML for ultralow-power AI on MCUs, and the integration of specialized hardware accelerators such as TPUs, NPUs, GPUs, and FPGAs to accelerate inference while minimizing power consumption (Hao et al., 2021). Advanced model optimization techniques such as quantization, pruning, and distillation are increasingly being used to make DL models lightweight for deployment at the edge. Federated and collaborative learning enables models to be trained on devices while maintaining data privacy, and real-time, context-aware AI is being applied in autonomous vehicles, robotics, and industrial automation (Fortino et al., 2021). In addition, edge AI benefits from 5G connectivity and hybrid edge-cloud architectures, AI-powered edge analytics, and cross-platform frameworks such as TensorFlow Lite, PyTorch Mobile, Edge Impulse, and OpenVINO, enabling seamless deployment across diverse devices. Advances in device design, battery technology, and network connectivity are essential to realizing the full potential of edge intelligence.

Conclusion

Edge AI represents a paradigm shift in how intelligent systems are deployed and operated, enabling real-time decision-making directly on resource-constrained devices. This chapter reviewed the fundamental computing paradigms (cloud, fog, and edge) and highlighted the unique advantages of edge computing for AI applications. It also explored a wide range of edge devices, from MCUs and microprocessors to advanced accelerators such as GPUs, TPUs, and FPGAs, each offering distinct capabilities for embedded intelligence.

Furthermore, the chapter examined key software frameworks and platforms that facilitate the deployment of AI models on edge devices, emphasizing their role in optimizing performance and scalability. As edge AI continues to evolve, its integration into devices will unlock new possibilities across various sectors, but it will also present challenges in energy efficiency, interoperability, and security. Addressing these challenges will be crucial in shaping the future of decentralized intelligent systems.

AI Disclosure

During the preparation of this work, the author(s) used ChatGPT in order to do proofreading. After using this tool/service, the author(s) reviewed and edited the content as needed and take(s) full responsibility for the content of the publication.

References

Adaptive socs and FPGAs, 2024. https://www.amd.com/en/products/adaptive-socs-and-fpgas/versal/ai-edge-series.html.

AI at the Edge. O'Reilly Media, Inc., (2023).

AI expansion pack for STM32CubeMX, 2018. https://www.st.com/en/embedded-software/x-cube-ai.html.

Arduino nano family, 2020. https://store.arduino.cc/collections/nano-family/products/nano-33-ble-rev2.

Bonomi, F., Milito, R., Zhu, J., Addepalli, S., 2022. Fog computing and its role in the internet of things. In Proceedings of the first edition of the MCC workshop on Mobile cloud computing. 13–16.

Buber, E., 2018. Performance analysis and CPU vs GPU comparison for deep learning. In 2018 6th International Conference on Control Engineering & Information Technology (CEIT) IEEE. 1–6.

Carrión, D.S., Prohaska, V., 2023. Exploration of TPUS for AI Applications. arXiv, Spain arXiv. Available at https://arxiv.org, 10.48550/arXiv.2309.08918.

Chi, T.K., Chen, T.Y., Lin, Y.C., Lin, T.L., Zhang, J.T., Lu, C.L., et al., 2024. An edge computing system with AMD Xilinx FPGA AI customer platform for advanced driver assistance system. Sensors 24 (10). Available at https://doi.org/10.3390/s24103098, http://www.mdpi.com/journal/sensors.

Computer Organization and Design ARM Edition: The Hardware Software Interface, 2016. 6, Morgan kaufmann, Morgan kaufmann.

Coral dev board datasheet, 2020. https://coral.ai/static/files/Coral-Dev-Board-datasheet.pdf.

Cuda-c-programming-guide, 2007. https://docs.nvidia.com/cuda/cuda-c-programming-guide/.

Dally, W.J., Keckler, S.W., Kirk, D.B., 2021. Evolution of the graphics processing unit (GPU). IEEE Micro 41 (6), 42–51. Available at https://doi.org/10.1109/MM.2021.3113475, https://ieeexplore.ieee.org/servlet/opac?punumber=40.

Dolui, K., Datta, S.K., 2017. Comparison of edge computing implementations: Fog computing, cloudlet and mobile edge computing. In 2017 Global Internet of Things Summit (GIoTS) IEEE. pp. 1–6.

Edge AI market report, 2025. https://www.gminsights.com/industry-analysis/edge-ai-market.

Edge Impulse, 2019. https://edgeimpulse.com/.

ELL, 2017. https://microsoft.github.io/ELL/.

ESP32, 2016. https://www.espressif.com/en/products/devkits/esp32-devkitc/overview.

Fortino, G., Zhou, M.C., Hassan, M.M., Pathan, M., Karnouskos, S., 2021. Pushing Artificial Intelligence to the edge: emerging trends, issues and challenges. Engineering Applications of Artificial Intelligence 103, 104298. Available at https://doi.org/10.1016/j.engappai.2021.104298.

Gill, S.S., Golec, M., Hu, J., Xu, M., Du, J., Wu, H., et al., 2025. Edge AI: a taxonomy, systematic review and future directions. Cluster Computing 28 (1). Available at https://doi.org/10.1007/s10586-024-04686-y, https://www.springer.com/journal/10586.

GPU parallel program development using CUDA, 2018. CRC Press.

Gridling, G. 2007. Introduction to microcontrollers. Vienna University of Technology Institute of Computer Engineering Embedded Computing Systems Group, 25.

Hao, C., Dotzel, J., Xiong, J., Benini, L., Zhang, Z., Chen, D., 2021. Enabling design methodologies and future trends for edge AI: specialization and codesign. IEEE Design and Test 38 (4), 7–26.

Hymel, S., Banbury, C., Situnayake, D., Elium, A., Ward, C., Kelcey, M. et al., 2022. Edge impulse: An Mlops platform for tiny machine learning. arXiv, United States. Available from https://arxiv.org.

Jayanth, R., Gupta, N., & Prasanna, V. 2024. Benchmarking Edge AI Platforms for High-Performance ML Inference IEEE High Performance Extreme Computing Conference, HPEC 2024. 10.1109/HPEC62836.2024.10938499, http://ieeexplore.ieee.org/xpl/mostRecentIssue.jsp?punumber=10938401.

Jouini, O., Sethom, K., Namoun, A., Aljohani, N., Alanazi, M.H., Alanazi, M.N., 2024. A survey of machine learning in edge computing: techniques, frameworks, applications, issues, and research directions. Technologies 12 (6). Available at https://doi.org/10.3390/technologies12060081, https://www.mdpi.com/journal/technologies.

Jouppi, N.P., Young, C., Patil, N., Patterson, D., Agrawal, G., Bajwa, R., et al., 2017. In-datacenter performance analysis of a tensor processing unit Proceedings - International Symposium on Computer Architecture, 28643, 1–121. 10.1145/3079856.3080246.

Lattice Intelligent Edge AI and FPGA-Solutions, 2018. https://www.latticesemi.com/Products/Lattice-Intelligent-Edge-AI-and-FPGA-Solutions.

Lee, K.J., 2021. Architecture of neural processing unit for deep neural networks. Advances in Computers 122, 217–245. Available at https://doi.org/10.1016/bs.adcom.2020.11.001, http://www.elsevier.com/wps/find/bookdescription.cws_home/705045/description#description.

Li, R., Li, Z. 2025. Dataflow & Tiling strategies in edge-AI FPGA accelerators: a comprehensive literature review. arXiv, United States. Available from: https://arxiv.org, 10.48550/arXiv.2505.08992.

Li, Z., Zhang, Y., Wang, J., Lai, J., 2020. A survey of FPGA design for AI era. Journal of Semiconductors 41 (2), 021402. Available at https://doi.org/10.1088/1674-4926/41/2/021402.

Meuser, T., Loven, L., Bhuyan, M., Patil, S.G., Dustdar, S., Aral, A., et al., 2024. Revisiting edge AI: opportunities and challenges. IEEE Internet Computing 28 (4), 49–59. Available at https://doi.org/10.1109/MIC.2024.3383758, https://ieeexplore.ieee.org/servlet/opac?punumber=4236.

Nicla vision, 2022. https://store.arduino.cc/collections/nicla-family.

NPU, 2017. https://www.utmel.com/blog/categories/integrated%20circuit/neural-processing-unit-npu-explained.

nucleo-l476rg, 2016. https://www.st.com/en/evaluation-tools/nucleo-l476rg.html.

ONNX, 2017. https://onnx.ai/onnx/intro/.

ONNX: Helping Developers Choose the Right Framework, 2019. https://www.opensourceforu.com/2019/06/onnx-helping-developers-choose-the-right-framework/.

Pham-Quoc, C., 2023. FPGA/AI-powered data security for IoT edge computing platforms: a survey and open issues. In International Conference on Intelligence of Things. Cham: Springer Nature Switzerland. pp. 3–14.

Portenta family, 2020. https://store.arduino.cc/collections/portenta-family.

Pytorch, 2016. https://docs.pytorch.org/audio/main/pipelines.html.

Raghunathan, K.R., 2021. History of microcontrollers: First 50 years. IEEE Computer Society, undefined IEEE Micro 41 (6), 97–104, Available from: https://ieeexplore.ieee.org/servlet/opac?punumber=40, 10.1109/MM.2021.3114754.

Raspberry Pi 5, 2023. https://www.raspberrypi.com/products/raspberry-pi-5/.

Raspberry-pi-pico, 2021. https://www.raspberrypi.com/products/raspberry-pi-pico/.

Reay, D.S., 2015. Digital Signal Processing Using the ARM® CORTEX®-M4. Wiley. Available at 10.1002/9781119078227.

Satyanarayanan, M., Bahl, P., Caceres, R., Davies, N., 2009. The case for VM-based cloudlets in mobile computing. IEEE Pervasive Computing 8 (4), 14–23. Available at https://doi.org/10.1109/MPRV.2009.82.

Seeed XIAO, 2019. https://www.seeedstudio.com/XIAO-nRF52840.html.

Shankar, V., 2024. Edge AI: A Comprehensive Survey of Technologies, Applications, and Challenges 1st International Conference on Advanced Computing and Emerging Technologies, ACET 2024 10.1109/ACET61898.2024.10730112, http://ieeexplore.ieee.org/xpl/mostRecentIssue.jsp?punumber=10729826.

Shi, W., Cao, J., Zhang, Q., Li, Y., Xu, L., 2016. Edge computing: vision and challenges. IEEE Internet of Things Journal 3 (5), 637–646. Available at https://doi.org/10.1109/JIOT.2016.2579198, http://ieeexplore.ieee.org/servlet/opac?punumber=6488907.

Singh, R., Gill, S.S., 2023. Edge AI: a survey. Internet of Things and Cyber-Physical Systems 3, 71–92. Available at https://doi.org/10.1016/j.iotcps.2023.02.004, https://www.sciencedirect.com/science/journal/26673452.

Sipeed-Maix, 2018. https://www.seeedstudio.com/sipeed-maix.html?srsltid=AfmBOooCS5RZywnHCjZjs0dZvxZnEgjrnooDwgvZpMpwIg_Vrb59dDUX.

Sipola, T., Alatalo, J., Kokkonen, T., Rantonen, M., 2022. Artificial Intelligence in the IoT Era: A Review of Edge AI Hardware and Software Conference of Open Innovation Association, FRUCT 320-331 http://ieeexplore.ieee.org/xpl/mostRecentIssue.jsp?punumber=6731008, 10.23919/FRUCT54823.2022.9770931.

Soro, S., 2021. TinyML for ubiquitous edge AI. arXiv, undefined arXiv. Available from: https://arxiv.org, 10.48550/arxiv.2102.01255.

Steadman, P., Rathore, P.G.R.S., Hewage, C., 2024. Challenges in Implementing Artificial Intelligence on the Raspberry Pi 4, 5 and 5. In Contributions Presented at The International Conference on Computing, Communication, Cybersecurity and AI, July 3–4, 2024, London, UK: The C3AI 2024 Springer Nature. 884.

Tank, B., Gandhi, V., 2023. A comparative study on cloud computing, edge computing and fog computing. In Recent Developments in Electronics and Communication Systems. pp. 665–670.

Tensorflow, 2017. https://www.tensorflow.org/api_docs/python/tf/lite.

TinyML: Getting Started with STM32 X-CUBE-AI, 2020. https://www.digikey.com/en/maker/projects/tinyml-getting-started-with-stm32-x-cube-ai/f94e1c8bfc1e4b6291d0f672d780d2c0.

Tinyml: Machine learning with tensorflow lite on arduino and ultra-low-power microcontrollers, 2019. O'Reilly Media.

Top 8 TinyML Frameworks and Compatible Hardware Platforms, 2024. https://www.dfrobot.com/blog-13921.html?srsltid=AfmBOoqzdMqqrOLg1HYA3yFa0uTO-yJz1dZ18I4_h4MzwjKmMd9_QJsu.

uTensor, 2018. https://github.com/uTensor/uTensor.

Wang, L., Tao, J., Kunze, M., Castellanos, A.C., Kramer, D., Karl, W., 2008. September. Scientific cloud computing: Early definition and experience. In 2008 10th ieee international conference on high performance computing and communications, pp. 825–830.

Wang, X., Han, Y., Leung, V.C.M., Niyato, D., Yan, X., Chen, X., 2020. Edge AI: Convergence of Edge Computing and Artificial Intelligence Edge AI: Convergence of Edge Computing and Artificial Intelligence. Springer Singapore, China, pp. 1–149. Available at https://link.springer.com/book/10.1007/978-981-15-6186-3, 10.1007/978-981-15-6186-3.

Welcome to the ExecuTorch Documentation, 2025. https://docs.pytorch.org/executorch/stable/index.html.

What Is Cloud Computing? Definition, Benefits, Types, and Trends, 2021. https://www.spiceworks.com/tech/cloud/articles/what-is-cloud-computing/.

Yi, S., Hao, Z., Qin, Z., Li, Q., 2015. Fog computing: Platform and applications. In 2015 Third IEEE workshop on hot topics in web systems and technologies IEEE, 73–78.

Chapter 4

Internet of things, artificial intelligence of things and digital twins

Adel Mellit[1], Abdelhamid Rabhi[2], and Marco Zennaro[3]
[1]*Faculty of Sciences and Technology, University of Jijel, Jijel, Algeria,* [2]*University of Picardie Jules Verne, Amiens Cedex 1, France,* [3]*ICTP, Trieste, Italy*

Introduction

The Internet of Things (IoT) has fundamentally transformed how physical devices sense, communicate, and interact with their environment by enabling widespread connectivity between sensors, actuators, embedded systems, and cloud computing services. By bridging the physical and digital worlds, IoT supports continuous data collection, real-time monitoring, and intelligent control across a wide range of application areas, including smart cities, healthcare, industrial automation, transportation, and energy systems (Khanna and Kaur, 2020). This transformation has spurred advancements in wireless communications, low-power electronics, embedded computing, and data analytics, enabling the deployment of scalable, diverse, and resource-constrained networks. This chapter provides a comprehensive overview of IoT technology, beginning with its definition, historical development, and multilayered architecture. It then reviews the communication protocols that enable reliable data exchange across IoT networks, including the physical, data link, network, transport, and application layers, and offers guidance on selecting appropriate protocols for specific use cases (Al-Sarawi et al., 2017). Based on this foundation, this chapter addresses IoT platforms and embedded devices, with a particular focus on microcontrollers (MCUs) as key enablers of embedded artificial intelligence.

The integration of artificial intelligence (AI) into IoT systems, commonly referred to as AI of Things (AIoT), is highlighted as a major technological advancement that enables real-time decision-making, adaptive edge intelligence, and increased system autonomy (Merenda et al., 2020). In this context, the chapter introduces digital twin technology, which creates virtual representations of physical assets or processes to support monitoring, simulation,

Embedded Artificial Intelligence and the Internet of Things for Photovoltaic Systems.
DOI: https://doi.org/10.1016/B978-0-443-44835-5.00004-7

and optimization across their lifecycle. By combining IoT sensing, AI-driven analytics, and digital twin models, intelligent systems can achieve enhanced performance through predictive maintenance, anomaly detection, and system-level optimization in diverse application domains.

The chapter also addresses the role of blockchain technology in overcoming key challenges related to trust, transparency, and security in IoT networks. Its decentralized and tamper-resistant architecture facilitates secure data exchange, device authentication, and data integrity, making blockchain a valuable complement to IoT and Industrial IoT (IIoT) systems(Awaisi et al., 2024). The final sections discuss IIoT architectures, the growing relevance of edge AI, and the opportunities and challenges associated with deploying AI in distributed environments. By linking foundational concepts with emerging technologies such as digital twins and blockchain, this chapter provides a cohesive perspective on the evolution of IoT and AI toward more secure, autonomous, and intelligent systems.

Internet of Things

Definition

Kevin Ashton first proposed the concept of the IoT in 1999, and referred to the IoT as interconnected, interoperable objects that can be uniquely identified using radio frequency identification (RFID) technology (Weber and Weber, 2010; Li et al., 2015). The IoT is a network of physical objects, or things, equipped with sensors, software, and other technologies that enable them to connect and exchange data with other devices and systems over the Internet. These "smart" devices range from everyday household appliances to complex industrial machines, facilitating communication and automation by collecting and exchanging data (Zhong et al., 2022).

In general, the IoT covers many areas, from enabling technologies and components to the many mechanisms for integrating these low-level components effectively. Software is a distinguishing factor for IoT systems. IoT operating systems are designed to operate on small-scale components in the most efficient manner possible, while simultaneously providing essential functionality to simplify and support the global IoT ecosystem in its goals and objectives (Minerva et al., 2015). Middleware, programmability, in terms of application programming interfaces (APIs), and data management appear to be key factors for building a successful system in the IoT world.

The concept of the IoT is shown in Fig. 4.1. These components can include electronic devices, wireless sensors, and smart layers (intelligent processing and decision-making components), which are integrated within smart platforms to provide advanced services and applications. By connecting these diverse components, the IoT enables real-time monitoring, automation, and intelligent decision-making, creating a seamless interaction between the physical and digital world (Minerva et al., 2015).

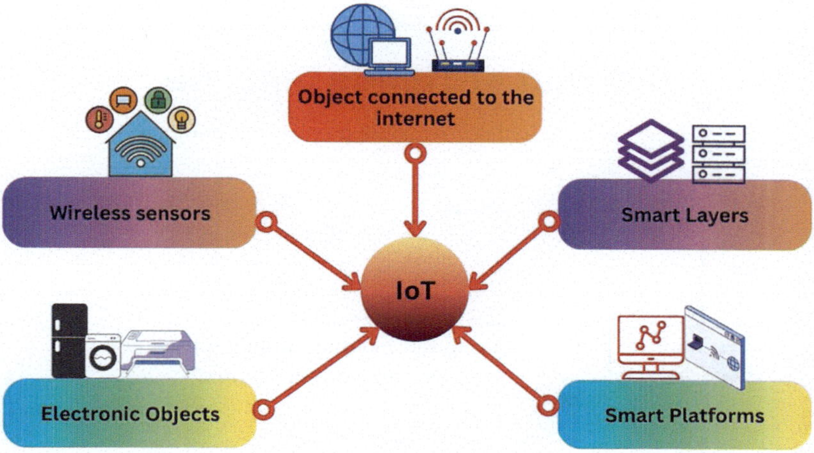

FIGURE 4.1 **The concept of the IoT.** The figure illustrates the basic concept of the Internet of Things (*IoT*), where physical devices are connected and communicate through the Internet. It shows how data is collected, shared, and processed to enable smart decision-making.

Brief history of Internet of Things

The history of the IoT can be traced back to the early days of computing and Machine-to-Machine communications, with major developments occurring in the late 20th and early 21st centuries. Although the official term Internet of Things was not coined until 1999, the underlying concepts and technologies had been developed and explored long before that (Sharma and Sharma, 2022; In Springer Handbook of Internet of Things, 2023). Fig. 4.2 shows the timeline of the IoT.

The idea of connecting objects to the Internet began in the 1980s, but the term IoT was coined in 1999 by Kevin Ashton. Early developments included a networked Coca-Cola vending machine at Carnegie Mellon University. The 2010s saw rapid IoT adoption with smartphones and affordable sensors, while the Industrial IoT expanded in 2015. Today, IoT is widely integrated into smart homes, cities, industry, and healthcare, with continued growth expected.

Internet of Things architecture

The basic IoT architecture comprises the following four layers (See Fig. 4.3) (Al-Sarawi et al., 2017; IoT Architecture Guide, 2018):

Perception layer

Sensors and actuators collect physical data from the environment. The primary purpose of the perception layer is to convert between digital and analog signals. This layer forms the foundation of the IoT infrastructure, collecting

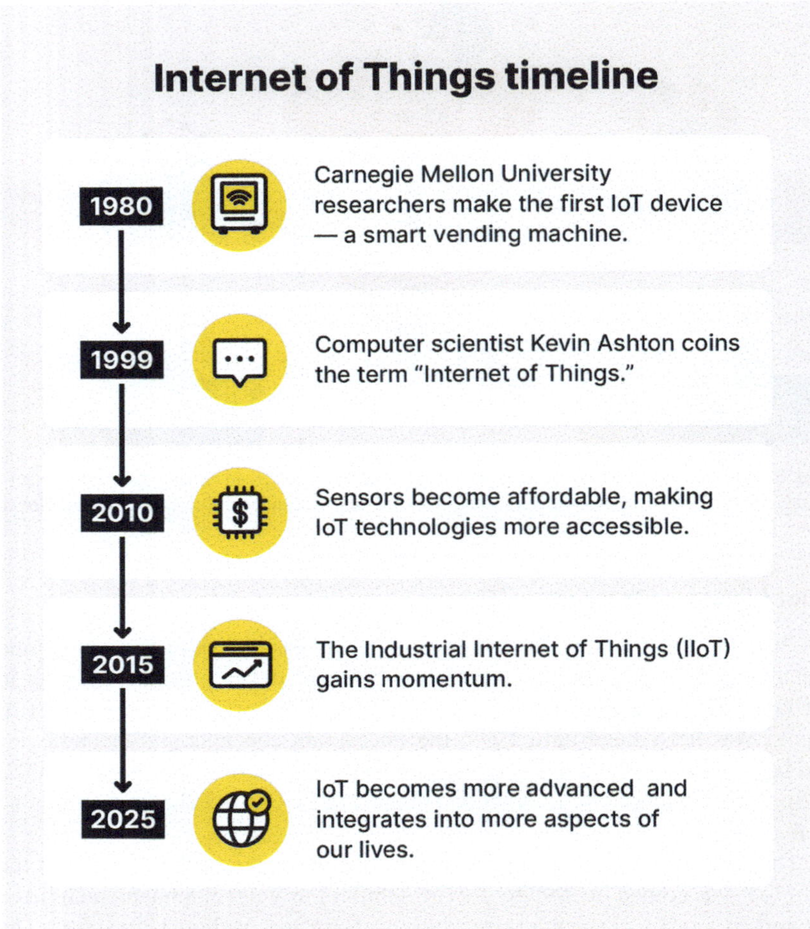

FIGURE 4.2 **Timeline of IoT.** This figure presents a timeline showing the key milestones in the evolution of the Internet of Things. It highlights how IoT technologies have developed and expanded over time.

data and potentially performing operations based on it. Internet-connected devices or endpoints represent the link between the digital and physical worlds.

Network layer or connectivity

Transmits data using various connectivity technologies, including Wi-Fi, Bluetooth, Ethernet, and 5G. It consists of managing communication between devices, networks, and cloud systems required by the IoT solution architecture. Communications between the perception layer and the processing layer are

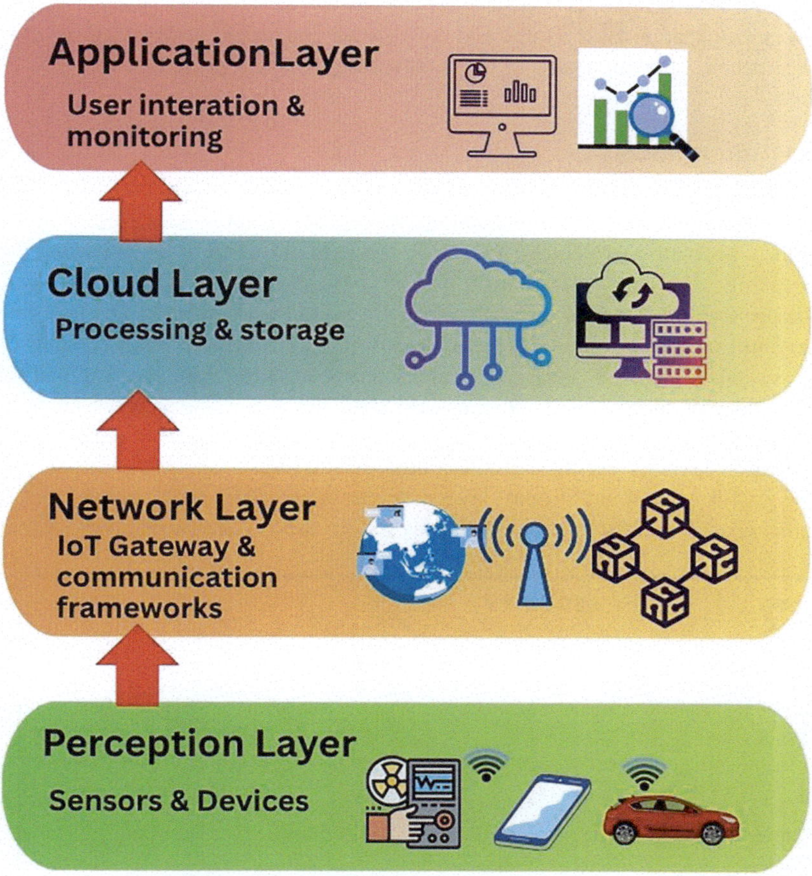

FIGURE 4.3 **Basic IoT architecture.** This figure shows the basic architecture of an IoT system. It highlights the main layers: sensing, communication, data processing, and application (*IoT*, Internet of Things).

established using the Transmission Control Protocol/Internet Protocol (TCP/IP) or User Datagram Protocol (UDP/IP) protocol stack. It can also be implemented using gateways that convert signals into different protocols. Gateways also perform encryption and decryption of data sent across the IoT system.

Processing layer

Processes and analyzes data locally to minimize latency and reduce network congestion. The processing layer is where data generated in the perception layer and sent through the communication layer is collected, stored, and processed. Data goes through two main stages in the processing layer: (1) the data accumulation stage and (2) the data abstraction stage. The processing layer also reformats the data returned from the application layer so that end devices can understand it. These stages hide device details, allowing teams to focus on solving business problems.

Application layer

Provides user interfaces and services through apps, dashboards, and APIs. The application layer in the IoT infrastructure is where data is analyzed to solve business problems or achieve goals. Software tools transform data from the processing layer into useful information for users or automated processes. APIs are used to integrate IoT software with middleware. Some IoT platforms offer built-in software development capabilities that provide features such as data visualization and analysis. Users interact directly with the application layer and can monitor IoT devices and data through various tools such as dashboards and mobile applications. Software tools also use data in the application layer to perform a wide range of activities.

Fig. 4.4 shows additional layers to the main IoT architecture (edge, business, and security layers). These layers are used to more precisely define

FIGURE 4.4 Additional layers to the main IoT architecture. This figure illustrates the extended IoT architecture with additional layers beyond the basic structure. It shows how these extra layers enhance data management, security, and application integration (*IoT*, Internet of Things). *Adopted from IoT Architecture Guide. Major and Additional Layers of the IoT System. (2018). https://www.helpwire.app/blog/iot-architecture/.*

the purpose of the main layers and to improve the performance of the overall IoT infrastructure (IoT Architecture Guide, 2018).

Edge or fog computing layer

The edge (or fog) computing layer processes and stores data close to its source, reducing latency and leveraging 5G capabilities, minimizing the need to send information to the cloud.

- Business layer
  The business layer derives decisions and insights from the application layer, guiding IoT-driven strategies and investments.
- Security layer

End-to-end security is vital in IoT, protecting sensitive data and critical infrastructure, and spans all layers to ensure system safety and reliability. Security can be generally classified into three categories: (1) device, (2) cloud, and (3) connection security.

Communication protocols

IoT communication protocols define the rules for data exchange between devices in the IoT. These protocols ensure that devices communicate efficiently, enabling diverse applications such as smart homes, industrial automation, and healthcare monitoring (Al-Sarawi et al., 2017). They can be broadly classified into network protocols (for communication) and data protocols (for data exchange).

Physical and data link layer protocols

An IoT network protocol can be used to connect a device over a network. These are commonly used on the internet. Some of the popular Standard IoT communication protocols are (Al-Sarawi et al., 2017; Cheng et al., 2018; Bayılmış et al., 2022):

Bluetooth Low Energy

A short-range, low-cost wireless technology that enables data exchange between devices within a personal area network (PAN). It operates at the 2.4 GHz frequency and supports transmission speeds of up to 3 Mbps over a range of 50–150 m. It is widely available in smartphones and is typically used in wearable and mobile applications (See Fig. 4.5). Bluetooth Low Energy (BLE) uses the generic attribute for structured data exchange, ensuring reliable and efficient communication between devices.

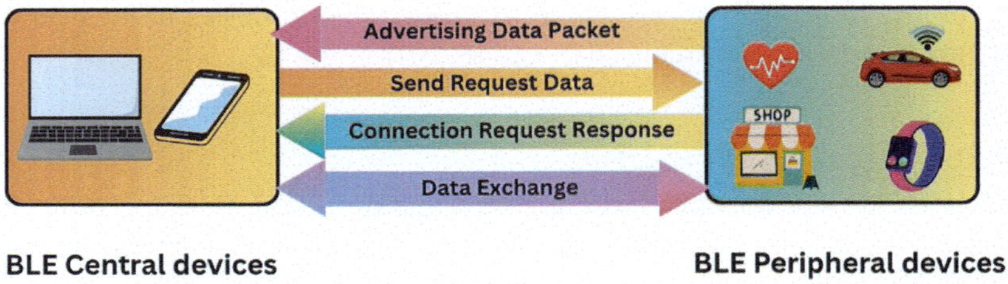

FIGURE 4.5 **BLE concept.** This figure illustrates the basic concept of Bluetooth Low Energy (*BLE*). It shows how BLE enables low-power wireless communication between devices over short distances.

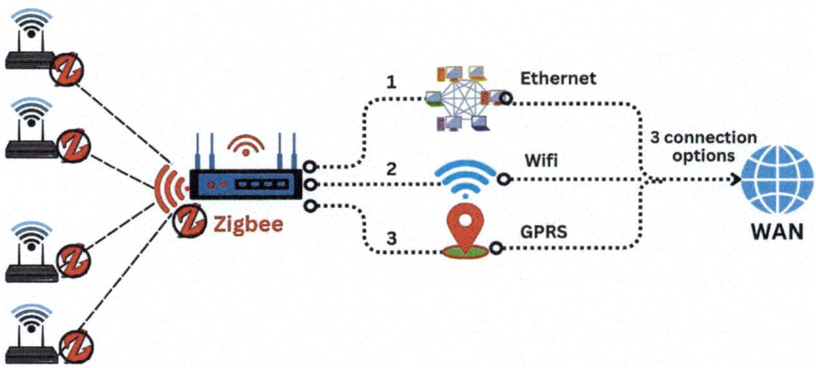

FIGURE 4.6 **ZigBee concept.** This figure shows the concept of ZigBee, a low-power wireless communication protocol designed for sensor networks. It highlights how ZigBee enables reliable data exchange between multiple devices in a mesh network.

ZigBee

Zigbee is a low-power wireless communications technology operating on the 2.4 GHz frequency, similar to Bluetooth. This technology is cost-effective and is primarily used in commercial and industrial applications. With a range of 10–100 m, it supports star and mesh networking, with mesh networking being one of its key features (See Fig. 4.6) and scalable networks with up to 65,000 devices. It uses AES-128 encryption for secure data transmission, ensuring secure and efficient communication in IoT environments.

Wireless fidelity

WiFi is a wireless local area network (LAN) technology standardized according to the IEEE 802.11 standard, introduced by the Wi-Fi Alliance. It provides internet connectivity within a range of 60–100 feet using high-frequency radio signals, with data rates ranging from 2 Mbps to 1.73 Gbps.

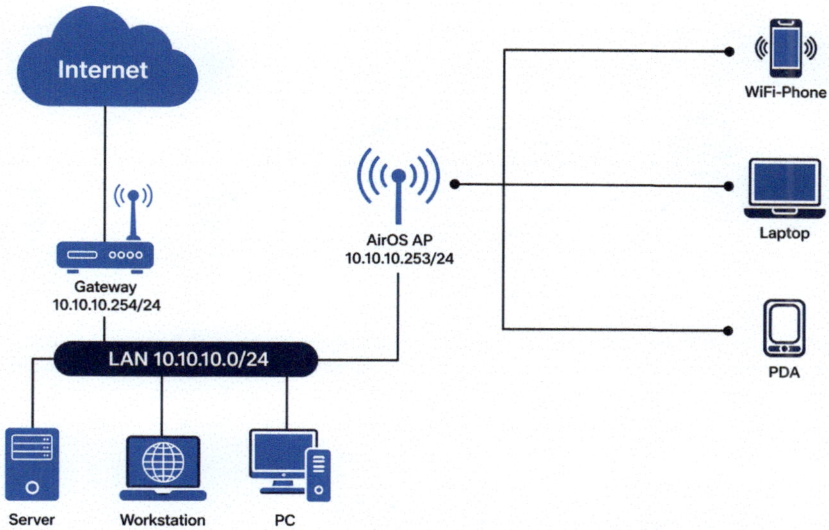

FIGURE 4.7 **Wifi concept.** This figure illustrates the concept of Wi-Fi communication. It shows how devices connect wirelessly to a network for internet access and data exchange (*Wi-Fi*, Wireless fidelity).

In the IoT, this technology supports PAN, LAN, and wide area network (WAN) configurations, and routing can extend its coverage (See Fig. 4.7). It provides WPA3 encryption for secure connections, making it ideal for consumer and industrial IoT applications that require high data throughput.

Z-Wave

Z-Wave is a low-power wireless protocol operating in the sub-1 GHz band (typically 908.42 MHz in the US and 868.42 MHz in the EU), mainly used for home automation. It is a more secure technology. It forms secure mesh networks with ranges of 30–100 m and supports data rates up to 100 Kbps, offering longer range but lower speed compared to other IoT technologies (See Fig. 4.8). Z-Wave uses AES-128 encryption, ensuring secure communication in home automation environments.

Radio frequency identification technology

This technology uses radio frequency waves to exchange data between a reader and a tagged object for identification and tracking. This technology operates without physical contact and relies on three main components: the RFID tag, the reader, and the antenna. The tags operate across various frequency bands, with 13.56 MHz being the most commonly used (See Fig. 4.9).

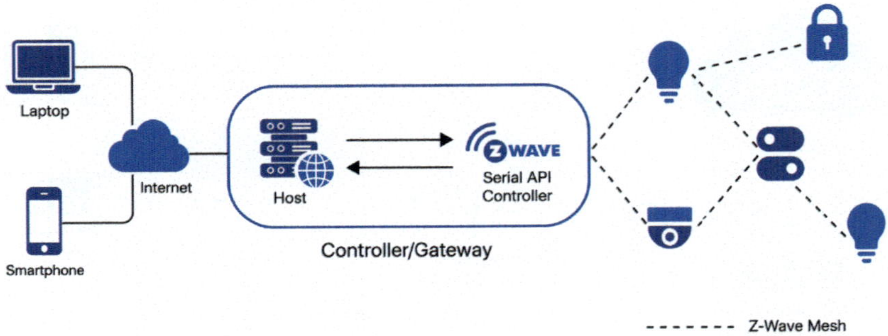

FIGURE 4.8 **Z-Wave concept.** This figure illustrates the concept of Z-Wave, a low-power wireless communication protocol used mainly in home automation. It shows how Z-Wave enables devices to form a mesh network for reliable and energy-efficient data exchange.

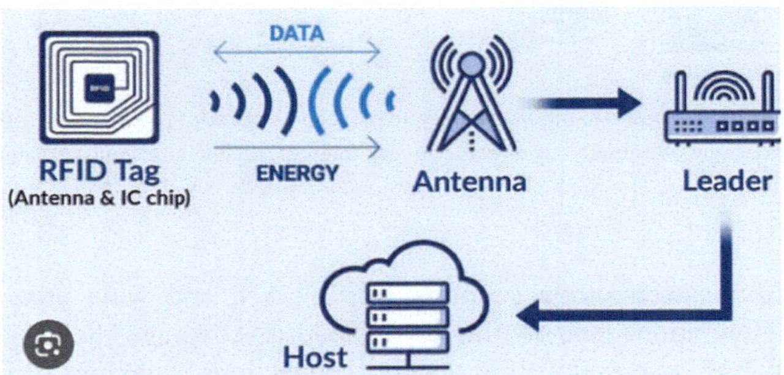

FIGURE 4.9 **RFID concept.** This figure illustrates the RFID concept, where a reader communicates wirelessly with an RFID tag using radio waves to identify or track an object. The reader sends a signal to the tag, which responds by transmitting stored data back to the reader for processing (*RFID*, Radio frequency identification).

Cellular

Cellular connectivity (Global System for Mobile Communications "GSM"/3G/4G/5G) enables data transmission over long distances. While it's convenient for large volumes, it's often used in the IoT to send small amounts of data over the Internet. Because telecom companies manage the infrastructure, users avoid setup and maintenance costs (See Fig. 4.10).

Long range wide area network

A low-power WAN protocol designed for large-scale IoT deployments. This protocol enables secure, mobile communications across public networks and supports millions of devices. It also allows for two-way communication over long distances, exceeding 15 km, making it ideal for wide-area network

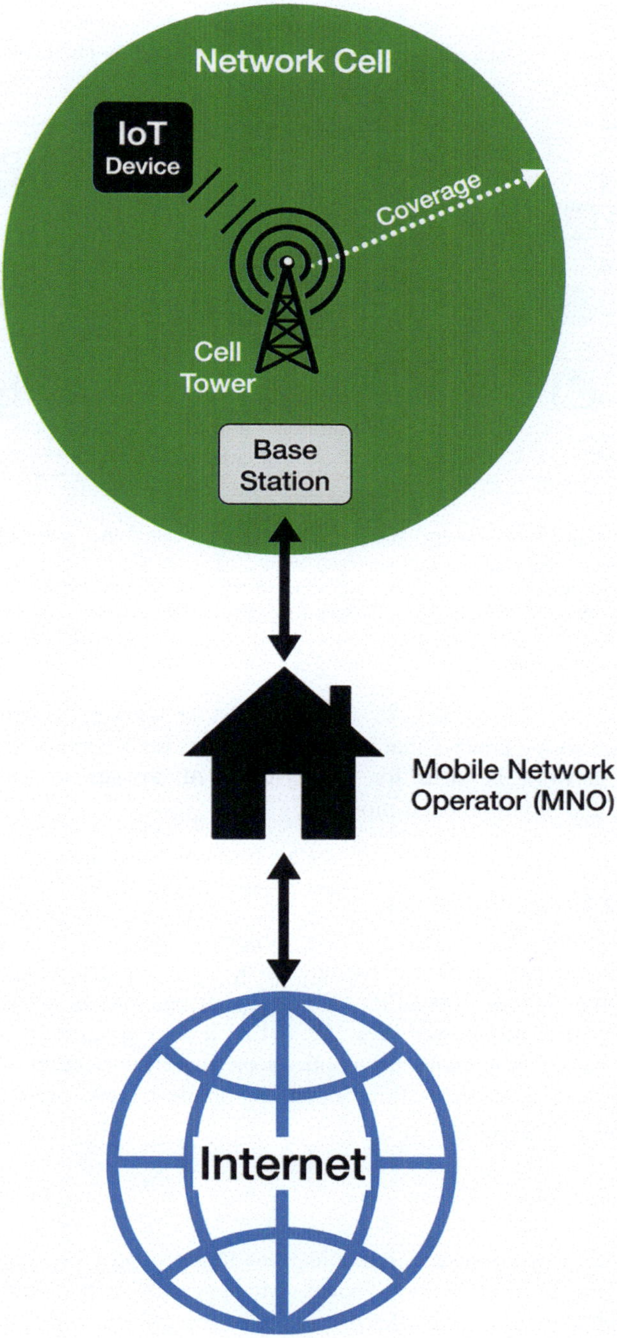

FIGURE 4.10 Cellular concept. This figure illustrates the cellular communication concept, where devices connect to a nearby cellular tower to transmit data over a wide-area network. The tower relays the information to the core network, enabling long-range communication between remote devices or systems.

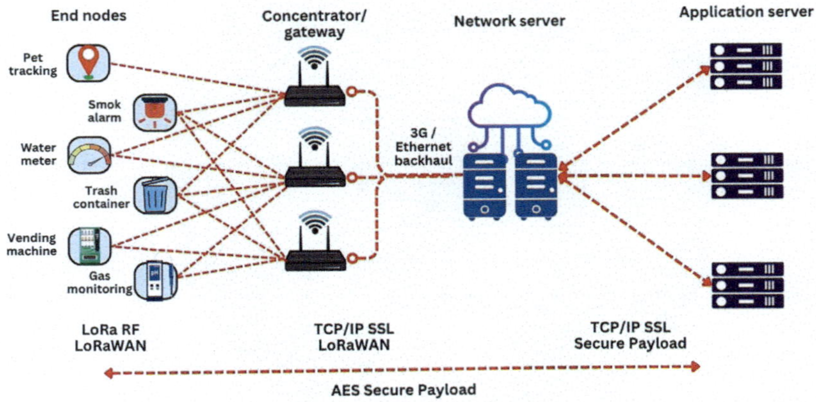

FIGURE 4.11 **LoRaWAN concept.** This figure illustrates the LoRaWAN concept, where low-power IoT sensors send data to a LoRa gateway using long-range wireless communication. The gateway then forwards the collected data to a cloud server or network server for processing and visualization (*LoRaWAN*, Long range wide area network).

applications (See Fig. 4.11). It supports start technology, where devices communicate with gateways that forward data to network servers. Long-range wide area network is ideal for battery-operated devices in smart cities, agriculture, and environmental monitoring.

Near Field Communication

An IoT technology that makes it easier to connect devices. This technology provides short-range wireless communication, but its transmission is slower than Bluetooth. It relies on RFID technology. It operates at a frequency of 13.56 MHz and is low-power (See Fig. 4.12). It uses encryption and secure channels, making it suitable for applications requiring secure, short-range communications. It supports three operating modes: reader/writer, peer-to-peer, and card emulation.

Sigfox

Sigox is a form of wireless communications that provides low-power, long-range wireless connectivity to devices. Messages are transmitted over the Sigfox global network. It provides one of the largest IoT networks. It is similar to cellular networks that install antennas on towers. It is a low-power wide area network. It transmits small data packets with very low power consumption, making it ideally suited for sensor-based applications such as asset tracking, environmental monitoring, and smart metering (See Fig. 4.13).

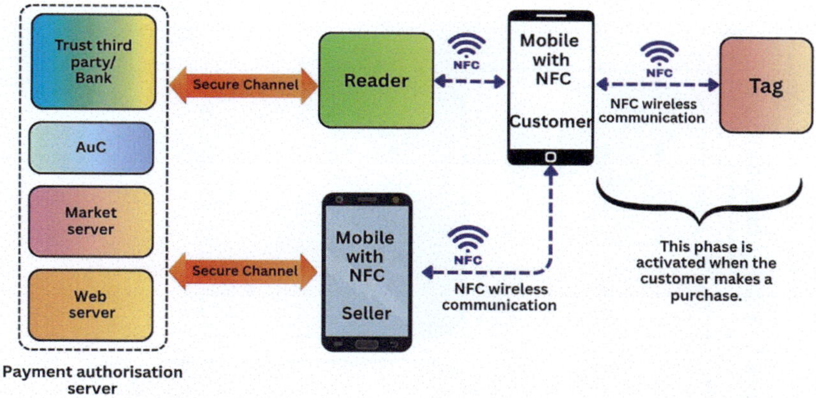

FIGURE 4.12 **NFC concept.** This figure illustrates the NFC concept, where two devices exchange data over very short distances using near-field electromagnetic coupling. Communication occurs only when the devices are brought close together, enabling secure and quick data transfer (*NFC*, Near field communication).

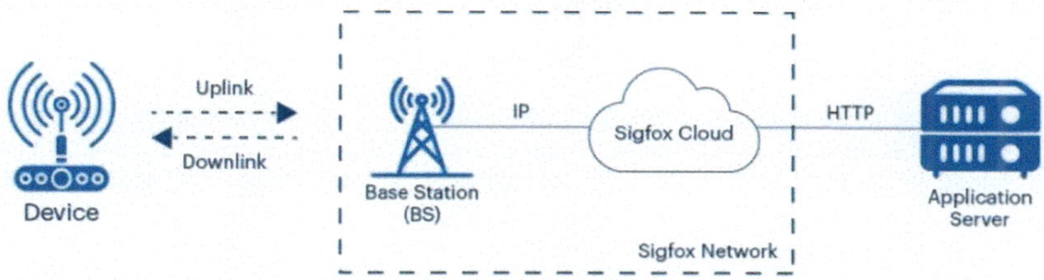

FIGURE 4.13 **SigFox concept.** This figure illustrates the SigFox concept, where low-power IoT devices send small packets of data to a SigFox base station using an ultra-narrowband network. The base station forwards the data to the SigFox cloud, where it is processed and made available to users or applications.

Comparison and selection of data link layer protocols

Selection depends on power consumption, data rate, communication range, network topology, scalability, and security (See Fig. 4.14) (Schizas et al., 2022). A comparison between various IoT link layer protocols is listed in Table 4.1.

The choice of IoT protocol for solar applications depends on range, power consumption, data rate, cost, complexity, and network architecture. LoRa and Sigfox are ideal for remote solar farms due to their long range, low power consumption, low cost, and ease of deployment, making them suitable for scattered sensor networks. Zigbee, Z-Wave, and BLE protocols work well for spot monitoring in residential or small-scale solar systems, providing low-power, moderate-cost, mesh networks for reliable coverage of inverters, batteries, or nearby sensors. Wi-Fi and cellular networks offer high data rates and broad coverage but come with higher power, cost, and complexity, making

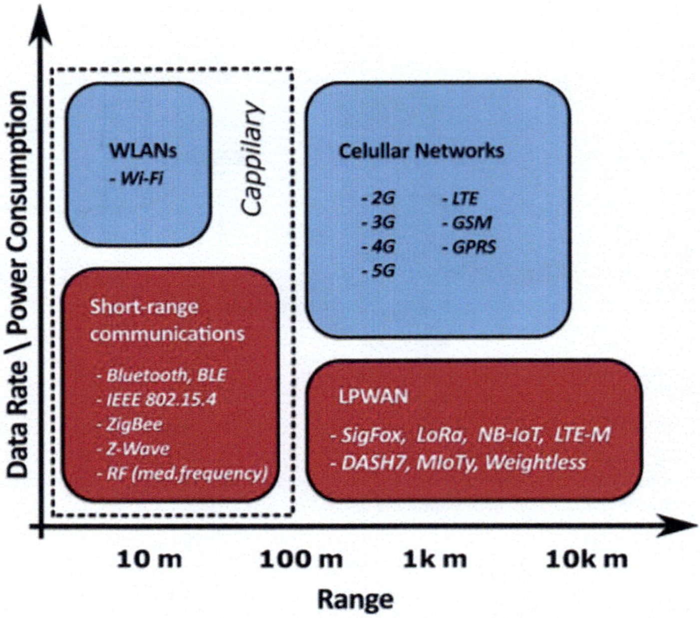

FIGURE 4.14 **Comparison of data link layer protocols.** This figure compares different data link layer protocols in terms of communication range, power consumption, data rate, and typical IoT applications. It visually highlights the trade-offs between short-range high-speed protocols and long-range low-power technologies (*IoT*, Internet of Things).

them suitable for applications requiring real-time monitoring or analytics. RFID and near field communication technologies are primarily used for asset tracking rather than continuous sensor monitoring. In general, long-range, low-power protocols serve large-scale solar farms, while short-range mesh protocols are best for spot or residential settings.

Network layer protocols

The main network layer protocols in the IoT include the Internet Protocol (IP), specifically IPv6, and 6LoWPAN (Internet Protocol version six over Low-Power Wireless Personal Area Networks). IPv6 provides addressing and routing to enable seamless Internet connectivity, while 6LoWPAN adapts IP communications to the low-power, resource-constrained wireless networks typically used in the IoT. Together, these two protocols ensure efficient and secure transmission of data packets between devices and across heterogeneous networks (Al-Sarawi et al., 2017; Cheng et al., 2018). Data is bundled into packets, or in the case of very large data sets, divided into smaller subpackets.

TABLE 4.1 Comparison between various IoT link-layer protocols.

Protocol	Range	Power consumption	Data rate	Network topology	Cost	Complexity	Strengths in IoT
Wi-Fi	50–100 m	High	High (Mbps)	Star	Medium-High	Medium-High	High-speed data transfer, easy integration
Zigbee	10–100 m	Low	Low–Medium (kbps)	Mesh	Low-Medium	Medium	Low-power, reliable, scalable
LoRa	>15 km	Very Low	Low (kbps)	Star	Low	Low-Medium	Long-range, low-power
NFC	<0.1 m	Very Low	Low	Point-to-point	Very Low	Very Low	Secure short-range communication
Sigfox	10–50 km	Very Low	Very Low (bps)	Star	Low–Medium	Low	Ultra-low power, long-range
Cellular	Several km	High	High	Star	High	High	Wide coverage, high-speed, mobile
RFID	<10 m	Very Low	Low	Point-to-point	Low	Very Low	Identification and tracking
Z-Wave	30–100 m	Low	Low	Mesh	Medium	Medium	Home automation, reliable
Bluetooth (BLE)	10–100 m	Low	Low	Star-Mesh	Low	Medium	Low-power, widely supported

IoT, Internet of Things; *NFC*, near field communication; *RFID*, radio frequency identification.

Each protocol used has its own characteristics and advantages (A Comprehensive Guide to Communication Protocols in IoT, 2025).

- IPv4/IPv6: These provide unique addressing and routing across the Internet. IPv6 is more suitable for large-scale IoT due to its large address space.
- 6LoWPAN: IPv6 is adapted to low-power, resource-constrained wireless networks, such as Zigbee and Thread.
- Routing Protocol for Low-Power and Lossy Networks (RPL): Designed specifically for the IoT, it enables efficient routing in dynamic and resource-constrained environments.
- Internet Protocol Security: Ensures secure data transmission at the network layer through encryption and authentication.
- Internet Control Message Protocol: Supports diagnostics, fault handling, and neighbor discovery in IoT networks.

IPv6 with 6LoWPAN and RPL is the most widely adopted combination for IoT systems, as it balances scalability, low power usage, and interoperability.

Transport layer protocols

Transport layer protocols provide comprehensive communication services for applications, ensuring efficient and reliable data communication between processes on different devices. These protocols operate above the network layer, managing essential functions such as reliability, flow control, congestion control, and multiplexing/demultiplexing. Main transport layer protocols used in the IoT and their key features (IoT Architecture Guide, 2018; Cheng et al., 2018):

Transmission Control Protocol

A connection-oriented protocol that ensures reliable and orderly delivery of data packets. It is suitable for applications that require accuracy and completeness, but is less suitable for limited IoT devices due to its high overhead and power consumption.

User Datagram Protocol

A connectionless, lightweight, and low-latency protocol. It provides faster transmission with minimal overhead, making it suitable for real-time IoT applications such as sensor data streaming, although it does not guarantee delivery.

MQTT for sensor networks

Operates over UDP instead of TCP and is optimized for limited devices and low-power wireless networks.

Constrained application protocol

Built on UDP, it is designed for resource-constrained devices. It supports request/response communications like HTTP but with lower overhead, making it highly suitable for the IoT.

Quick Internet connections over UDP

A modern transport protocol based on UDP, offering reliability, multicast, and encryption. It is increasingly important for Internet of Things applications that require efficiency and security.

In the Internet of Things, UDP-based protocols (Constrained application protocol, MQTT for sensor networks) are preferred for constrained devices and instant communications, while TCP (with MQTT or HTTP) is used when reliability and orderly delivery are critical.

Application layer protocols

Application layer protocols are communication protocols that operate at the upper layer of the open systems interconnection and TCP/IP models. These protocols define how data is exchanged and how applications interact across devices. In the IoT, these protocols provide the interface between IoT devices (sensors, actuators, and gateways) and user applications or cloud platforms. These protocols ensure that data collected by IoT devices is formatted, transmitted, and correctly interpreted by the receiving application. Common application layer protocols in the IoT (IoT Architecture Guide, 2018; Bayılmış et al., 2022):

- HTTP/HTTPS (Hypertext Transfer Protocol)
  A standard web communication protocol; HTTPS adds encryption for security. It is used when IoT devices interact with web servers or cloud APIs.
- Message queue telemetry (MQTT)
  A lightweight, publish/subscribe protocol optimized for low-bandwidth and unreliable networks, widely used in the IoT.
- Constrained application protocol (CoAP)
  Designed for constrained devices, it is built on UDP, enabling efficient request-response communication, similar to HTTP.
- Advanced message queuing protocol
  A reliable and secure message-oriented broker protocol widely used in enterprise IoT.
- Data distribution service
  A scalable, real-time publish/subscribe protocol suitable for mission-critical IoT systems.
- Extensible messaging and presence protocol

An XML-based messaging protocol is sometimes used in IoT for device communications.

Internet of Things platforms and devices

Internet of Things platforms

An IoT platform is a comprehensive software system that enables data from IoT devices to be connected, managed, and analyzed. An IoT platform provides a central hub for its various components, offering functions such as device management, data collection, processing, storage, and application development. At its core, an IoT platform represents a bridge between the physical world of connected devices and the digital world of actionable insights (Network Layer Protocols, 2025). It consists typically of four interconnected components: the client user interface or software solution, the cloud, the gateway, and one or more connected smart devices (See Fig. 4.15) (Babun et al., 2021). The high-level solution architecture illustrates these interconnected solutions:

- An end device: Is an end device that provides sensing services, control services, or both. A thermostat senses temperature, reports that data, and provides the ability to change the setpoint to control heating, ventilation, and air conditioning or other climate systems.
- A gateway: A gateway is a hardware or software component that facilitates communication between smart devices, smart objects, or controllable objects and the cloud-based systems that support them. In mesh communication networks like ZigBee, all devices transmit communication to the desired gateway.
- The cloud: This is commonly referred to as the IoT platform.
- Customer user interface: The user's interface with their device or services occurs through a customer user interface, often a mobile or responsive app.

There are mainly five types of IoT platforms (The 5 Types of IoT Platforms, 2024; Fortino et al., 2022):

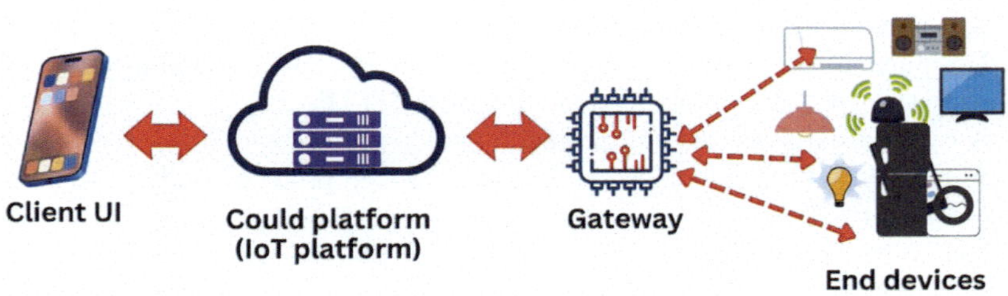

FIGURE 4.15 A basic configuration of an IoT platform. This figure shows that a typical IoT platform consists of connected smart devices that collect data, communication networks that transmit the data securely, a cloud/edge layer that stores and processes it, and a user interface that enables monitoring, analytics, and control actions (*IoT*, Internet of Things).

Internet of Things connectivity management platforms

IoT connectivity management platforms, as the name suggests, focus on the networking component of IoT systems. They provide users with the software, connectivity hardware, and data routing needed to keep their devices connected to the internet. Their networks generally rely on existing carrier services and Wi-Fi networks, configuring connectivity in a way that makes IoT setup easy.

Internet of Things device management platforms

IoT device management platforms specialize in the hard work associated with IoT devices. They ensure everything is connected and secure, and keep you up-to-date on the status of your devices. Device management platforms update firmware, notify you of changes to your devices, report metrics, and patch security updates. This type of IoT platform will help you with routine tasks associated with your devices, regardless of their number.

Cloud Internet of Things platforms

Cloud IoT platforms provide users with the infrastructure needed to build a cohesive IoT system. They represent a central location for all your back-end operations and data to operate. One of the most important benefits of cloud platforms is their scalability; regardless of the size of your project, the cloud platform can grow with you and your IoT system.

Internet of Things application enablement platforms

IoT application enablement platforms offer a one-stop shop approach, providing users with everything they need to launch an IoT system. They provide hardware, software, and IoT system development and deployment. They provide a one-stop destination for getting your system up and running, eliminating the need to manage developers, configure the network, and engineer the hardware yourself.

Advanced Internet of Things analytics platforms

Advanced IoT analytics platforms are a great solution for data-driven IoT systems. Users looking for sophisticated IoT systems that utilize machine learning (ML), artificial intelligence, statistical modeling, and big data can use these platforms to interpret and act on the collected data. IoT systems that primarily ingest data, rather than perform tasks, will benefit most from these platforms. Choosing the right IoT platform is critical, as the subsequent migration becomes costly and complex. The platform must meet your needs while ensuring security, reliability, and ease of use. It must simplify system management, protect sensitive data, and provide reliable service, as choosing a low-cost but unreliable platform can lead to serious problems on a large scale. The most popular and prominent IoT platforms, especially for enterprise and industrial applications, include Amazon Web Services (AWS), Microsoft Azure, Oracle, and Google Cloud. A brief comparison is shown in Table 4.2.

Amazon Web Services (AWS), Microsoft Azure, Oracle Cloud, and Google Cloud each offer distinct strengths in IoT, AI, and edge computing. AWS leads the market with the largest global network, a broad IoT/AI ecosystem, and robust edge solutions, although its pricing and complexity can be challenging. Microsoft Azure offers enterprise integration and hybrid cloud solutions, making it attractive to organizations already invested in the Microsoft ecosystem, but its services can be expensive. Oracle Cloud focuses on database-driven enterprises, offering cost-effective solutions and Roving Edge devices, but it lags in global coverage and ecosystem breadth. Google Cloud, with its AI-focused strategy and powerful tools like Vertex AI, TensorFlow, and Edge Tensor Processing Unit (TPU), is best suited for businesses that rely heavily on analytics and machine learning, but its adoption and enterprise scale still lag behind (AWS) and Azure. In short, AWS and Azure dominate the enterprise market, Google Cloud leads AI innovation, and Oracle Cloud serves niche sectors focused on databases.

Internet of Things devices

IoT devices are physical objects embedded with sensors, software, and other technologies that enable them to connect and exchange data over the internet. They are transforming various sectors, from smart homes to industrial manufacturing, by automating processes, enhancing efficiency, and providing valuable data insights. In this subsection, we will focus more on hardware devices, including MCUs used in IoT applications. Key features of IoT MCUs are:

- Low power consumption: Essential for battery-powered or energy-harvesting devices.
- Processing power: Processing sensor data, running lightweight algorithms, and supporting communication protocols.
- Memory: Sufficient flash memory and RAM for storing firmware and running applications.
- Communication support: Interfaces such as Universal Asynchronous Receiver/Transmitter (UART), Serial Peripheral Interface (SPI), Inter-Integrated Circuit (I^2C), and wireless modules (Wi-Fi, BLE, LoRa, etc.).
- Peripheral integration: Built-in Analog-to-Digital converter, timers, Pulse Width Modulation (PWM), and General-Purpose Input/Output (GPIO) ports for interacting with sensors and actuators.

Table 4.3 lists available MCUs used in IoT applications. These MCUs are compared in terms of storage, memory, frequency, and supported communication protocols. Fig. 4.16 shows a real photo of each MCU.

TABLE 4.2 Comparison between some IoT platforms.

Provider	IoT services	AI and ML services	Edge computing	Strengths	Weaknesses	Global coverage	Target users
Amazon Web Services (AWS)	AWS IoT Core, Greengrass, IoT Analytics, SiteWise, IoT Device Defender	Amazon SageMaker, Rekognition, Lex, Comprehend, Polly, Forecast	AWS Greengrass, AWS Snow Family (Snowcone, Snowball, Snowmobile)	Market leader, broad ecosystem, strong ML/AI integration, global reliability	Complex pricing, steep learning curve	Largest global network	Enterprises, startups, IoT developers, ML researchers
Microsoft Azure	Azure IoT Hub, IoT Edge, Azure Digital Twins, Time Series Insights	Azure Machine Learning, Cognitive Services, Bot Service, Applied AI	Azure IoT Edge, Azure Stack Edge, Azure Arc	Strong enterprise integration, hybrid cloud leader, developer tools	Expensive, some services are less mature	Very wide coverage, close to AWS	Enterprises with a Microsoft ecosystem, hybrid cloud users
Oracle Cloud	Oracle IoT Cloud Service, Oracle Edge Services	Oracle AI Services (Vision, Speech, Anomaly Detection, Language), Data Science Service	Oracle Roving Edge Infrastructure	Optimized for Oracle DB workloads, strong enterprise focus, cost-effective for DB-heavy apps	Smaller market share, fewer developer tools, limited ecosystem	Limited compared to AWS/Azure	Enterprises using Oracle DB/ERP, regulated industries
Google Cloud	(Cloud IoT Core retired 2023; partner ecosystem), Edge TPU, IoT solutions via partners	Vertex AI, TensorFlow Enterprise, AutoML, Cloud AI APIs	Google Distributed Cloud (Edge, Hosted), Edge TPU devices	AI-first, strong analytics/ML, open-source leadership (Kubernetes, TensorFlow)	Smaller enterprise adoption, weaker global reach than AWS/Azure	Moderate, growing but behind AWS/Azure	AI/ML-driven companies, startups, analytics-heavy businesses

TABLE 4.3 Some available IoT-based MCUs.

MCU/SoC family	Key features	ROM	RAM	Frequency	Supported communication protocols
ESP32	Dual-core, Wi-Fi and BLE, versatile GPIO, moderate power, low cost, Integrated Wi-Fi + BLE SoC	4–16 MB	512–520 KB	80–240 MHz	I^2C, SPI, UART, 1-Wire, CAN (via software), Ethernet
ESP8266	Single-core, Wi-Fi, compact, low-power, Integrated Wi-Fi SoC	1–4 MB	80–160 KB	80–160 MHz	^2C, SPI, UART, 1-Wire
Arduino Nano BLE 33 IoT	ARM Cortex-M4, BLE, Wi-Fi via NINA module, small form factor, low-power, u-blox NINA-W102	1 MB	256 KB	64 MHz	I^2C, SPI, UART
STM32Wx series	ARM Cortex-M cores, integrated Wi-Fi, low-power modes, rich peripherals, and integrated Wi-Fi/Sub-GHz RF	1–2 MB	256–512 KB	64–120 MHz	I^2C, SPI, UART, CAN, 1-Wire, Modbus (via software), Ethernet (with external PHY)
Seeed Studio XIAO nRF52840	Ultra-compact, ARM Cortex-M4, BLE, Wi-Fi via module, low-power, Wi-Fi via external module.	2 MB	264 KB	133 MHz	I^2C, SPI, UART
Raspberry Pi Pico W	Dual-core Cortex-M0+, Wi-Fi, low-cost, low-power, flexible GPIO, Infineon CYW43439 Wi-Fi chip	2 MB	264 KB	133 MHz	I^2C, SPI, UART
Portenta H7 Lite Connect	Dual-core ARM Cortex-M7/M4, Murata 1DX Wi-Fi + BLE module	16 MB	8 MB	480 MHz / 240 MHz	I^2C, SPI, UART, CAN

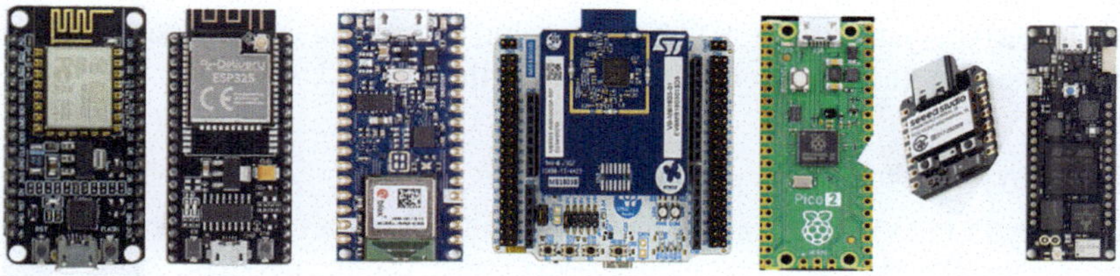

FIGURE 4.16 IoT devices based on MCUs: (A) ESP32, (B) ESP8866, (C) Nano BLE IoT 33, (D) STM32W, (E) Raspberry Pi Pico W, (F) Seeed XIOA, and (G) Portenta H7 Lite connected. This figure shows examples of popular MCUs used in IoT applications include: (A) ESP32, (B) ESP8266, (C) Arduino Nano 33 BLE IoT, (D) STM32W series, (E) Raspberry Pi Pico W, (F) Seeed XIAO IoT, and (G) Portenta H7 Lite Connected.

ESP32

The ESP32 is a single-chip, 2.4 GHz integrated Wi-Fi and Bluetooth module, built on TSMC's low-power 40 nm process technology. Designed for optimal power and RF performance, it is robust, versatile, and reliable across a wide range of applications and power scenarios (ESP32 Series, 2025).

ESP8266

The ESP8266 is a small, low-cost Wi-Fi chip with built-in TCP/IP networking software and a MCU, produced by Espressif Systems in Shanghai, China. ESP8266. ESP8266-IC. Manufacturer: Espressif Systems (ESP8266EX, 2025).

Arduino Nano 33 BLE Sense

The Arduino Nano 33 BLE Sense combines a tiny form factor, different environment sensors, and the possibility to run AI using TinyML and TensorFlow Lite. Whether you are looking at creating your first embedded ML application or you want to use Bluetooth Low Energy to connect your project to your phone, the Nano 33 BLE Sense will make that journey easy (Arduino Nano 33 BLE Sense, 2019).

STM32Wx

The STM32Wx MCUs enable wireless connectivity. They support both the sub-GHz and 2.4 GHz frequency ranges. These modules are highly integrated and reliable, meeting a wide range of industrial and consumer applications. The STM32Wx solutions are compatible with multiple protocols, from point-to-point and mesh networks to wide-area networks. They are also energy-efficient and offer built-in security features. The STM32Wx wireless MCUs feature a dual-core "two-in-one" architecture. Built around a microcontroller and a wireless transceiver, these modules come in a highly integrated and cost-effective system. The STM32Wx solutions enable instant performance while ensuring energy efficiency (STM32 wireless MCUs, 2025).

Raspberry Pi Pico W

Raspberry Pi Pico W has been designed to be a low-cost yet flexible development platform for RP2040, with a 2.4 GHz (Pico-series Microcontrollers, 2023).

Seeed Studio XIAO (nRF52840)

The Seeed Studio XIAO nRF54L15 is a compact, high-performance development board powered by the cutting-edge Nordic nRF54L15 chipset. This next-generation system-on-chip integrates an ultra-low-power 2.4 GHz multiprotocol radio with a MCU containing a 128 MHz Arm Cortex-M33 processor and an Arm Cortex-M0+ processor for advanced power management. It offers up to 1.5 MB of expandable Non-Volatile Memory (NVM) memory and 256 KB of RAM, and an ultra-low-power internal design that significantly extends battery life (Getting Started with Seeed Studio XIAO nRF54L15Sense, 2024).

Portenta H7 Lite connect

The Arduino Portenta H7 Lite connected brings the computing power of the Portenta H7 to AI applications and low-latency control solutions, plus a module designed to be more cost-effective by removing the HD video interface feature. Additionally, it can run TensorFlow Lite for machine learning at the edge, making it suitable for AI-powered IoT solutions. Compared to the full Portenta H7, the Lite Connected version is more affordable while maintaining core features for IIoT, smart energy, predictive maintenance, and automation (Arduino Portenta H7, 2023).

Supported Internet of Things devices and platforms

Wi-Fi-compatible MCUs and systems on a chip (SoCs) are selected based on a balance of processing power, memory, connectivity, and power efficiency. Low-power devices such as the ESP32, ESP8266, Arduino Nano BLE 33 IoT, Seeed Studio XIAO nRF54L15, and Raspberry Pi Pico W are ideal for lightweight IoT applications, supporting protocols such as I^2C, SPI, and UART for sensor and peripheral communication. The STM32Wx series offers built-in Wi-Fi connectivity, rich peripherals, and support for industrial protocols such as CAN and Modbus, making it suitable for more complex or large-scale IoT systems. The Arduino Portenta H7 Lite Connected enhances its capabilities by combining dual-core processing, advanced wireless connectivity, and expanded memory, enabling edge AI, real-time analytics, and industrial IoT solutions. Choosing the right MCU or SoC ensures efficient, reliable, and scalable operation across a wide range of IoT applications.

Most popular IoT platforms for the above mentioned MCUs are: ThingSpeak (Thingspeak, 2010), Blynk (Blynk, 2020), Aafruit IO (The internet of things for everyone, 2025), Particle (Particule, 2025), IBM Watson IoT (IBM cloud, 2016), and Arduino Cloud (Cloud Arduino, 2019). Table 4.4 reports a short comparison between these IoT platforms in terms of free availability, available Phone App, supported MCUs, complexity, and typical applications.

Among the compared IoT platforms, Adafruit IO and Blynk stand out as low-complexity solutions, ideal for beginners, manufacturers, and rapid prototyping. Blynk focuses heavily on mobile applications, while ThingSpeak adds moderate complexity due to the simplicity of basic data logging, while advanced analytics require MATLAB® expertise, making it popular in academia. Arduino Cloud also offers a medium-complexity solution, offering seamless integration with Arduino boards, an intuitive web and mobile interface, and cloud data management, making it ideal for hobbyists and educational projects. Particle offers higher complexity with software development kits, cloud integration, and fleet management features, making it suitable for professional and industrial IoT deployments. At the top of the list is IBM Watson IoT, which is more complex and designed for large-scale enterprise applications with strong AI and machine learning integration, but requires significant cloud and systems expertise. In general, complexity is proportional to the intended use: simple platforms are suitable for education and prototyping, while advanced platforms target enterprise and industrial IoT.

Artificial intelligence of things

Definition

The AIoT is a new paradigm that emerged from the combination of AI and the IoT (Siam et al., 2025), which permits the creation of intelligent systems capable of analyzing data, learning, making decisions, and automating actions without human intervention. By applying AI techniques, such as machine learning, to the vast data collected by interconnected IoT devices, the IoT enables more efficient operations, enhanced user experiences, and innovative applications in various areas (Jiyi et al., 2021; What is AIoT, 2025). AIoT enables more efficient operations, predictive maintenance, and enhanced human–machine interactions across various sectors like smart homes, manufacturing, healthcare, and smart cities. AIoT is based on three fundamental elements, as shown in Fig. 4.17: sensing, processing, networking, and communication (Siam et al., 2025).

Specifically, the AIoT collects information from the physical environment using a range of embedded sensors, including cameras, microphones, motion sensors, and physiological sensors. AI algorithms process the collected sensor data for multiple purposes, including anomaly detection, classification, and

TABLE 4.4 A brief comparison between IoT platforms that support MCUs.

Platform	Cost model	Open source	Complexity	Typical applications	Supported MCUs/ Boards	Mobile app availability	Comments
ThingSpeak	Free for noncommercial use (limited channels), paid licenses for higher limits	Proprietary, cloud-based	Low	Data logging, research, academic projects, IoT analytics	Arduino, ESP8266/ ESP32, Raspberry Pi, STM32, etc.	No	Strong MATLAB integration for analytics & ML
Blynk	Free tier + paid Pro plans (for commercial/scale)	Closed-source, cloud + app	Low	Smart home, remote monitoring, rapid IoT prototyping	Arduino, ESP8266/ ESP32, Raspberry Pi, STM32, TI, etc.	Yes	Focus on app-centric IoT with drag-and-drop UI
Adafruit IO	Free (up to limits), $10/month for Plus	Cloud-based, but MQTT APIs are open	Low	Hobbyist IoT, DIY electronics, small data logging	Adafruit Feather, Arduino, ESP8266/ ESP32, Raspberry Pi	No	Beginner-friendly, great for makers/ education

Particle	Paid subscription (free developer tier, enterprise plans for fleets)	Proprietary cloud, SDKs available)	Medium	Industrial IoT, logistics, asset tracking, fleet/ device mgmt.	Particle hardware (Photon, Boron, Argon), ESP32 (via SDK)	Yes	Strong focus on commercial IoT deployments
IBM Watson	Enterprise subscription (pay-as-you-go via IBM Cloud)	Proprietary, enterprise-focused)	High	Industrial IoT, predictive maintenance, large-scale enterprise apps	Broad support: Arduino, Raspberry Pi, ESP32, industrial gateways (via MQTT/ HTTP/REST)	No	Advanced AI/ ML and enterprise integration
Arduino cloud	Free basic tier, paid Maker/Pro plans (more devices, dashboards)	Proprietary, cloud-based	Low	Smart home, DIY IoT, education, prototyping	Arduino boards, ESP8266/ESP32, MKR series, Nano 33 IoT	Yes	Seamlessly integrated with the Arduino ecosystem

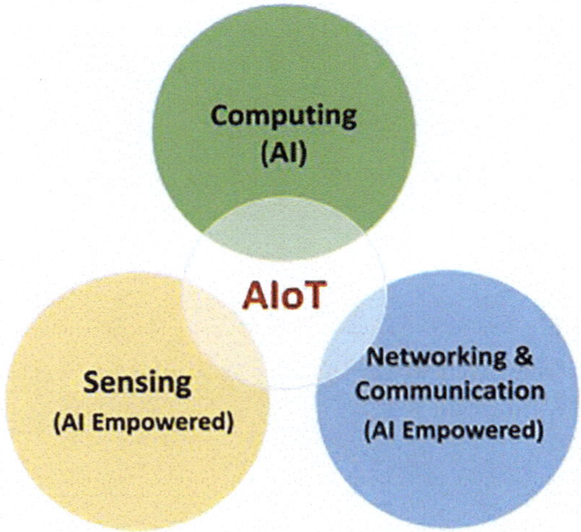

FIGURE 4.17 **AIoT paradigm.** The figure illustrates the AIoT paradigm, where IoT devices collect data from the physical environment and transmit it to edge or cloud platforms, where artificial intelligence algorithms analyze the data and enable intelligent decision-making and automation (*AIOT*, Artificial intelligence of things; *IoT*, Internet of Things).

location, among others. Finally, the networking and communications component of the AIoT ensures that sensor data and/or calculated results are reliably transmitted to the cloud, edge devices, or other nearby AIoT devices. The computing component of the AIoT focuses on AI-oriented computing workloads, unlike the traditional IoT. Furthermore, AI powers the sensing, networking, and communications components of the AIoT (Siam et al., 2025).

AIoT systems are primarily configured in either cloud or edge settings: (1) Cloud AIoT and (2) edge AIoT.

Cloud

AIoT involves managing and processing data from IoT devices using cloud computing platforms. Connecting IoT devices to the cloud is crucial, as this is where data is stored, processed, and accessed by various applications and services. The cloud AIoT architecture consists of four layers: the device layer, the connectivity layer, the cloud layer, and the user connectivity layer.

Edge artificial intelligence of things

AIoT data can also be processed at the edge, meaning that data from IoT devices is processed as close to these devices as possible to reduce the bandwidth required for data transmission and avoid potential data analysis delays. The edge AIoT architecture consists of three layers: the aggregation layer, the connectivity layer, and the edge layer.

Artificial intelligence of things architecture

With the integration of AI models into IoT devices and edge computing platforms, these new smart devices are bringing new levels of optimization and efficiency to many edge computing applications. AIoT architectures typically include (Jiyi et al., 2021; What is AIoT, 2025):

- IoT devices: Sensors, cameras, actuators, and edge devices.
- Connectivity: for data transmission
- Edge processing: AI models deployed locally to make low-latency decisions.
- Cloud intelligence: Centralized AI systems for model training and coordination between devices.
- Data management: Systems that ensure data is stored and shared securely, scalable, and compliant.
- User interfaces: Dashboards or mobile apps for displaying insights and alerts.

Edge computing is critical in the AIoT to reduce latency and improve response times for critical applications. Security measure is another key important factor to keep the AIoT system safe.

Edge artificial intelligence in Internet of Things

Edge AI in IoT is the process of running AI algorithms directly on local IoT devices, enabling real-time data processing and decision-making at the source. This approach, also known as edge computing, reduces reliance on cloud infrastructure, resulting in lower latency, reduced bandwidth usage, enhanced security and privacy, and improved efficiency in applications such as smart homes, self-driving vehicles, and industrial automation (Zou et al., 2019; Rathi et al., 2021; Merenda et al., 2020; Cheng et al., 2024). The future trajectory of AI and the IoT is intertwined with advances in edge AI. Edge computing enables faster distribution of information across the enterprise, reducing stagnation and increasing efficiency. Advances in AI algorithms will ensure transparency and responsible deployment.

Edge AI shifts computing from centralized cloud systems to local devices, such as sensors, smart cameras, and other IoT devices. By deploying AI models and algorithms at the edge, data can be analysed in real-time at the point of origin, enabling immediate responses and more efficient decision-making. This decentralized approach distributes AI across multiple devices, resulting in a more flexible, responsive, and scalable system. Key benefits of edge AI include reduced latency through local data processing, reduced bandwidth consumption through reduced cloud data transfers, and improved efficiency through more energy-efficient operations. AI also enhances security and privacy by storing sensitive data on-device, while ensuring reliable offline performance when network connectivity is weak.

Key applications of AIoT are:

- Smart cities and homes: Analysis of sensor and camera data for real-time traffic management, security, and home automation.
- Industrial Internet of Things (IIoT): Monitoring production lines to detect defects or inefficiencies, enable predictive maintenance, and improve safety.
- Autonomous vehicles: Real-time processing of sensor data (such as lidar and radar) to make critical driving decisions.
- Healthcare (IoMT): Analysis of medical device data for real-time diagnosis and patient monitoring.

Advantages and challenges in artificial intelligence of things

The benefits of the Artificial Internet of Things include (Awaisi et al., 2024; Matin et al., 2023):

- Real-time decision making: AI models can analyze streaming data and trigger immediate actions.
- Predictive maintenance: AI can identify patterns in sensor data that signal failure before it happens.
- Autonomous operations: AI-enabled devices can act without human input, reducing errors and improving efficiency.
- Scalability: AIoT systems adapt and learn, making them more efficient over time.
- Increased operational efficiency. IoT devices integrated with AI can analyze data to uncover patterns and insights, and fine-tune system processes for increased efficiency.
- Real-time adaptability. Data can be generated and analyzed to identify vulnerabilities, enabling the system to make necessary adjustments.
- Data analytics. Employees don't need to spend significant time monitoring IoT devices, saving money.
- Transformative technology. The Artificial Internet of Things is transformative and beneficial to both types of technology. AI adds value to IoT through machine learning capabilities and improved decision-making. IoT also adds value to AI through communication, signals, and data exchange. The AIoT can improve businesses and services by creating greater value from the data generated by IoT.

While AIoT offers significant advantages, implementation comes with challenges (Security Challenges and Solutions in AIOT Systems, 2024; Chang et al., 2021):

- Cybersecurity issues. The increasing number of AIoT-connected devices increases the risk of cyberattacks and security breaches.
- Complexity. Integrating AI and IoT technologies can be challenging and requires specialized knowledge and capabilities.

- Data management issues. Processing data collected from various sensors requires effective data management strategies.
- High cost. Implementing AIoT technologies can be expensive due to the need for specialized equipment, software, and personnel.
- Privacy issues. There are concerns about how data collected by AIoT devices is processed and stored, which could lead to privacy issues and breaches.
- Data security: More endpoints mean a larger attack surface for cyberthreats.
- Interoperability: Devices from different vendors must work together.
- Skills gap: Implementing and managing AIoT demands expertise in both fields.

The AIoT is evolving rapidly thanks to advances in generative AI, 5G networks, and edge computing. As networks become faster and more devices become connected to the internet, AIoT systems will enable entirely new experiences, from self-repairing factories to autonomous retail stores and smart public infrastructure. The global AI and IoT market, estimated at $126.1 billion in 2023, is projected to grow to $1319.4 billion by 2032, fueled by the rapid adoption of IoT devices, which are expected to surpass 30.9 billion units by 2025 (Artificial Intelligence of Things AIoT: Current State and Future Path, 2025).

The AIoT is ushering in a new era of technological advancement and opportunities across various sectors. As it continues to evolve, it plays a pivotal role in driving digital transformation, promoting sustainability, and enabling innovative applications.

The concept of secure artificial intelligence of things

Advances in IoT and AI technologies, particularly edge computing and advances in machine learning and deep learning, have significantly accelerated the growth of the AIoT. However, many challenges remain, most notably data privacy and security risks posed by the massive amount of sensitive information generated. Furthermore, interoperability across diverse IoT systems and ethical concerns, such as bias in AI algorithms, remain major obstacles (Artificial Intelligence of Things AIoT: Current State and Future Path, 2025). This data is often transmitted over networks and processed by AI systems, making it vulnerable to cyberattacks, breaches, and unauthorized access. Security challenges in AIoT systems stem from the complex interactions between data streams, intelligent algorithms, and interconnected devices. Given their high vulnerability to cyberattacks, these systems face significant risks. However, AI technologies are being used to process IoT data and extract valuable insights, enhancing their resilience and efficiency.

Several regulatory frameworks guide the deployment and security of AIoT systems. The EU AI Act (2024) establishes a comprehensive framework for AI, including AIoT applications, while the EU Cyber Resilience Act (2024) focuses on implementing effective cybersecurity measures for AI-based products. The General Data Protection Regulation (GDPR, 2016) protects the collection and processing of personal data, and the Network and Information Security Directive 2 (NIS2, 2023) provides cybersecurity guidance for various digital systems, including IoT networks. In the United States, the California IoT Security Act (2020) requires unique security features and passwords for consumer IoT devices, and the US IoT Cybersecurity Improvement Act (2020) sets cybersecurity standards for IoT devices used by federal agencies.

Major risks in artificial intelligence of things

AIoT security falls under cybersecurity, which focuses on protecting cloud-connected devices, also known as IoT devices, and the networks connected to them (Thapliyal et al., 2024). The risk of unauthorized access to sensitive data is one of the most serious vulnerabilities in AIoT deployments. Another major concern is the possibility of AI algorithms being tampered with or corrupted, which could compromise the system's accuracy and reliability. Additionally, denial-of-service attacks highlight the consequences of weak cybersecurity: attackers can flood AIoT systems with excessive traffic, rendering them inaccessible or unresponsive (Li et al., 2023). To address these risks, companies must place cybersecurity at the core of their AIoT strategies. This requires implementing robust ecosystem-wide protection measures, including device protection, network security, and data integrity and confidentiality (Securing AIoT, 2025; Khurshid et al., 2025; Megala et al., 2025). Key strategies for securing AIoT are (Megala et al., 2025; Qureshi and Newe, 2024):

Authentication mechanisms

Implementing strong authentication mechanisms, such as two-factor authentication (2FA), is critical to protecting IIoT devices and networks from unauthorized access. 2FA requires users to provide two types of credentials, typically something they know (password or PIN) and something they have (smartphone, token, or biometric data). This additional layer of security significantly reduces the risk of a breach, even if one of the credentials is compromised, and helps mitigate the vulnerabilities of easily guessed passwords. Various 2FA methods can be chosen, including SMS codes, authentication apps, hardware tokens, and biometric data, based on specific system requirements. In addition to security, 2FA can also enhance user confidence, as many users are already familiar with it from online services, making its integration into IIoT systems practical and easy to use.

Data encryption

Protecting AIoT data requires strong authentication and robust encryption. Encrypting data at rest, stored in databases or devices, ensures that even if unauthorized access is gained, the information remains unreadable without the correct encryption key. Similarly, encrypting data in transit protects against interception as it moves between devices or networks, allowing only the intended recipients to decrypt its content. Advanced algorithms, such as advanced encryption standard, provide high levels of confidentiality and integrity, while effective key management—including secure storage, controlled access, and regular key rotation—ensures that encryption remains reliable and resistant to attacks.

Monitoring and anomaly detection

Continuous monitoring and anomaly detection are essential for identifying and responding to security threats in AIoT systems in real-time. By tracking device behavior, network activity, and data flows, businesses can detect unusual patterns, such as unauthorized access attempts, abnormal network traffic, or unexpected device behavior, before they escalate into serious incidents. Identifying underlying behaviors and comparing them with real-time data enables rapid anomaly identification, enabling immediate mitigation actions such as isolating affected devices or blocking suspicious traffic. These mechanisms not only improve incident response but also provide valuable insights into recurring threats and system vulnerabilities. Leveraging tools such as intrusion detection systems, log monitoring solutions, network traffic analyzers, and Security Information and Event Management (SIEM) platforms, along with advanced analytics and machine learning, enhances AIoT security, helping businesses proactively protect their infrastructure and maintain the resilience of their systems.

Segmentation of the network

Segmenting AIoT networks is a key strategy for enhancing system security. By dividing the network into isolated segments with dedicated security measures, companies can contain potential breaches and prevent an attack in one segment from compromising the entire ecosystem. Segmentation also simplifies the management of complex AIoT infrastructures, allowing organizations to monitor and secure each segment individually. Critical segments that house sensitive data or essential operations can receive stronger protection, while controls on less critical areas can be relaxed, improving resource allocation. Effective network segmentation should be accompanied by robust access controls, continuous monitoring, and intrusion detection systems to quickly and effectively identify and mitigate unauthorized access or suspicious activity.

Regular software updates and patch management

Regularly updating AIoT hardware and software is essential to address known vulnerabilities and protect against evolving cyberthreats. Security patches and firmware updates fix discovered flaws, improve system performance, and often add new security features to address emerging threats. Neglecting updates can compromise AIoT systems, putting sensitive data and critical infrastructure at risk. Given the complexity of AIoT environments, especially with large-scale or diverse deployments, companies should implement an effective patch management process. This includes testing updates in controlled settings, scheduling installations to minimize disruption, and monitoring the update process to ensure successful deployment. By keeping hardware and software up to date, organizations can enhance security, maintain system integrity, and stay safe from potential threats.

Blockchain

Blockchain is a distributed digital ledger system that records transactions across multiple nodes in a transparent, secure, and tamper-resistant manner. Each record, or block, is linked to the previous one, forming a chain that ensures data consistency and prevents unauthorized modifications. Unlike traditional databases managed by a central entity, blockchain relies on decentralization, which enhances trust and accountability (Arora and Makani, 2024). In the IoT, blockchain plays a significant role in addressing challenges such as data integrity, privacy, and secure communication. IoT devices constantly generate massive amounts of sensitive information, and blockchain provides a mechanism for protecting this data, verifying its source, and maintaining traceability. It can be used, for example, to authenticate connected devices, manage permissions, and ensure reliable communication between machines (Li et al., 2024).

In AIoT, blockchain enhances its value by ensuring the reliability of data used in AI-based analytics and decision-making processes. Since AI systems rely heavily on accurate and reliable data, blockchain technology prevents tampering and ensures the integrity of information-feeding algorithms. Furthermore, through the use of smart contracts, blockchain enables the automation of Industrial Internet of Things (AIoT) systems—for example, coordinating devices, regulating energy exchanges in smart grids, or enhancing transparency in supply chains (Lakhan et al., 2025).

Blockchain technology has the potential to significantly enhance AIoT networks by providing a decentralized approach to storing and managing the massive data streams generated by connected devices. Its applications in the AIoT include secure data storage, decentralized identity management, and automated transactions. By recording information in an immutable, tamper-resistant ledger, blockchain ensures the accuracy and integrity of data generated by devices while preventing unauthorized access. Additionally, it enables the creation of decentralized digital identities for IoT devices, allowing them to authenticate and interact securely without relying on a central authority.

Fig. 4.18 shows a conceptual model of an integrated IoT ecosystem with blockchain technology. The left side highlights the blockchain layer, along

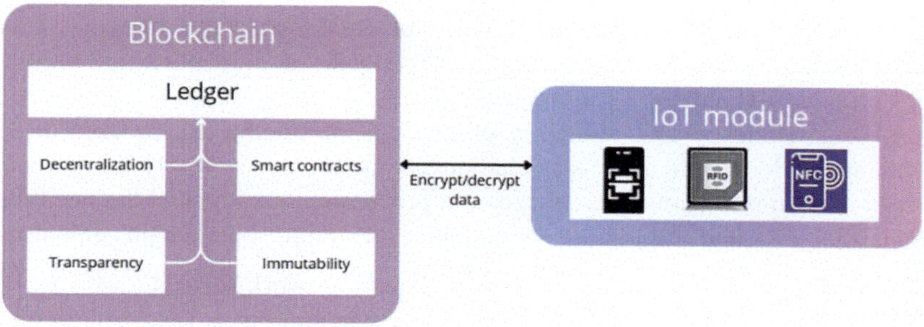

FIGURE 4.18 Blockchain-IoT conceptual diagram. The diagram shows how blockchain integrates with IoT systems: IoT devices collect and transmit data, which is recorded in secure, decentralized blocks on the blockchain network to ensure transparency, immutability, and trusted data sharing among stakeholders (*IoT*, Internet of Things).

with key features that add value to IoT networks, while the right side highlights commonly used IoT technologies (Almarri and Aljughaiman, 2024).

Key features of blockchain technology (Arora and Makani, 2024):

- *Programmable:* Blockchains can be programmed (e.g., using Solidity) to create smart contracts that execute automatically.
- *Secure:* Encryption, hashing, and Merkle trees ensure data integrity, making tampering or unauthorized access virtually impossible.
- *Distributed:* Managed as a peer-to-peer ledger, all authorized nodes share the same copy, ensuring transparency and reliability.
- *Anonymous:* Participant identities are concealed, preventing others from identifying or tracing transactions.
- *Timestamp:* Each block is assigned a timestamp upon its creation, ensuring the freshness of the data and allowing entities to verify when the record was stored.
- *Immutability:* Data in blockchain blocks cannot be modified because they are linked by hash chains. Any attempt to modify a block requires changing all subsequent hashes, making it virtually impossible to tamper with.
- *Consensus:* A new block is only added after a consensus mechanism is implemented. Mining nodes (such as cloud servers) must agree—usually by a majority (e.g., 80%)—before the block can be validated and added, ensuring collective consensus and trust.

Blockchain technology enhances the IoT by improving data privacy, access control, and cybersecurity, while supporting artificial intelligence with higher system reliability and efficiency. Its integration enables AIoT systems to deliver more stable and advanced services. It provides a secure, transparent, and decentralized framework for managing data and transactions in digital systems. It fosters trust by ensuring the tamper-proof and verifiable nature of records, which is crucial in distributed networks such as the IoT and AIoT.

Blockchain facilitates secure device authentication, data integrity, and traceability, reducing the risks of fraud, cyberattacks, and unauthorized access. Furthermore, it supports automated processes through smart contracts, improving efficiency and reducing reliance on intermediaries. Despite these advantages, blockchain also faces challenges, including scalability limitations, energy consumption concerns, transaction processing latency, and the need for standardization and regulatory compliance. Overall, blockchain enhances security, transparency, and accountability while carefully addressing its technical and operational limitations.

Digital twins

Definition and principle

A digital twin is a digital representation of a physical object, process, or service. It is a computer program that uses real data to create simulations that predict the performance of a product or process. The goal is to access all the information about a product without physically possessing it (See Fig. 4.19).

The operation of digital twins relies on close integration with the IoT and data analytics systems. This data is then processed by AI algorithms to model accurate predictions of the state of physical systems. This modeling includes simulations of different scenarios to test the system's reactions before real-world application, thereby increasing efficiency.

History and evolution of digital twins

Digital twin technology was first proposed by NASA during the Apollo missions in the 1960s (Glaessgen and Stargel, 2012). NASA established the first definition of a digital twin as an integrated, multiphysics, multiscale, probabilistic simulation of a system. Digital twin technology underwent a significant shift, first introduced by Michael Grieves in a presentation on

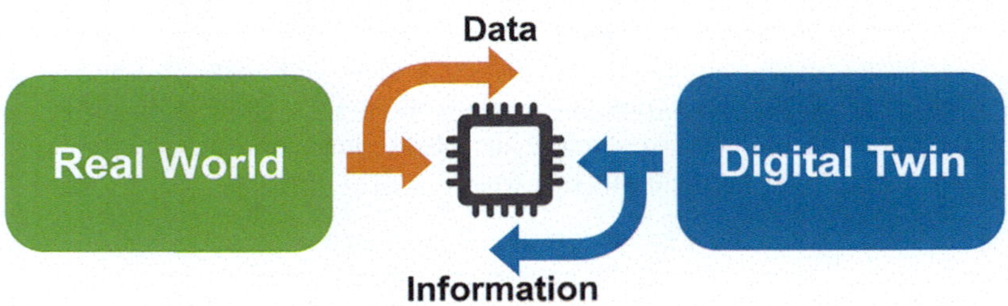

FIGURE 4.19 **Digital twin concept.** The figure illustrates the concept of a digital twin, showing a virtual model that mirrors the behavior and performance of a physical system in real time. Data flows continuously between the physical asset and its digital counterpart, enabling monitoring, simulation, and predictive decision-making.

product lifecycle management (PLM) (Grieves, 2005). The digital twin is considered a key element of Industry 4.0. He initially called his idea Conceptual Ideal for PLM, before renaming it the Mirror Spaces Model, and later the Information Mirror Model. Mr. Grieves defined three essential elements of a digital twin (The Digital twins, 2023):

- A virtual twin (digital representation)
- A physical counterpart (the real object)
- A data flow cycle linking the physical and virtual entities

These three elements remain at the core of how the term digital twin is understood today. However, it is important to note that researchers have not yet reached a consensus on a single, unified definition of a digital twin.

Structure and functioning of digital twins

A digital twin is a virtual representation designed to accurately reflect a physical object (e.g., a system). It covers the object's lifecycle, is updated with real-time data, and uses simulation, machine learning, and reasoning to facilitate decision-making. A digital twin is only truly useful if it enables the transformation of data streams into concrete actions. To achieve this, the implementation is structured around four key steps:

Data collection

Information and data models, along with real-world data, form the core of the digital twin. Data quality is crucial to the success or failure of the digital twin. The data must be accurate and precise for successful use. Both stock data and data generated by sensors are recorded and prepared. They form the backbone for creating the virtual model.

Virtual model

The virtual model is created based on the collected data. It offers a digital representation of the processes and facilities, and provides an initial overview of the processes. From this stage onward, the digital twin is a valuable tool for selecting the right facility and conveyor system.

Simulation and analysis

The digital twin simulates the processes and tests different scenarios using artificial intelligence algorithms. This allows companies to optimize logistics processes in the virtual environment before implementing them in real-world operations.

Monitoring and optimization

The digital twin offers advantages that go beyond simple trouble-free implementation. It continuously monitors logistics processes and compares actual performance to the virtual model. If discrepancies are found, the necessary optimization measures can be implemented immediately. Similarly, the digital twin allows for the evaluation of process changes alongside ongoing operations. This enables digital testing of effectiveness and efficiency.

Key benefits and challenges addressed by the digital twin

Digital twin technology addresses several key challenges in the design, operation, and maintenance of complex systems. It enhances real-time visibility by creating a virtual copy of physical assets, enabling continuous monitoring and a deeper understanding of system behavior. Digital twins improve maintenance by shifting from reactive or time-based approaches to predictive strategies, allowing for early fault detection and optimal intervention. They also support performance optimization by simulating various operational scenarios and identifying potential improvements. Furthermore, digital twins provide a secure environment for testing and validating new designs or control strategies, reduce data fragmentation by integrating multiple data sources into a unified model, and enhance forecasting and planning by assessing future conditions. Overall, digital twins enable more accurate diagnostics, improved decision-making, and better asset lifecycle management.

Conclusion

The AIoT offers tremendous potential for innovation and efficiency, but it also introduces new cybersecurity risks. Companies must implement strategies such as strong authentication, data encryption, regular updates, network segmentation, and continuous monitoring to protect their systems. Cybersecurity is an ongoing process that requires regular audits, risk assessments, and employee training. Looking ahead, AI-powered defences, privacy-preserving technologies, blockchain-enabled trust mechanisms, and evolving regulations will shape the future of AIoT security. Therefore, taking proactive measures is essential to protect operations, data, and stakeholder trust. Popular and emerging AIoT trends include edge computing, which processes data near its source to reduce latency and network congestion; swarm intelligence, which enables decentralized and self-organized optimization of IoT devices; and 5G connectivity, which supports faster data transfer with higher bandwidth and lower latency. The AIoT is also improving operational efficiency across various sectors, streamlining supply chains and human capital management, while computer vision enables machines to interpret visual data for automation, monitoring, and quality control.

Further developments include the use of digital twins to create virtual copies of physical systems for improved simulation, predictive maintenance, and performance enhancement, as well as the adoption of blockchain technology to enhance data integrity, transparency, and device authentication in distributed IoT environments. Moreover, AI-powered cybersecurity tools support proactive threat detection and mitigation, while environmental sustainability initiatives utilize smart systems to reduce energy consumption and optimize resource utilization. These trends collectively highlight the transformative role of AIoT in building smarter, safer, and more sustainable technology ecosystems.

AI Disclosure

During the preparation of this work, the author(s) used ChatGpt in order to Proofreading. After using this tool/service, the author(s) reviewed and edited the content as needed and take(s) full responsibility for the content of the publication.

References

A Comprehensive Guide to Communication Protocols in IoT 2025. Available at https://www.cavliwireless.com/blog/nerdiest-of-things/communication-protocols-in-iot.

Almarri, S., Aljughaiman, A., 2024. Blockchain technology for IoT security and trust: a comprehensive SLR. Sustainability 16 (23), 10177. Available at https://doi.org/10.3390/su162310177.

Al-Sarawi, S., Anbar, M., Alieyan, K., Alzubaidi, M., 2017. Internet of Things (IoT) communication protocols. In: 2017 8th International conference on information technology (ICIT) IEEE, pp. 685–690.

Arduino Nano 33 BLE Sense 2019. Available at https://docs.arduino.cc/hardware/nano-33-ble-sense/.

Arduino Portenta H7 2023. Available at https://docs.arduino.cc/resources/datasheets/ABX00042-ABX00045-ABX00046-datasheet.pdf.

Arora, P., Makani, R., 2024. Blockchain integration with AIoT data security and privacy for sustainability. Journal of Artificial Intelligence and Systems 6 (1), 112–123. Available at https://doi.org/10.33969/ais.2024060108.

Artificial Intelligence of Things (AIoT): Current State and Future Path 2025. Available at https://www.rinf.tech/artificial-intelligence-of-things-aiot-current-state-and-future-path/.

Awaisi, K.S., Ye, Q., Sampalli, S., 2024. A survey of industrial AIoT: opportunities, challenges, and directions. IEEE Access 12, 96946–96996. Available at https://doi.org/10.1109/ACCESS.2024.3426279, http://ieeexplore.ieee.org/xpl/RecentIssue.jsp?punumber=6287639.

Babun, L., Denney, K., Celik, Z.B., McDaniel, P., Uluagac, A.S., 2021. A survey on IoT platforms: communication, security, and privacy perspectives. Computer Networks 192.

Bayılmış, C., Ebleme, M.A., Çavuşoğlu, Ü., Küçük, K., Sevin, A., 2022. A survey on communication protocols and performance evaluations for Internet of Things. Digital Communications and Networks 8 (6), 1094–1104. Available at https://doi.org/10.1016/j.dcan.2022.03.013, https://www.sciencedirect.com/science/journal/23528648.

Blynk 2020 https://blynk.io/getting-started.

Chang, Z., Liu, S., Xiong, X., Cai, Z., Tu, G., 2021. A survey of recent advances in edge-computing-powered artificial intelligence of things. IEEE Internet of Things Journal 8 (18), 13849–13875. Available at https://doi.org/10.1109/JIOT.2021.3088875, http://ieeexplore.ieee.org/servlet/opac?punumber=6488907.

Cheng, Y., Zhang, H., Huang, Y., 2018. Overview of communication protocols in internet of things: Architecture, development and future trends. In: 2018 IEEE/WIC/ACM International Conference on Web Intelligence (WI) IEEE, pp. 627–630.

Cheng, L., Gu, Y., Liu, Q., Yang, L., Liu, C., Wang, Y., 2024. Advancements in accelerating deep neural network inference on AIoT devices: a survey. IEEE Transactions on Sustainable Computing 9 (6), 830–847. Available at https://doi.org/10.1109/TSUSC.2024.3353176, http://ieeexplore.ieee.org/xpl/RecentIssue.jsp?punumber=7274860.

Cloud Arduino 2019. Available at https://cloud.arduino.cc/.

ESP32 Series 2025. Available at https://www.espressif.com/sites/default/files/documentation/esp32_datasheet_en.pdf.

ESP8266EX 2025. Available at https://www.espressif.com/sites/default/files/documentation/0a-esp8266ex_datasheet_en.pdf.

Fortino, G., Guerrieri, A., Pace, P., Savaglio, C., Spezzano, G., 2022. IoT platforms and security: an analysis of the leading industrial/commercial solutions. Sensors 22 (6), 2196. Available at https://doi.org/10.3390/s22062196.

Getting Started with Seeed Studio XIAO nRF54L15(Sense) 2024. Available at https://wiki.seeedstudio.com/xiao_nrf54l15_sense_getting_started/.

Glaessgen, E., Stargel, D., 2012. The digital twin paradigm for future nasa and us air force vehicles. In: 53rd AIAA/ASME/ASCE/AHS/ASC structures, structural dynamics and materials conference 20th AIAA/ASME/AHS adaptive structures conference 14th AIAA.

Grieves, M.W., 2005. Product lifecycle management: the new paradigm for enterprises. International Journal of Product Development 2 (2).

IBM cloud 2016. Available at https://www.ibm.com/cloud/internet-of-things.

In Springer Handbook of Internet of Things. Springer International Publishing, (2023).

IoT Architecture Guide. 2018 Major and Additional Layers of IoT System https://www.helpwire.app/blog/iot-architecture/.

Jiyi, W.U., Wenjuan, L.I., Jian, C.O., Shiyou, Q.I.A.N., 2021. AIoT: a taxonomy, review and future directions. Telecommunications Science 37 (8).

Khanna, A., Kaur, S., 2020. Internet of things (IoT), applications and challenges: a comprehensive review. Wireless Personal Communications 114 (2).

Khurshid, K., Khurshid, K., Usman Hadi, M., Al Bataineh, M., Saeed, N., 2025. Securing AIoT surveillance: techniques, challenges, and solutions. IEEE Open Journal of the Communications Society 6, 6517–6550. Available at https://doi.org/10.1109/OJCOMS.2025.3593311, https://ieeexplore.ieee.org/xpl/RecentIssue.jsp?punumber=8782661&_ga=2.260517901.1496039865.1593798416-2073569860.1593798416.

Lakhan, A., Alyasseri, Z.A.A., Mohammed, M.A., Al-Attar, B., Nedoma, J., Alubady, R., et al., 2025. Sustainable secure blockchain assisted AIoT and green multiconstraints supply chain system. IEEE Internet of Things Journal 12 (19), 39392–39406. Available at https://doi.org/10.1109/JIOT.2025.3548037, http://ieeexplore.ieee.org/servlet/opac?punumber=6488907.

Li, S., Xu, L.D., Zhao, S., 2015. The internet of things: a survey. Information Systems Frontiers 17 (2), 243–259. Available at https://doi.org/10.1007/s10796-014-9492-7, http://www.springeronline.com/sgw/cda/frontpage/0,11855,4-170-70-35673075-0,00.html?detailsPage&contentItemPagecontentItemId=140346&CIPageCounter=CI_FOR_AUTHORS_AND_EDITORS_PAGE1.

Li, J., Herdem, M.S., Nathwani, J., Wen, J.Z., 2023. Methods and applications for Artificial Intelligence, Big Data, Internet of Things, and Blockchain in smart energy management. Energy and AI 11. Available at https://doi.org/10.1016/j.egyai.2022.100208, www.journals.elsevier.com/energy-and-ai.

Li, Y., Shen, J., Ji, S., Lai, Y.H., 2024. Blockchain-based data integrity verification scheme in AIoT cloud-edge computing environment. IEEE Transactions on Engineering Management 71, 12556–12565. Available at https://doi.org/10.1109/TEM.2023.3262678, https://ieeexplore.ieee.org/servlet/opac?punumber=17.

Matin, A., Islam, M.R., Wang, X., Huo, H., Xu, G., 2023. AIoT for sustainable manufacturing: overview, challenges, and opportunities. Internet of Things 24, 100901. Available at https://doi.org/10.1016/j.iot.2023.100901.

Megala, P., Prabhu, R., Seethalakshmi, V., Sathiyapriya, T., Thilagavathi, S., Gurumoorthy, K.B., 2025. Privacy and Security in AIoT. Informa UK Limited, pp. 34–56. Available at https://doi.org/10.1201/9781003482338-3.

Merenda, M., Porcaro, C., Iero, D., 2020. Edge machine learning for AI-enabled IoT devices: a review. Sensors 20 (9), 2533. Available at https://doi.org/10.3390/s20092533.

Minerva, R., Biru, A., Rotondi, D., 2015. Towards a definition of the Internet of Things (IoT). IEEE Internet Initiative 1 (1).

Network Layer Protocols, Available at https://www.geeksforgeeks.org/computer-networks/network-layer-protocols/.

Particule 2025. Available at https://www.particle.io/.

Pico-series Microcontrollers 2023. Available at https://www.raspberrypi.com/documentation/microcontrollers/pico-series.html.

Qureshi, K.N., Newe, T., 2024. Artificial Intelligence of Things (AIoT): New Standards, Technologies and Communication Systems. CRC Press. Available at https://doi.org/10.1201/9781003430018.

Rathi, V.K., Rajput, N.K., Mishra, S., Grover, B.A., Tiwari, P., Jaiswal, A.K., et al., 2021. An edge AI-enabled IoT healthcare monitoring system for smart cities. Computers and Electrical Engineering 96. Available at https://doi.org/10.1016/j.compeleceng.2021.107524, https://www.journals.elsevier.com/computers-and-electrical-engineering.

Schizas, N., Karras, A., Karras, C., Sioutas, S., 2022. TinyML for ultra-low power AI and large scale IoT deployments: a systematic review. Future Internet 14 (12), 363. Available at https://doi.org/10.3390/fi14120363.

Securing AIoT 2025 https://www.kajeet.com/en/blog/securing-aiot-the-importance-of-cybersecurity.

Security Challenges and Solutions in AIOT Systems 2024. Available at https://medium.com/@apurva.sorte/security-challenges-and-solutions-in-aiot-systems-4d76cb9305b7.

Sharma, R., & Sharma, D. (Eds.). (2022). New trends and applications in internet of things (IoT) and big data analytics (Vol. 14, p. 271). Springer.

Siam, S.I., Ahn, H., Liu, L., Alam, S., Shen, H., Cao, Z., et al., 2025. Artificial intelligence of things: a survey. Association for computing machinery. ACM Transactions on Sensor Networks 21 (1). Available at https://doi.org/10.1145/3690639, http://dl.acm.org/citation.cfm?id=J981.

STM32 wireless MCUs 2025. Available at https://www.st.com/en/microcontrollers-microprocessors/stm32-wireless-mcus.html.

Thapliyal, S., Wazid, M., Singh, D.P., Chauhan, R., Mishra, A.K., Das, A.K., 2024. Secure Artificial Intelligence of Things (AIoT)-enabled authenticated key agreement technique for smart living environment. Computers and Electrical Engineering 118. Available at https://doi.org/10.1016/j.compeleceng.2024.109353, https://www.sciencedirect.com/science/journal/00457906.

The 5 Types of IoT Platforms 2024. Available at https://blog.velosiot.com/the-5-types-of-iot-platforms.

The Digital twins. Springer International Publishing, (2023).

The internet of things for everyone 2025. Available at https://io.adafruit.com/.

Thingspeak 2010. Available at https://thingspeak.mathworks.com/.

Weber, R.H., Weber, R., 2010. Internet of things. Heidelberg 12.

What is AIoT 2025. Available at https://blog.tmcnet.com/blog/rich-tehrani/ai/what-is-aiot-understanding-the-artificial-intelligence-of-things.html.

Zhong, Y., Chen, L., Dan, C., Rezaeipanah, A., 2022. A systematic survey of data mining and big data analysis in internet of things. The Journal of Supercomputing 78 (17), 18405–18453. Available at https://doi.org/10.1007/s11227-022-04594-1.

Zou, Z., Jin, Y., Nevalainen, Y., Huan, Y., Heikkonen, J., Westerlund, T., 2019. Edge and fog computing enabled AI for IoT-an overview. In: 2019 IEEE International Conference on Artificial Intelligence Circuits and Systems (AICAS) IEEE, pp. 51–56.

Chapter 5

Data augmentation and synthetic techniques for photovoltaics

Nadia Drir[1], Adel Mellit[2], and Mahmoud Dhimish[3]
[1]*Faculty of Electrical Engineering, USTB University, Algiers, Algeria,* [2]*Faculty of Sciences and Technology, University of Jijel, Jijel, Algeria,* [3]*Department of Photonics Engineering, Technical University of Denmark, Roskilde, Denmark*

Introduction

The increasing penetration of photovoltaic (PV) systems into global energy infrastructures has established them as a key technology for the transition toward sustainable electricity production (Rocha et al., 2023). However, their efficiency and long-term reliability are compromised by defects such as microcracks, hotspots, potential-induced degradation, and soiling, which reduce energy yield and increase maintenance costs if not diagnosed early (Drir et al., 2025a). Consequently, accurate, timely, and automated fault diagnosis is essential to ensure optimal performance and reduce operational expenses. Traditional diagnostic methods, including electrical characterization and I–V curve analysis, provide useful insights but are limited in scalability and spatial precision. By contrast, image-based techniques such as electroluminescence (EL), infrared (IR) thermography, and visible imaging allow high-resolution, noninvasive fault detection without interrupting energy production (Karimi et al., 2020; Su et al., 2019).

The integration of artificial intelligence (AI), particularly deep learning (DL), has further advanced this field by enabling automatic feature extraction and classification (Mellit and Kalogirou, 2021; Tang et al., 2024). This shift from manual inspection to AI-driven diagnosis has opened the path toward real-time, intelligent monitoring of PV systems. A persistent challenge, however, lies in the issue of data imbalance: datasets are often dominated by healthy modules, whereas defective samples, particularly those corresponding to rare fault types, remain underrepresented (Drir et al., 2025b). Such an imbalance biases deep learning models toward the majority classes, thereby diminishing diagnostic accuracy for infrequent yet critical defects. To mitigate this problem, advanced data augmentation and synthetic data generation

Embedded Artificial Intelligence and the Internet of Things for Photovoltaic Systems.
DOI: https://doi.org/10.1016/B978-0-443-44835-5.00013-8

techniques are essential for enriching minority classes and enhancing overall model generalization (Goodfellow et al., 2020). Basic augmentation methods (e.g., rotations, flips, and noise injection) offer partial improvements but cannot capture the intrinsic variability of PV defects. To overcome these limitations, more sophisticated generative approaches such as generative adversarial networks (GANs), variational autoencoders (VAEs), and hybrid architectures have been introduced (Radford et al., 2016). These techniques generate realistic synthetic images that not only mitigate class imbalance but also enhance the robustness and scalability of AI-driven diagnostics (Arjovsky et al., 2017; Gulrajani et al., 2017). The growing body of research in this domain highlights the central role of synthetic data in advancing reliable PV fault detection by producing diverse and representative training samples.

This chapter investigates data augmentation and synthetic data generation to enhance PV defect diagnosis using EL and IR thermal imaging. It first reviews conventional augmentation techniques, including geometric and optical transformations, and their role in mitigating dataset imbalance. The chapter then compares advanced synthetic data generation methods, such as GAN-based models, feature-guided adversarial frameworks, and conditional variational autoencoders (CVAEs), for producing realistic PV images. An evaluation framework combining histogram-, distribution-, and statistical-based metrics is introduced to assess the quality of the generated data. The strengths and limitations of different augmentation strategies are discussed, highlighting how synthetic data can complement real datasets to improve model robustness. Finally, open-source PV datasets are presented, and future directions for integrating synthetic data pipelines into scalable PV diagnostic systems are outlined.

Definition of data augmentation and synthetic data

Data augmentation refers to a set of techniques that artificially expand the volume and diversity of training datasets by applying label-preserving transformations to existing samples. In the context of EL images of cell-level manufacturing defects and IR images of installed PV panels, these transformations may include geometric modifications (e.g., rotation, flipping, translation), photometric adjustments (e.g., contrast enhancement, noise injection), or more advanced spatial alterations. Such techniques are crucial for DL models, as they simulate real-world variability such as changes in panel orientation, lighting, or thermal behavior, thereby enhancing the model's generalization ability and reducing overfitting. However, data augmentation is inherently limited by the information present in the original dataset. It does not generate entirely new content or unseen defect patterns; rather, it diversifies existing examples while preserving their semantic integrity (Tan et al., 2024; Tang et al., 2023).

Synthetic data, by contrast, consists of artificially generated samples produced using mathematical models, physics-based simulations, or deep generative architectures such as GANs and VAEs. In PV applications, synthetic data allows the creation of entirely new EL or IR images representing rare, extreme, or safety-critical defect scenarios such as microcracks, hot spots, delamination, or soldering faults that are often underrepresented or missing from field-collected datasets. This approach is particularly valuable when acquiring real images is expensive, impractical, or limited by operational constraints. When properly validated, synthetic data serves as a powerful training resource for robust and high-performance DL models, thereby enabling more accurate fault detection, classification, and predictive maintenance in PV systems.

Dataset description

To establish a reliable foundation for the development of data-driven augmentation and synthetic generation techniques, two complementary datasets of PV imagery were used in this chapter. The first dataset consists of high-resolution EL images made publicly available by Hebei University of Technology and Beijing University of Aeronautics and Astronautics (PVEL-AD, 2024). It contains 12 clearly defined fault categories and covers a broad spectrum of degradation modes encountered in PV modules. While this dataset is highly detailed and diverse, it presents a substantial imbalance across classes: common defects such as Finger and Crack are largely overrepresented, whereas critical categories such as Corner and Scratch are represented by only a few samples.

Fig. 5.1 presents a visual overview of the twelve defect categories represented in the EL dataset, whereas Table 5.1 summarizes their distribution by detailing the number of images available in each class. Together, these resources illustrate both the richness of the dataset and the imbalance challenge that underscores the need for data augmentation and synthetic data generation.

The second dataset consists of 20,006 IR thermography images captured by Unmanned Aerial Vehicle (UAV) platforms using medium and long-wave infrared sensors (Millendorf et al., 2020). The images have a resolution of 24×40 pixels and cover a similar set of 12 anomalous conditions as well as normal operating states. Although this dataset is well-suited for hotspot fault analysis, it suffers from the same imbalance problem, with the "no anomaly" class dominating the distribution and rare classes such as diode-multi and soiling containing only a limited number of samples.

Fig. 5.2 shows representative IR images of the anomaly categories, whereas Table 5.2 details their distribution. This presentation highlights the predominance of normal states and the underrepresentation of critical faults.

Given this imbalance, implementing data augmentation and synthetic data generation techniques is essential to mitigate uneven class distributions and

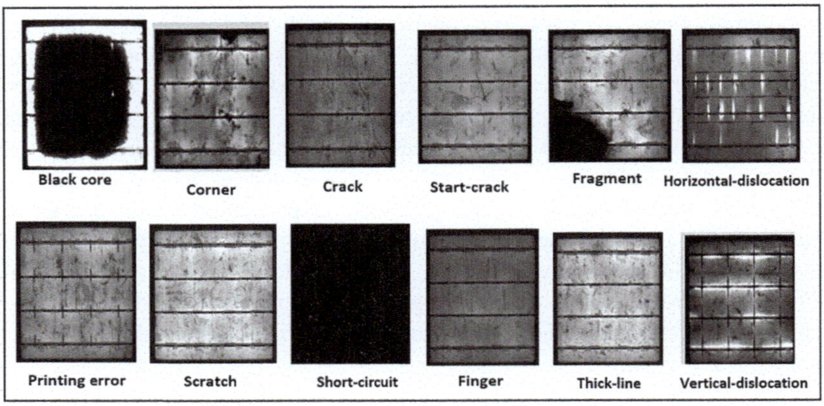

FIGURE 5.1 Illustration of the 12 defect types in the EL dataset. The figure illustrates the 12 defect types included in the EL dataset, providing visual examples of each category. It highlights the diversity of defects encountered in photovoltaic modules, which are used for training and evaluating diagnostic and classification models (*EL*, Electroluminescence). *https://github.com/binyisu/PVEL-AD.*

TABLE 5.1 Defect categories and corresponding image counts in the EL dataset.

Faults	Labels	Number images
Black_core	0	1028
Corner	1	9
Crack	2	1260
Star_crack	3	135
Fragment	4	7
Horizontal_dislocation	5	798
Printing_error	6	32
Scratch	7	5
Short-circuit	8	492
Finger	9	2958
Thick_line	10	981
Vertical_dislocation	11	137
Total		7842

EL, Electroluminescence.

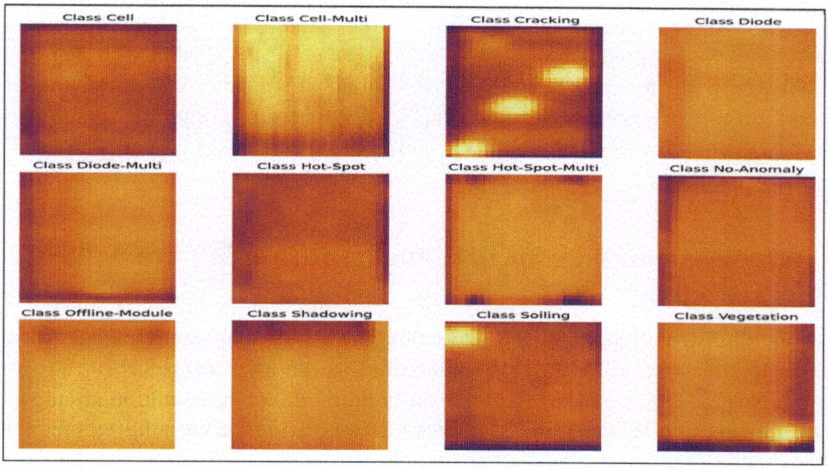

FIGURE 5.2 Visual examples of the 12 fault categories in the IR dataset. The figure presents visual examples of the 12 fault categories in the IR dataset, showing representative infrared images of each defect type. It demonstrates the variety of thermal anomalies used for training and testing PV module fault detection models (*IR*, Infrared; *PV*, photovoltaic). *https://github.com/RaptorMaps/InfraredSolarModules.*

TABLE 5.2 Fault classes and image distribution in the IR dataset.

Faults	Labels	Number of images
Cell	0	1877
Cell-Multi	1	1288
Cracking	2	941
Diode	3	1499
Diode-Multi	4	175
Hot-Spot	5	251
Hot-Spot-Multi	6	247
No-Anomaly	7	10,000
Offline-Module	8	828
Shadowing	9	1056
Soiling	10	205
Vegetation	11	1639
Total		20,006

IR, Infrared.

improve the representativeness of minority faults. The following section introduces the adopted augmentation strategy, specifically designed to increase the diversity of rare categories while preserving the intrinsic characteristics of each fault type. Accordingly, an overview is provided of commonly employed data augmentation approaches in PV defect imaging, encompassing both EL and IR modalities.

Common techniques for data augmentation in photovoltaic defect imaging

In PV defect detection, both EL and IR imaging function as powerful diagnostic modalities. Despite differences in acquisition conditions and defective visibility, both modalities benefit greatly from tailored data augmentation strategies that improve model generalization and robustness. The following techniques are among the most commonly applied (Wang and Cheng, 2025):

Geometric transformations

- Rotation: Helps the model learn rotational invariance, especially when the orientation of cells or panels varies in the dataset.
- Flipping (horizontal/vertical): Increases robustness to symmetry, which is especially useful for module-level IR images.
- Scaling and resizing: Simulates changes in camera distance or resolution.
- Cropping and random zooming: Focuses the model's attention on local defect regions, which is critical in high-resolution EL imagery.
- Translation and shifting: Accounts for sensor misalignment or module displacement.

Photometric transformations

- Brightness and contrast adjustment: Useful for simulating variability in lighting or signal intensity, especially relevant for EL images captured under different excitation currents.
- Noise injection (Gaussian, Salt-and-Pepper): Enables the model to become robust to sensor or environmental noise, a phenomenon common to both IR and EL acquisition.
- Blurring and sharpening: Simulates focus variability, enhancing the model's tolerance to imperfect imaging conditions.

These techniques are particularly effective when chosen carefully to align with modality-specific characteristics: EL images benefit from contrast enhancement and precise cropping, whereas IR images often require noise mitigation and geometric robustness. Thoughtful application of data augmentation is therefore critical for training robust and generalizable models for PV fault classification. While EL and IR images share several applicable techniques, optimal augmentation must consider the imaging modality, defect type, and environmental acquisition conditions. Balancing fundamental transformations with more advanced augmentation strategies ensures a rich and representative training set, thereby improving fault localization, classification accuracy, and ultimately the reliability of PV plant monitoring systems.

Algorithm's and diagrams

In the context of PV defect detection using EL and IR imagery, the data augmentation pipeline is carefully structured to emulate real-world variability while preserving the semantic integrity of the original image content. To achieve this, both geometric and photometric operations are applied in a randomized yet sequential manner to each image (See Fig. 5.3). This approach helps mitigate overfitting and promotes stronger model generalization, especially when the available datasets are limited or imbalanced. The augmentation workflow can be summarized as follows (Zhang et al., 2024):

Step #1 Random image selection

A single image is randomly selected from the original dataset for each augmentation operation. This randomness ensures a broad and unbiased sampling of the available data, reducing the risk of overrepresenting specific modules or defect types. By diversifying the input samples at the start of the process, the pipeline encourages greater variability in the resulting augmented dataset.

Step #2 Preprocessing of the selected image

If the selected image is in Red-Green-Blue (RGB) format, it is converted to grayscale to maintain uniformity across all subsequent transformations and analyses. The grayscale image is then normalized to the range [0, 1], ensuring

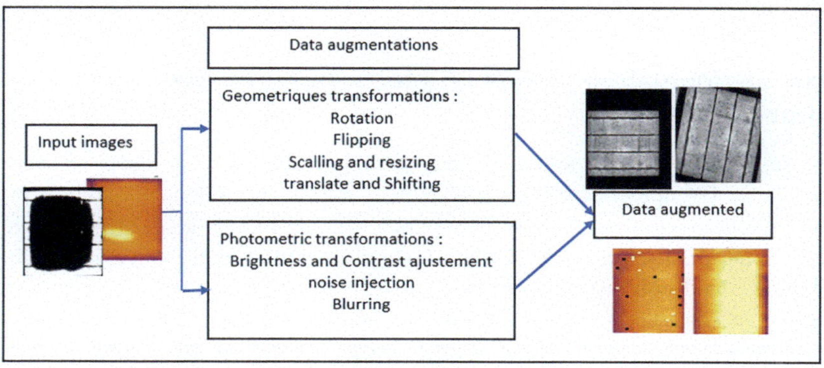

FIGURE 5.3 Data augmentation process. The figure illustrates the data augmentation process, showing how original images are transformed through techniques such as rotation, scaling, flipping, and color adjustments. These augmentations increase dataset diversity, improve model generalization, and enhance the performance of machine learning algorithms.

consistent pixel intensity scaling. Normalization improves numerical stability during training and enhances visual clarity when applying color maps to emphasize structural differences.

Step #3 Definition of the augmentation module

A comprehensive augmentation module is defined, comprising eight commonly employed geometric and photometric transformations: Rotation; Horizontal and vertical flipping; Scaling and resizing; Cropping combined with zooming; Translation; Brightness and contrast adjustment; Noise injection; and Blurring or sharpening filters. Each transformation has been selected to reproduce specific variations that occur in real PV monitoring scenarios, such as changes in viewing angle, illumination, focus, and sensor performance.

Step #4 Independent transformation application

Each transformation is applied independently to a newly selected random image, ensuring that every augmented sample originates from a distinct source. This strategy prevents augmentation-induced correlations between images and significantly enhances intraclass variability. As a result, the trained model becomes more resilient to unseen defect configurations.

Step #5 Postprocessing of augmented images

Following augmentation, the resulting images are standardized for consistency. If required, they are converted back to grayscale and renormalized to the [0, 1] range prior to visualization or model input. This step guarantees uniform pixel scaling across all augmented datasets and prepares the images for effective visual inspection.

Step #6 Visualization using colormap

To highlight subtle structural features and emphasize defect regions, the normalized augmented images are visualized using the Inferno colormap. This visualization enhances contrast in both dark and bright regions, making defect patterns more perceptible and improving interpretability during qualitative analysis.

Data augmentation results on electroluminescence and infrared datasets

To improve the robustness and generalization capability of deep learning models for PV defect detection, a systematic and well-defined data augmentation pipeline is applied. Starting from each original image, a set of geometric (e.g., rotation, flipping, translation) and photometric (e.g., noise injection, brightness and contrast adjustment) transformations is randomly chosen and applied in sequence to generate multiple augmented instances. These transformations emulate realistic variations in image acquisition conditions, such as changes in illumination, viewing angle, and sensor response, while preserving the original semantic label of each sample. By increasing the diversity of the training dataset, the augmentation strategy effectively mitigates class imbalance and enhances the model's ability to identify rare fault patterns that frequently occur in real-world PV inspection scenarios (da Silveira Junior et al., 2025).

The implemented transformations include:

- Rotation: Random rotation by 90°, 180°, or 270° to introduce rotational invariance, particularly relevant for modules captured at varying orientations.
- Flipping: Horizontal, vertical, or diagonal flips to increase robustness against mirrored configurations, especially common in UAV-based IR imaging.
- Scaling and resizing: Random scaling within the range of 0.8, 1.2× followed by resizing to the original dimensions, simulating variations in sensor distance or resolution.
- Cropping and zooming: Random crops covering 70%–90% of the original area, resized back to the original size, enabling the network to focus on localized defects.
- Translation: Random affine shifts of up to ±20 pixels along both axes to account for sensor misalignment or UAV drift.
- Brightness and contrast adjustment: Random intensity scaling (0.5 or 1.5×) to simulate illumination variability in EL images or sensor gain variations in IR data.
- Noise injection: Gaussian noise ($\sigma \approx 15$) and salt-and-pepper noise (2%-pixel corruption) were introduced to increase robustness against sensor or environmental noise.
- Blurring and sharpening: Gaussian blur (5×5 kernel) simulated focus variability, while sharpening enhanced defect edges for improved feature learning.

Figs. 5.4 and 5.5 provide visual examples of these transformations applied to representative EL and IR images, illustrating the diversity achieved through augmentation. Table 5.3 summarizes the implementation details and the diagnostic benefits of each technique.

Advantages and limitations of data augmentation in photovoltaic imaging

Data augmentation techniques offer substantial benefits in the field of PV fault detection and diagnosis, particularly when dealing with limited or highly imbalanced datasets, a common challenge in both EL and IR image collections. Nevertheless, their implementation also introduces certain challenges, requiring careful tuning and validation to ensure their suitability for each imaging modality (Hijjawi et al., 2023). Table 5.4 provides a concise summary of the principal strengths and inherent drawbacks of applying data augmentation in PV defect detection using EL and IR imagery, emphasizing its contribution to improved model generalization while recognizing the remaining limitations related to dataset diversity and physical realism.

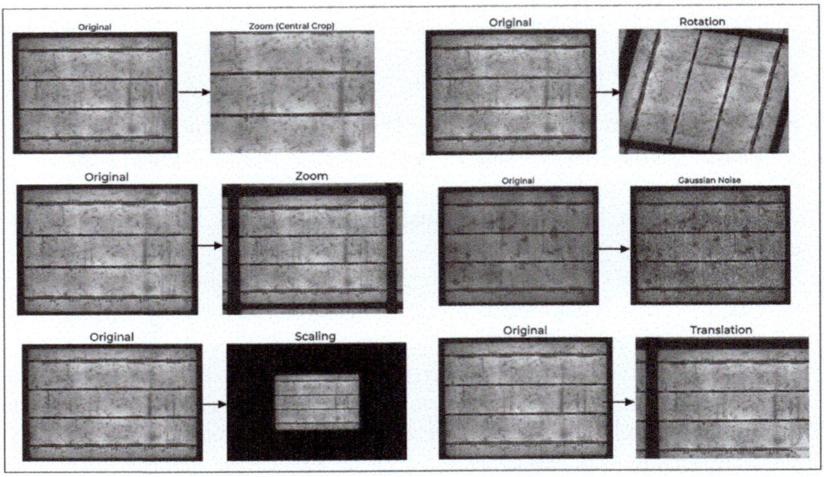

FIGURE 5.4 Examples of augmented EL images. The figure shows examples of augmented EL images, demonstrating transformations such as rotation, scaling, and flipping applied to the original dataset. These augmentations expand the dataset and help improve the robustness and accuracy of machine learning models for PV defect detection (*EL*, Electroluminescence; *PV*, photovoltaic).

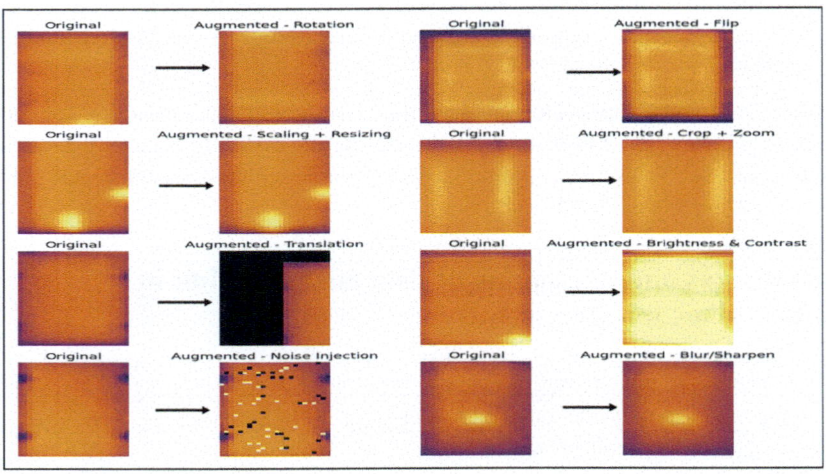

FIGURE 5.5 Examples of augmented IR images. The figure presents examples of augmented IR images, illustrating transformations such as rotation, flipping, and contrast adjustment applied to the original IR dataset. These augmentations increase data diversity and enhance the robustness of machine learning models in detecting PV module faults (*IR*, Infrared; *PV*, photovoltaic).

TABLE 5.3 Summary of basic data augmentation for PV defect imaging.

Techniques	Implementation	Purpose in PV imaging
Rotation (90°, 180°, 270°)	Random rotation	Increases rotational invariance
Flipping (H/V/Diagonal)	Random flips	Improves robustness to symmetry
Scaling and resizing (0.8–1.2×)	Rescale + resize	Simulates varying distances/resolution
Cropping and zooming (70–90%)	Random crop + resize	Focus on localized defect regions
Translation (±20 px)	Affine shift	Accounts for UAV drift or misalignment
Brightness/Contrast (0.5–1.5×)	Intensity adjustment	Simulates illumination/gain variability
Noise injection	Gaussian/Salt-and-Pepper	Improves robustness to acquisition noise
Blurring/Sharpening	Gaussian blur/Edge enhancement	Models focus on variability, highlighting edges

PV, Photovoltaic.

Synthetic data

Synthetic data refers to artificially generated information produced by computational models rather than collected from direct measurements or real-world observations. In the context of image-based fault detection in PV systems, such data are typically created to simulate realistic defect patterns or module conditions using generative approaches such as GANs (Ebied et al., 2025), VAEs (Harrou et al., 2024), or advanced physics-informed simulation frameworks. These artificial samples are designed to replicate the statistical distributions and visual features of real EL or IR images, providing a valuable complement to limited experimental datasets. The relevance of synthetic data has grown substantially as annotated real-world images remain scarce, imbalanced, or difficult to obtain, particularly for rare, early-stage, or safety-critical defects. By augmenting underrepresented classes with synthetic instances, researchers can construct more balanced datasets that enhance model robustness and fault detection accuracy.

Furthermore, synthetic data mitigates challenges related to data privacy, acquisition cost, and the extensive time required for manual labeling in industrial environments. In deep learning applications, synthetic data serves

TABLE 5.4 Advantages and limitations of data augmentation in PV fault detection.

Advantages	Limits
Artificially increasing the variability of the training samples enables the model to better generalize to unseen inputs. This is particularly important in PV applications where annotated datasets are often small and skewed toward normal operating conditions.	Data augmentation cannot generate defect characteristics that are entirely absent from the original dataset. Consequently, rare or subtle fault types remain underrepresented.
Transformations such as noise injection, brightness/contrast adjustment, and geometric operations make the model more resilient to real-world variations in acquisition settings (e.g., illumination changes or camera misalignment).	When applied in excess or without domain awareness, augmentations may produce unrealistic patterns that are not representative of actual PV behavior, which can mislead the model.
Since augmented samples retain the original semantic labels, the technique avoids the need for costly and time-consuming manual annotation conducted by PV experts.	After a certain level of augmentation, additional transformations lead to marginal improvements and may even deteriorate model specificity due to overregularization.
Appropriate photometric and spatial transformations can simulate variability across module types, operational conditions, and sensor configurations, thus improving transferability between PV systems.	Certain transformations (e.g., horizontal flipping of EL images) may introduce spatial inconsistencies that are physically implausible in the context of PV fault localization.

PV, Photovoltaic.

not only to expand training datasets but also to introduce controlled variability that improves model generalization. For example, in PV module inspection, synthetic EL images can be generated to represent cracks, hotspots, or delamination defects under varying illumination and noise conditions, thereby enabling models to learn invariant features for more reliable real-world performance. However, despite these advantages, achieving sufficient realism and diversity remains a critical challenge. If not properly modeled, synthetic samples may introduce nonphysical artifacts or distributional biases, limiting their ability to generalize to real-world data and potentially leading to overfitting on unrealistic features.

Techniques for generating synthetic data

Generating synthetic data for PV defect imaging involves a variety of techniques, ranging from traditional image processing to advanced deep generative models. Each method offers distinct advantages depending on the complexity of the data, the level of realism required, and the intended application, whether for classification, detection, or segmentation tasks.

Generative adversarial networks

GANs are among the most powerful and widely adopted approaches for synthetic image generation. A GAN comprises two competing neural networks, the generator and the discriminator, trained simultaneously in an adversarial framework. The generator learns to synthesize realistic PV defect images, while the discriminator evaluates whether each image is genuine or artificially produced (Kasi et al., 2023). Through this adversarial training process, the generator progressively improves its ability to produce high-fidelity images that capture the textural, spatial, and spectral characteristics of real EL and IR data. Variants such as conditional GANs (cGANs) and deep convolutional GANs (DCGANs) have demonstrated particular effectiveness for controlled image generation conditioned on specific defect types or operational scenarios (Shorten and Khoshgoftaar, 2019).

The framework employs a DCGAN to synthesize defect images, addressing the critical challenge of data scarcity within minority defect categories. In PV system diagnostics, the uneven distribution of defect types often limits the development of robust and well-generalized classifiers. By leveraging DCGANs, the model produces high-fidelity synthetic samples conditioned on specific defect classes, thereby expanding the training dataset with diverse, realistic representations of underrepresented faults. This synthetic data augmentation enhances detection accuracy and strengthens model generalization.

The DCGAN architecture is composed of two primary neural networks: a generator (G) and a discriminator (D). The generator receives as input a latent noise vector (z) and a class label (y) and outputs a synthetic image, denoted $G(z, y)$, designed to reproduce the visual and structural characteristics of real defect images corresponding to class y.

Conversely, the discriminator functions as a conditional binary classifier, denoted $D(x, y)$, which estimates the probability that an input image x, conditioned on label y, is real rather than generated. The adversarial training process progresses through the iterative optimization of both networks: the generator seeks to deceive the discriminator by minimizing the likelihood that synthetic images are correctly identified, whereas D simultaneously strives to maximize its discriminative accuracy between real and generated samples. Formally, this adversarial interaction is expressed through the conditional minimax objective function defined in Eq. (5.1):

$$\min_G \max_D L_{GAN}(D, G) = E_{(x,y)}[\log(D(x, y)] - E_{(y,z)}[\log(1 - D(G(z, y), y))] \tag{5.1}$$

This implementation focuses on a DCGAN designed to generate synthetic defect images with dimensions 224×224×3, addressing class imbalance in defect datasets by conditioning the generation on discrete class labels. Network architectures process consists of:

Step# 1 Generator

The generator receives as input a latent noise vector z sampled from a standard normal distribution, concatenated with an embedding vector corresponding to the class label y. The label embedding layer maps each discrete class to a continuous 50-dimensional vector, enabling the generator to condition image synthesis on class-specific features.

The latent vector is projected and reshaped into a low-resolution feature map, which is then progressively sampled by a stack of transposed convolutional layers with batch normalization and ReLU activations, until reaching a 3×224×224 RGB image. The final layer uses Tanh activation to normalize pixel values between [−1,1].

Step #2 Discriminator

The discriminator accepts an image tensor 3×224×224 and the corresponding class label y. The image passes through four convolutional layers with kernel size 4, stride 2, and padding 1, each followed by LeakyReLU activations (negative slope 0.2). This reduces spatial resolution while increasing channel depth, extracting hierarchical features. The discriminator is a deep convolutional network composed of strided convolutional layers with LeakyReLU activations, which progressively reduce spatial resolution while increasing channel depth. A final fully connected layer with sigmoid activation outputs the probability that the input image is real or fake.

Step #3 Data preparation

Data images x and corresponding labels y are normalized to [-1,1] and resized 224×224 using torchvision transformations. A custom dataset class applies these transformations and prepares data tensors suitable for GPU processing.

Step #4 Training process

- Loss function: Binary cross-entropy loss (BCE) is employed for both generator and discriminator objectives, promoting the generator to produce realistic images and the discriminator to correctly classify images as real or fake.
- Optimization: Adam optimizers with different learning rates are used for the generator and discriminator, leveraging the Adam hyperparameters β1 =0.5 and β2=0.999 to stabilize training.
- Adversarial training loop: For each epoch and batch: Real images and labels are transferred to the device (CPU/GPU).
- The discriminator is updated first: It predicts real images as real (target label=1) and synthetic images (produced by the generator) as fake (target label=0). The discriminator loss is the sum of losses on real and fake predictions.

- The generator is updated next: It attempts to fool the discriminator by producing images that the discriminator classifies as real (target label=1).

Losses for both networks are backpropagated and their parameters updated accordingly. The adversarial nature ensures that the generator progressively improves to synthesize realistic, class-conditioned defect images.

Step #5 Image visualization

A utility function visualizes generated images by reversing normalization and displaying a batch of synthetic samples, aiding qualitative assessment during training. Table 5.5 summarizes the key stages of the proposed DCGAN framework.

TABLE 5.5 Summary of the DCGAN-based synthetic data generation framework.

Step	Technical description
Conditional generator	Combines a latent noise vector (dim=512) with a class label embedding (12 classes) to generate 3-channel 224×224 images normalized to [-1, 1].
CNN-based discriminator	Uses 4 convolutional layers with LeakyReLU activations; flattens features and concatenates one-hot encoded labels before a sigmoid output layer.
Custom Dataset	Loads normalized images resized to 224×224 with labels
Initialization	Models moved to GPU if available; uses BCE Loss; Adam optimizers with learning rates 1e-4 (generator) and 1e-5 (discriminator).
Discriminator training	Computes BCE loss on real images (label=1) and fake images (label=0); backpropagates to update discriminator parameters.
Generator training	Generates fake images; discriminator evaluates them; generator loss computed as BCE with real label (1) to fool discriminator; backpropagation applied.
Alternating optimization	Sequentially updates the discriminator, then the generator per batch for adversarial training. (G)Adam, lr=1e-4, betas=(0.5, 0.99), (D) Adam, lr=1e-5, betas=(0.5, 0.99)
Visualization	Function to denormalize and display generated images for qualitative monitoring.
Hyperparameters	Latent vector size: 512; Number of classes: 12; Batch size: 16; Total epochs: 500.
Objective	Generate realistic, class-conditional synthetic images to balance the dataset and improve defect detection performance.

DCGANs, Deep convolutional GANs.

The description encompasses the architecture of the generator and discriminator, dataset preparation, initialization, training process, and hyperparameter configuration. The overall objective is to generate realistic, class-conditional electroluminescence images and IR images that address dataset imbalance and support improved defect detection in photovoltaic modules.

Deep convolutional GANs-VGG16

To enhance the realism of the generated EL images, the DCGAN architecture was extended with a perceptual loss component derived from a pretrained VGG16 network. While the standard DCGAN framework relies solely on the adversarial interaction between generator and discriminator, this integration introduces a feature-level constraint that encourages the generator to capture structural and textural details more faithfully. Specifically, low-level convolutional layers of VGG16 are used as a fixed feature extractor, and the discrepancy between real and synthetic images is minimized in this feature space (Drir et al., 2025b).

Formally, let f_{gen} and f_{real} represent the feature maps of generated and real samples, respectively. The perceptual loss is expressed as the mean squared error (MSE) between the two feature sets:

$$L_{perc} = \frac{1}{N} \sum_{i=1}^{N} (f_{gen}^{(i)} - f_{real}^{(i)})^2 \tag{5.2}$$

where N denotes the number of extracted features. The overall generator loss then combines the classical adversarial term with the perceptual constraint as follows:

$$L_G^{total} = L_G + \lambda L_{perc} \tag{5.3}$$

with λ serving as a weighting factor that balances adversarial and perceptual objectives. This enhancement enables the generator to synthesize images that are not only adversarially plausible but also perceptually faithful to the fine-grained characteristics of real EL data. The integration of VGG16 perceptual guidance proved particularly effective in reproducing complex structural attributes of defects, thereby enriching the diversity and quality of the synthetic dataset and ultimately improving the robustness of defect classification models

The overall architecture of the proposed DCGAN–VGG16 framework, including the generator, discriminator, and perceptual loss module, is outlined as follows:

Step #1 Generator

The generator takes as input a latent noise vector z. This vector is projected and reshaped into a low-dimensional feature map, which is then progressively upsampled using transposed convolutional layers. Each upsampling step employs batch normalization and ReLU activations to stabilize training. The final layer applies a Tanh activation function to generate RGB images of size 224×224, with pixel values normalized to the range [−1,1].

Step #2 Discriminator

The discriminator receives an image tensor of size 3×224×224 and processes it through a stack of strided convolutional layers (4×4 kernels, stride 2, padding 1) with LeakyReLU (α=0.2). These layers progressively reduce spatial dimensions while extracting hierarchical features. The resulting feature map is flattened and passed through a fully connected layer with a sigmoid activation to output a scalar probability indicating whether the input image is real or generated.

Step #3 VGG16-Based perceptual loss

To guide the generator towards producing perceptually realistic images, a pretrained VGG16 network from torchvision models is used. The first convolutional blocks (features [:8]) are extracted and frozen to serve as a feature extractor. For each training batch, both real and generated images are passed through this truncated VGG16, and the perceptual loss is computed as the MSE between their feature representations.

Step #4 Training process

The discriminator is first updated by minimizing the BCE loss on real images (target=1) and fake images generated by the generator (target=0). The generator is then updated to minimize a combined objective (Eq. 5.3). A weighting factor λ=0.1 is applied, as in the implementation. This architecture ensures that the generator not only learns to deceive the discriminator but also produces images with structural and perceptual fidelity to real EL images.

Table 5.6 lists a stepwise description of the proposed DCGAN-VGG16 framework. The summary delineates the architectural components of the generator and discriminator, the incorporation of a perceptual loss derived from VGG16, and the sequential training procedure. Unlike conventional GAN formulations that rely solely on adversarial learning, the proposed design integrates low-level feature constraints from a pretrained VGG16 to preserve structural and textural fidelity.

Variational autoencoders

VAEs are probabilistic generative models that learn to represent input images within a continuous latent space from which new samples can be generated (Bian et al., 2019). The model consists of two primary components: an encoder and a decoder. The encoder compresses an input image into a latent representation characterized by a mean and variance, capturing the underlying probability distribution of the data. The decoder then reconstructs the original image, or generates a new variation, by sampling from this latent distribution and mapping it back into image space. Unlike GANs, VAEs provide a well-defined and interpretable latent space, which allows for smooth interpolation

TABLE 5.6 Summary of the DCGAN-VGG16-based synthetic data generation framework.

Step	Technical description
Latent vector sampling	A random latent vector Z (e.g., 256 dimensions, sampled from a Gaussian distribution) is fed into the generator.
Generator processing	Z is projected and reshaped into a low-dimensional feature map, then upsampled progressively via transposed convolutions, batch normalization, and ReLU activations. Final layer uses Tanh to generate a synthetic RGB image of size 224×224. I
Discriminator input	Both real EL images and generated images are fed to the discriminator.
Discriminator processing	The discriminator applies a sequence of strided convolutions (4×4 kernels, stride=2, padding=1) with LeakyReLU activations. Output is flattened and passed through a fully connected layer with sigmoid activation
Adversarial loss	Binary cross-entropy (BCE) loss is computed: real images → label 1, fake images → label 0. The discriminator is updated accordingly.
VGG16 feature extraction	Real and generated images are passed through frozen low-level convolutional layers of a pretrained VGG16 (features [0–7]) to obtain feature representations.
Perceptual loss calculation	The mean squared error (MSE) is computed between VGG16 feature maps of real and fake images.
Generator loss update	The generator is optimized using a combined loss with λ=0.1.
Hyperparameters	Latent vector size: 512; Number of classes: 12; Epochs=400; Batch size=16; Learning rate=1e-4 (G), 1e-5 (D), Adam betas= (0.5, 0.999).

and controlled manipulation between different defect types or severity levels. This feature is particularly advantageous in PV defect analysis, where researchers may explore transitions between healthy and faulty states to study degradation patterns.

Although VAEs typically generate blurrier images than GANs due to their reliance on probabilistic reconstruction objectives, they offer greater training stability and require less hyperparameter tuning. These characteristics make VAEs well-suited for exploratory data synthesis, defect state interpolation, and anomaly generation tasks in photovoltaic image analysis (See Fig. 5.6).

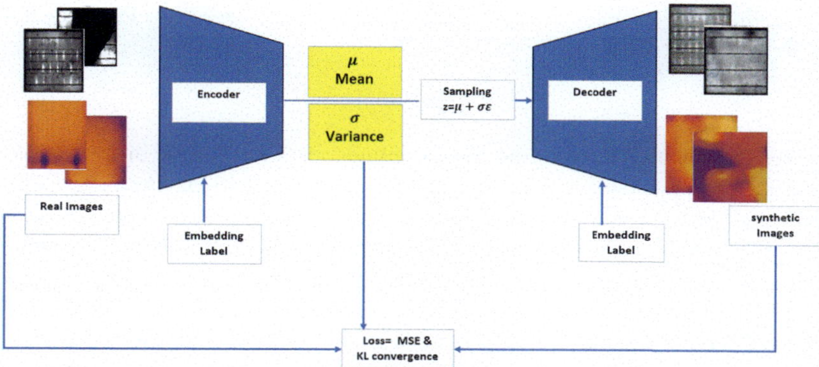

FIGURE 5.6 Variational autoencoders: probabilistic generative models. The figure illustrates the structure of a VAE, highlighting its encoder–decoder architecture and the probabilistic latent space. It shows how the model learns to generate new data samples by mapping inputs to a distribution rather than a fixed latent vector (*VAE*, Variational autoencoder).

Probabilistic model and formulation

The goal of a VAE is to learn a parametric generative model $P_\theta(x)$ capable of sampling new data instances x from a latent variable z drawn from a prior distribution:

$$P_\theta(x) = \int P_\theta(x \mid z)p(z)dz \text{ with} p(z) = N(0,\ 1) \tag{5.4}$$

Since the marginal likelihood $P_\theta(x)$ is generally intractable, VAEs employ variational inference and introduce an approximate posterior distribution $q_\varnothing(x, z)$(the encoder), parameterized by $\varnothing$, to approximate the true posterior $P_\theta(x, z)$.

Variational inference and learning objective: The model maximizes a variational lower bound (evidence lower bound) on the log-likelihood of the data:

$$L_{\theta,\varnothing,x} = E_{q(z,x)}[\log P_\theta(x, z)] - D_{KL}(q_\varnothing(z \mid x) \| p(z)) \tag{5.5}$$

where the first term promotes accurate data reconstruction from the latent code. The second term penalizes divergence of the learned posterior from the prior to encourage a smooth and regularized latent space.

Encoder (Inference Network): A NN that predicts the mean and variance of the approximate posterior distribution:

$$q_\varnothing(x, z) = N(z; \mu_\varnothing(x), diag(\sigma^2(x))) \tag{5.6}$$

Decoder (Generative Network): A NN that models the conditional likelihood $P_\theta(x, z)$ and maps latent variables z back to the data domain.

To allow backpropagation through the stochastic sampling process, the latent variable is reparametrized as follows:

$$z = \mu_{\emptyset}(x) + \sigma_{\emptyset}(x) * \varepsilon \varepsilon \sim N(0, I) \tag{5.7}$$

This formulation transforms the sampling operation into a differentiable transformation with respect to ∅ and θ, enabling end-to-end training using stochastic gradient-based optimization. The well-structured latent space learned by VAEs enables smooth interpolation between different defect states, facilitating the synthesis of new defect variants for training and testing. Despite producing outputs of lower visual fidelity compared to GANs, VAEs are preferred for their stable training dynamics and probabilistic interpretability, making them effective for anomaly generation and exploratory data augmentation in defect detection tasks.

Detailed description of the conditional variational autoencoder architecture

The implemented architecture is a CVAE designed to generate and reconstruct images conditioned on discrete class labels (Neloy and Turgeon, 2024; Li et al., 2024). This approach extends the traditional VAE by integrating label information directly into both the encoder and decoder, enabling class-specific generation and better control over the latent space representation. The CVAE models the conditional distribution $P_{\theta}(x|y)$, where x is an input image, and y is its corresponding class label. Conditioning on the label allows the latent space to be structured with class-aware embeddings, thus facilitating the generation of images belonging to specific defect categories or states.

Step #1 Input and conditioning

The model accepts RGB images of size 224×224×3.

Defect class labels are passed through an embedding layer, which maps each label to a dense vector of dimension 50. This embedding is used in two ways:

It is reshaped into a spatial label map and concatenated as an additional channel to the input image before encoding (resulting in an input tensor with four channels).

It is concatenated with the latent vector before decoding, ensuring that generation is explicitly conditioned on the class label.

Step #2 Encoder architecture

The 4-channel input passes through four convolutional layers, progressively increasing feature depth (32–256 filters) while down-sampling via a stride of two. The resulting feature map of size 256×14×14 is flattened. Two fully connected layers output the mean $\mu_{\emptyset}(x, y)$ and log-variance $\log(\sigma^2(x, y))$ of the posterior distribution over latent variables.

Step #3 Reparameterization trick

The latent variable z is sampled using:

$$z = \mu_{\varnothing}(x, y) + \sigma_{\varnothing}(x, y) * \varepsilon \varepsilon \sim N(0, I) \tag{5.8}$$

This formulation ensures differentiability and allows gradient backpropagation through stochastic sampling.

Step #4 Decoder architecture

The latent vector z is concatenated with the embedded class label. A fully connected layer projects this concatenated vector into a tensor of shape 256×14×14. A sequence of four transposed convolutional layers progressively up-samples the tensor to the original resolution. The final layer applies Tanh activation, producing pixel values normalized in [−1,1].

Step #5 Loss function

The model optimizes a composite loss combining:

$$L = ||X - \hat{X}||^2 + D_{KL}(q_{\varnothing}(z \mid x, y)||p(z)) \tag{5.9}$$

The first term corresponds to the MSE between the input image and its reconstruction. The second term represents the Kullback–Leibler divergence between the approximate posterior and the prior distribution $p(z) = N(0, I)$, encouraging a smooth and well-structured latent space

Step #6 Training details

- The model is trained using the Adam optimizer with a learning rate of 10^{-4}.
- Training proceeds for 900 epochs on mini-batches of size 16.
- Input images are normalized to [0,1] and then standardized with a mean of 0.5 and a standard deviation of 0.5.

Step #7 Image generation

Generally, latent vectors z are sampled from the prior$N(0, I)$, concatenated with the embedding of a desired class label, and fed into the decoder to produce class-conditioned synthetic images. This enables targeted synthesis of images belonging to specific defect classes, useful for data augmentation or anomaly detection tasks.

Table 5.7 provides a stepwise overview of the CVAE architecture. The summary outlines the main processing stages, including input conditioning, encoder–decoder design, reparameterization, and training configuration.

To highlight the distinctive characteristics of each generative model employed in this chapter, a comparative analysis is presented in Table 5.8. The table contrasts the baseline DCGAN, the enhanced DCGAN-VGG16 hybrid, and the CVAE across multiple aspects, including architectural design, conditioning mechanisms, latent space formulation, loss functions, and applicability to PV defect analysis. This comparison underscores how each approach contributes differently to addressing dataset imbalance and improving the quality of synthetic EL images for defect detection.

TABLE 5.7 Summary of the CVAE-based synthetic data generation framework.

Step	Technical description
Input conditioning	Images (224×224×3) and defect class labels are jointly encoded; labels are embedded (dim=50) and integrated into both encoder and decoder.
Encoder	Extracts latent parameters (mean and variance) from conditioned inputs to approximate the posterior distribution.
Reparameterization	Samples latent vector z (dimension=512) using the reparameterization trick to enable backpropagation.
Decoder	Reconstructs images from z concatenated with label embeddings, progressively upsampling to RGB output (224×224).
Loss function	Combines reconstruction loss (MSE) with KL divergence to enforce fidelity and a structured latent space
Training parameters	Adam optimizer, learning rate=10^{-4}, batch size=16, training epochs=900; input images normalized to [0,1].
Image generation	Synthetic images are generated by sampling z ~ N(0,I), concatenated with class embeddings, and decoded into class-conditional outputs

CVAE, Conditional variational autoencoder.

Qualitative results

Visual inspection reveals that the DCGAN generates defect images with higher sharpness and realistic texture variations across classes, effectively mimicking real PV module anomalies. Generated samples closely replicate key defect patterns such as microcracks and hotspots.

D convolutional GAN qualitative analyses electroluminescence images

Visual examination of the generated EL images (See Fig. 5.7) shows that the DCGAN is able to reproduce visually plausible defect patterns in certain classes. In particular, for Classes 8, 11, and 1, the synthetic images globally preserve the overall structure of the real defects, although the visual quality remains slightly lower. By contrast, for Classes 0, 2, 5, 6, 9, and 10, the generated EL samples show mediocre quality: the fine defect details are not faithfully reproduced and noticeable noise appears in several regions of the images. These results highlight the limited capacity of the DCGAN to capture complex EL defect structures across all classes.

TABLE 5.8 **Comparative analysis of generative architectures for PV defect data synthesis.**

Aspect	DCGAN	DCGAN -VGG16	CVAE
Architecture type	Standard deep convolutional GAN with generator and discriminator.	Hybrid architecture combining DCGAN with VGG16 as a feature extractor in the discriminator.	A conditional variational autoencoder integrating label embeddings into the encoder and decoder.
Input conditioning	Noise vector z sampled from a Gaussian distribution.	Noise vector z with class-conditioning via VGG16-based feature learning.	Input image (224×224×3) + class label embedding (dim=50).
Latent space	Random latent vector without explicit structure	Latent vector structured with VGG16-driven features.	Maps input + labels into the mean and variance of the latent Gaussian distribution.
Encoder role	Not applicable	Not applicable	Maps input + labels into the mean and variance of the latent Gaussian distribution.
Decoder/ Generator	Transposed convolutional layers generating synthetic images.	Generator similar to DCGAN but guided by VGG16 feature maps	Decoder reconstructs/ generates images from latent code + label embeddings
Loss function	Adversarial loss (min–max optimization between generator and discriminator).	Adversarial loss with enhanced discriminator using VGG16 features.	Reconstruction loss (MSE) + KL divergence (regularizing latent space).
Strengths	An effective baseline for synthetic image generation.	Improved image realism and class separability due to VGG16 features	Strong control over latent space; enables class-specific generation.
Limitations	Struggles with mode collapse and class conditioning.	Computationally heavier due to VGG16 integration.	Generated images may lack sharpness compared to GAN-based approaches

DCGAN, Deep convolutional GAN; *GAN*, generative adversarial network; *PV*, photovoltaic.

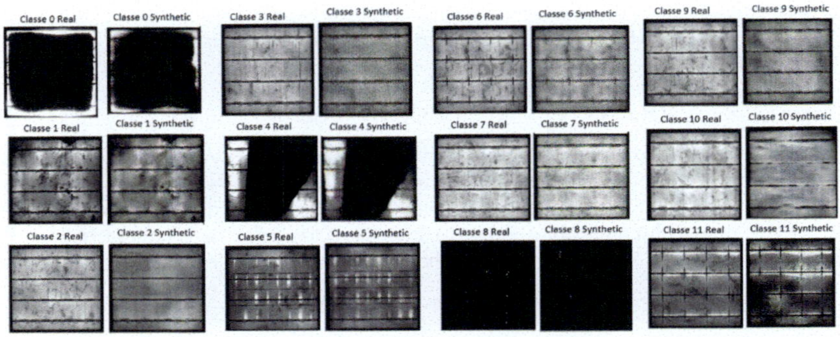

FIGURE 5.7 Qualitative comparison between real EL defect images and synthetic counterparts generated by DCGAN. The figure presents a qualitative comparison between real EL defect images and synthetic images generated by a DCGAN. It highlights the visual similarity between both sets, demonstrating the model's capability to produce realistic defect patterns for data augmentation and analysis (*DCGANs*, Deep convolutional GANs; *EL*, electroluminescence).

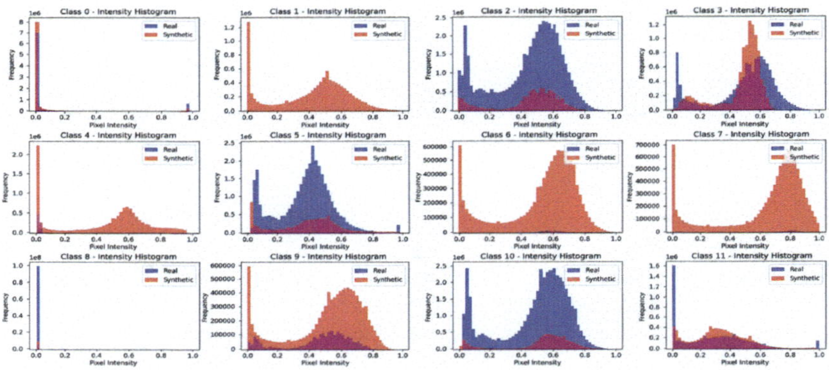

FIGURE 5.8 Histogram-based analysis of pixel intensity distributions in DCGAN-generated EL images. The figure shows a histogram-based analysis of pixel intensity distributions in EL images generated by the DCGAN. It compares the statistical characteristics of synthetic images with real EL data to evaluate how well the model reproduces realistic brightness and contrast patterns (*DCGANs*, Deep convolutional GANs; *EL*, electroluminescence).

The histogram analysis presented in Fig. 5.8 highlights a pronounced discrepancy between real and synthetic EL images, particularly in the low-intensity regions where synthetic samples appear unrealistically concentrated. For example, Classes 0 and 8 exhibit distinct synthetic peaks near zero-intensity values, whereas the corresponding real images demonstrate broader and more naturally distributed histograms, indicating limited generative fidelity in these categories. Likewise, Class 4 shows a noticeable divergence within the mid-intensity range, reflecting reduced variability and insufficient diversity in synthetic data. A moderate improvement is observed in Class 5, where partial overlap between real and synthetic histograms suggests that the generative model begins to capture some structural and textural characteristics of the original data. Nevertheless, the restricted overlap observed across most classes confirms that the current generative process still struggles to reproduce the intrinsic variability and realistic luminance distributions of authentic EL imagery.

The cumulative distribution functions (CDFs) of pixel intensities derived from the DCGAN-generated images reveal distinct deviations between real and synthetic distributions, particularly within mid-range defect classes (See Fig. 5.9). For example, Classes 2, 3, and 6 exhibit noticeable shifts in their CDF curves, where the synthetic data either underrepresents or overrepresents intermediate intensity values. This behavior indicates a limited ability of the model to reproduce subtle intraclass variations in luminance, which are crucial for accurately reflecting real EL characteristics. In contrast, simpler categories such as Classes 0 and 8 demonstrate closer correspondence between real and synthetic distributions, suggesting that the DCGAN framework performs more effectively when defect morphology and intensity gradients are less complex.

However, for defect types exhibiting higher structural heterogeneity and greater textural diversity, the increasing divergence between real and generated distributions implies that the baseline DCGAN architecture struggles to generalize across the full range of defect-specific intensity behaviors. This

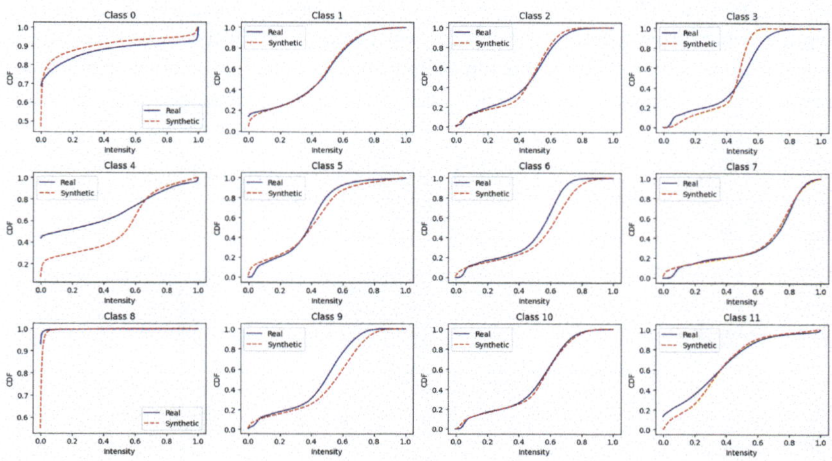

FIGURE 5.9 Cumulative distribution functions of pixel intensities for real and synthetic images. The figure presents the cumulative distribution functions (*CDFs*) of pixel intensities for both real and synthetic EL images. It illustrates how closely the synthetic data matches the statistical distribution of real images, serving as an indicator of generative model accuracy (*EL*, Electroluminescence).

limitation ultimately constrains its reliability for detailed and physically consistent defect reconstruction.

Deep convolutional GAN qualitative analyses infrared images

Visual inspection of IR images (see Fig. 5.10) shows that the model succeeds in preserving the overall temperature distribution and the class-specific signature for several defect categories. In particular, for Classes 1, 4, 10, and 11, the synthetic samples exhibit global patterns that closely resemble the real IR images and reproduce the position of the dominant hot-spot regions. Nevertheless, the generated samples tend to lose fine-grained details, leading to smoother textures and missing local fluctuations. This is particularly evident in Classes 2, 5, and 7, where the reconstructed images contain noticeable noise and artificial color gradients that are not present in the real samples. These results demonstrate that although the model can capture the macroscopic structure of IR defects, it still struggles to accurately reconstruct complex or subtle temperature variations.

As illustrated in Fig. 5.11, which presents results obtained using the DCGAN model, the analysis reveals clear discrepancies between the pixel-intensity distributions of real and synthetic images across multiple defect classes. For example, in Class 1, the synthetic samples fail to reproduce the multimodal structure evident in the real histogram, instead generating smoother, oversimplified distributions that underestimate localized intensity variations. Similarly, Class 7 exhibits an even greater divergence, indicating that the baseline DCGAN struggles to capture complex textural features and displays instability when modeling intraclass variability. Overall, these findings underscore the limited capability of standard adversarial training to preserve the detailed statistical and structural characteristics of real

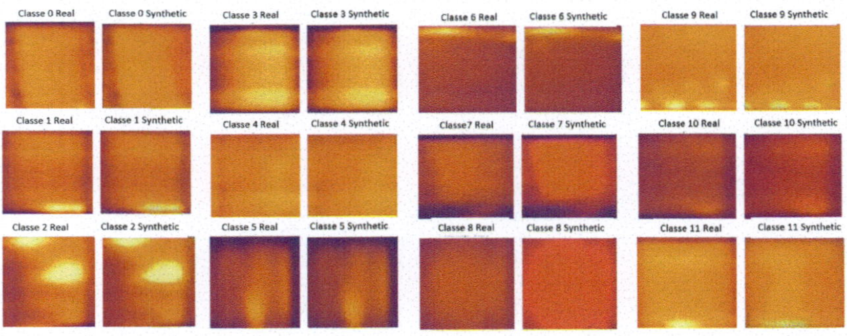

FIGURE 5.10 Qualitative comparison between real IR defect images and synthetic counterparts generated by DCGAN. The figure shows a qualitative comparison between real IR defect images and synthetic versions generated using a DCGAN. It highlights the visual similarity between the two sets, demonstrating the model's ability to replicate realistic patterns associated with PV faults (*DCGANs*, Deep convolutional GANs; *IR*, infrared; *PV*, photovoltaic).

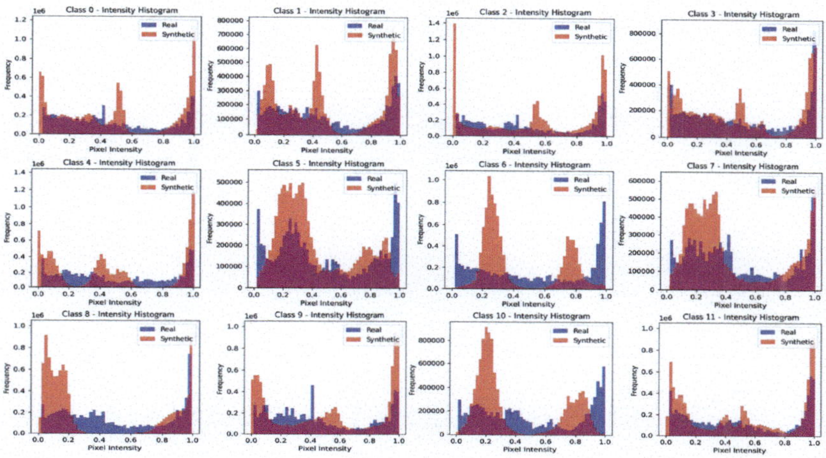

FIGURE 5.11 Histogram-based analysis of pixel intensity distributions in DCGAN-generated IR images. The figure displays a histogram-based analysis of pixel intensity distributions in IR images generated by the DCGAN. It compares these distributions with those of real IR data to assess how effectively the model captures the thermal intensity characteristics of PV defects (*DCGANs*, Deep convolutional GANs; *IR*, infrared; *PV*, photovoltaic).

electroluminescence data, highlighting the need for enhanced generative frameworks capable of maintaining both global fidelity and local texture realism.

As shown in Fig. 5.12, the CDFs generated by the DCGAN model reveal systematic discrepancies between the real and synthetic datasets across several defect classes. For instance, in Class 5, the synthetic CDF diverges markedly from the real curve within the mid-intensity range, indicating that the DCGAN fails to preserve the complete distributional behavior of pixel intensities. Likewise, in Class 9, the synthetic curve rises earlier than its real counterpart, signifying an overrepresentation of low-intensity values in the generated images. Collectively, these deviations demonstrate that the baseline DCGAN introduces consistent biases in pixel-intensity distributions, a finding that aligns with the distortions previously observed in the histogram-based analysis (See Fig. 5.11). This behavior emphasizes the model's limited capacity to reproduce the fine-grained luminance characteristics essential for accurate electroluminescence defect modeling.

Deep convolutional GAN–VGG16 qualitative analysis on electroluminescence images

Compared to the baseline DCGAN, the integration of VGG16 feature extraction results in a clear improvement in the generation of EL images across several classes. In particular, for Classes 0, 4, 5, 6, and 11, the generated samples exhibit more realistic contrast distributions and begin to reproduce

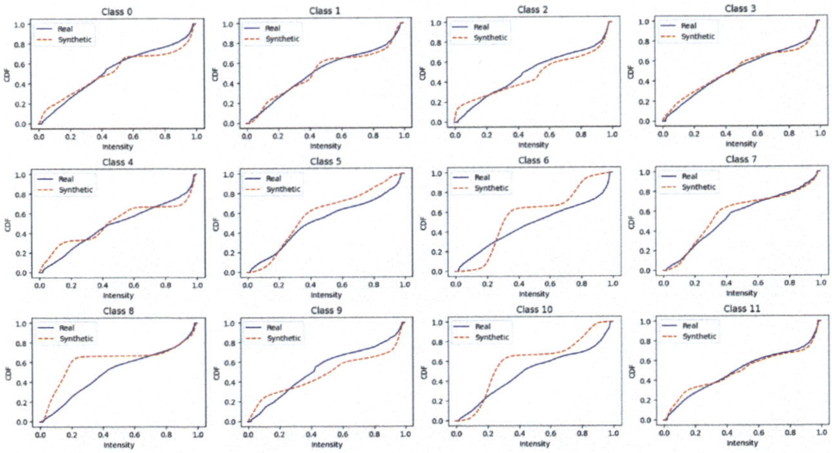

FIGURE 5.12 Cumulative distribution functions of pixel intensities for real and synthetic images in DCGAN-generated IR images. The figure presents the cumulative distribution functions (*CDFs*) of pixel intensities for real and synthetic IR images generated by the DCGAN. It illustrates how closely the synthetic images replicate the statistical intensity patterns of real IR data, providing insight into the model's generative accuracy (*DCGANs*, Deep convolutional GANs; *IR*, infrared).

specific defect-related structures that were completely missing in the previous model. In these classes, some fine defect features such as micro-crack borders or partially degraded cells become visible, even if a small amount of noise is still present in the background. Conversely, for Classes 2, 3, 9, and 10, the network still experiences difficulties in accurately reconstructing fine-grained details. In these cases, the main defect area is often roughly located, but the geometry and texture of the fault patterns remain blurred, indicating that the model does not yet fully capture the intraclass variability at the EL modality level (see Fig. 5.13).

The histogram analysis (see Fig. 5.14) comparing real and synthetic images generated by the DCGAN–VGG16 framework shows a substantial improvement in alignment between the two distributions across several defect classes. For instance, Classes 1 and 6 exhibit synthetic histograms that closely follow the intensity spread of real images, demonstrating a stronger capacity to reproduce structural variability and fine texture gradients. Similarly, Class 9 displays an almost complete overlap between synthetic and real distributions, indicating a significant reduction in distributional divergence and improved generative fidelity. However, notable discrepancies persist in Classes 3 and 7, where synthetic peaks are visibly shifted relative to their real counterparts, and in Class 11, where the synthetic data fail to capture the bimodal intensity structure observed in the real distribution. These findings suggest that, although the DCGAN–VGG16 model effectively captures global intensity trends and enhances overall realism, further optimization is required

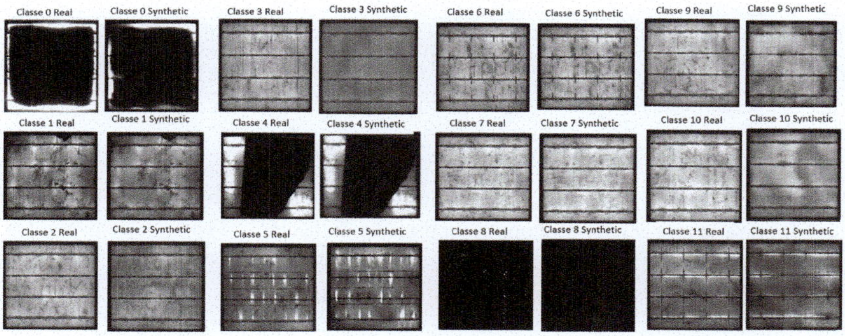

FIGURE 5.13 Qualitative comparison between real EL defect images and synthetic counterparts generated by DCGAN–VGG16. The figure shows a qualitative comparison between real EL defect images and synthetic counterparts generated by the DCGAN-VGG16 model. It highlights the enhanced visual fidelity of the synthetic images, demonstrating the model's ability to reproduce fine defect details and structural patterns found in real EL data (*DCGANs*, Deep convolutional GANs; *EL*, electroluminescence).

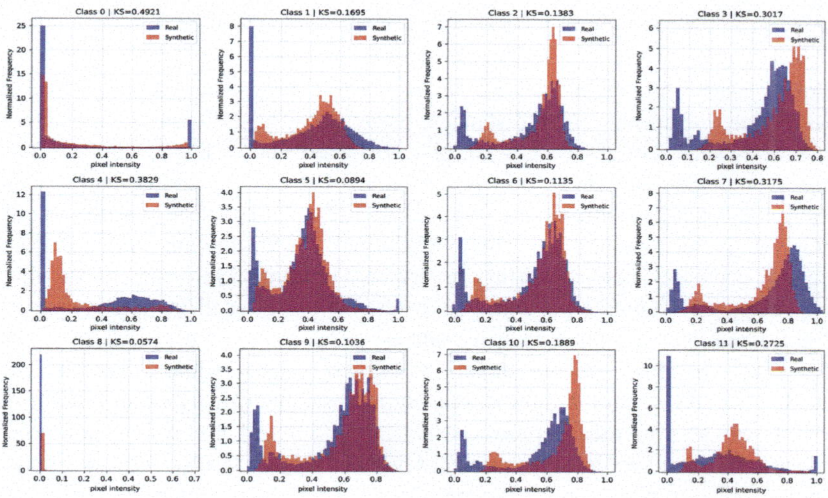

FIGURE 5.14 Histogram-based analysis of pixel intensity distributions in DCGAN-VGG16 generated EL images. The figure presents a histogram-based analysis of pixel intensity distributions in EL images generated by the DCGAN-VGG16 model. It compares the statistical characteristics of synthetic images with real EL data to evaluate the model's ability to reproduce realistic brightness and contrast patterns (*DCGANs*, Deep convolutional GANs; *EL*, electroluminescence).

to improve class-specific fidelity and replicate complex intraclass luminance variations.

The integration of the VGG16 network into the DCGAN discriminator markedly enhances the alignment between real and synthetic EL distributions

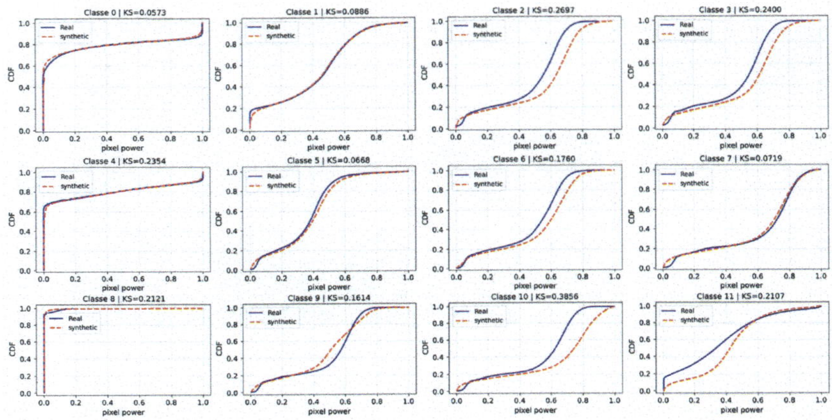

FIGURE 5.15 Cumulative distribution functions of pixel intensities for real and synthetic images in DCGAN-VGG16 generated EL images. The figure shows the CDFs of pixel intensities for real and synthetic EL images generated by the DCGAN-VGG16 model. It illustrates how closely the synthetic images match the statistical intensity patterns of the real EL data, highlighting the model's effectiveness in generating realistic defect images (*CDF*, Cumulative distribution function; *DCGANs*, deep convolutional GANs; *EL*, electroluminescence).

across nearly all defect classes. The CDFs show consistently reduced Kolmogorov–Smirnov (KS) distances, particularly in mid-complexity defect categories such as Classes 1, 2, 6, and 10, where the synthetic curves more closely replicate the shape and progression of their real counterparts. This improvement stems from the enhanced feature extraction capability of VGG16, which strengthens class-conditional discrimination and enforces greater structural fidelity during adversarial training. However, minor discrepancies persist in Classes 2 and 10, where the synthetic distributions display slight deviations at higher intensity ranges.

These residual inconsistencies suggest that while the hybrid DCGAN–VGG16 model significantly improves global distributional fidelity, it still struggles to fully reproduce subtle textural and luminance variations present in complex defect patterns. Overall, the DCGAN–VGG16 hybrid demonstrates superior generative performance, producing more realistic and statistically consistent EL images, particularly in classes where the baseline DCGAN exhibited weaker representation (See Fig. 5.15).

Deep convolutional GAN–VGG16 qualitative analysis on infrared images

For the infrared modality (See Fig. 5.16), the DCGAN–VGG16 architecture provides a clear qualitative improvement across all defect classes. Unlike the baseline model, the generated IR images allow a clear distinction of the main defect regions and correctly reproduce the thermal signature associated with

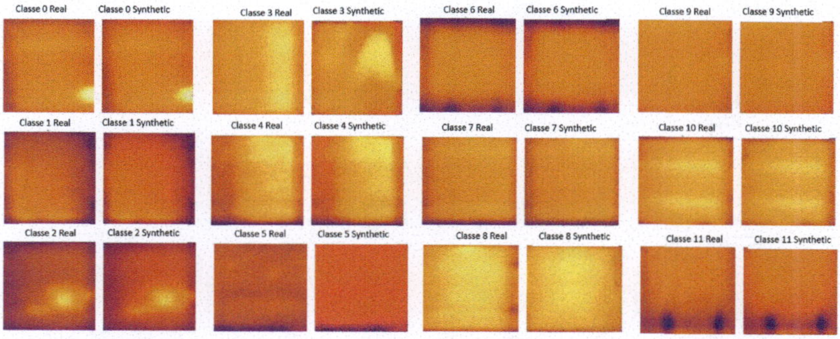

FIGURE 5.16 Qualitative comparison between real IR defect images and synthetic counterparts generated by DCGAN-VGG16. The figure presents a qualitative comparison between real IR defect images and synthetic counterparts generated by the DCGAN-VGG16 model. It highlights the improved visual realism of the synthetic images, demonstrating the model's capability to replicate detailed thermal patterns observed in real IR data (*DCGANs*, Deep convolutional GANs; *IR*, infrared).

each category. In most classes, the synthetic samples capture the spatial localization of hotspot areas and the global temperature distribution observed in the real images. However, despite this improvement, the visual quality remains lower than that of the real data.

The generated images still include noticeable noise and do not fully reproduce some of the fine-scale thermal variations, particularly in complex defect regions. This indicates that while the model has learned to represent the macroscopic structure of IR defects, the accurate reconstruction of subtle temperature gradients still represents a challenge.

In contrast, CGAN enhanced with VGG16 perceptual guidance demonstrates a clear improvement in distributional alignment (See Fig. 5.17). For example, in Class 2, the synthetic histogram exhibits sharper peaks that are more consistent with the real data compared to the baseline, indicating a more faithful reproduction of underlying structures. Likewise, in Class 10, the reduced distance between the real and synthetic distributions shows that this hybrid approach is better suited to capturing high-frequency details. These observations suggest that incorporating perceptual-level constraints helps the generative model preserve both global and local features, thereby producing data that more closely resembles the statistical characteristics of real samples.

Fig. 5.18 presents the results of DCGAN guided with VGG16 features, where the alignment between real and synthetic CDFs is visibly improved. In Class 2, the synthetic curve tracks the real distribution more closely across the full intensity range, demonstrating the effect of perceptual constraints in guiding the model toward realistic distributions. A similar improvement can be observed in Class 7, where the synthetic and real curves nearly overlap, with only minor deviations at higher intensities. The integration of perceptual loss

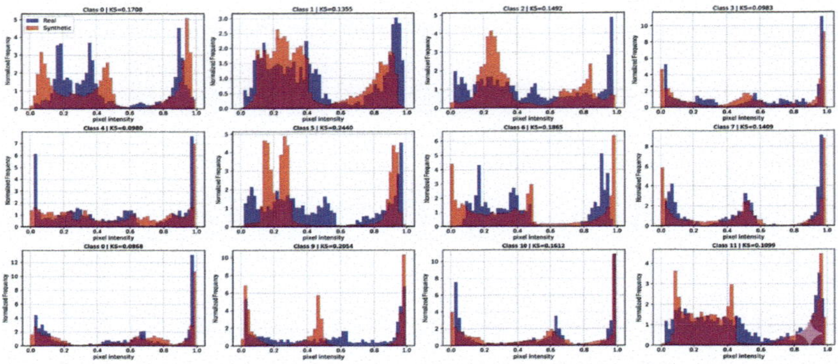

FIGURE 5.17 Histogram-based analysis of pixel intensity distributions in DCGAN–VGG16 generated IR images. The figure displays a histogram-based analysis of pixel intensity distributions in IR images generated by the DCGAN–VGG16 model. It compares these distributions with those of real IR data to evaluate the model's ability to replicate realistic thermal intensity patterns (*DCGANs*, Deep convolutional GANs; *IR*, infrared).

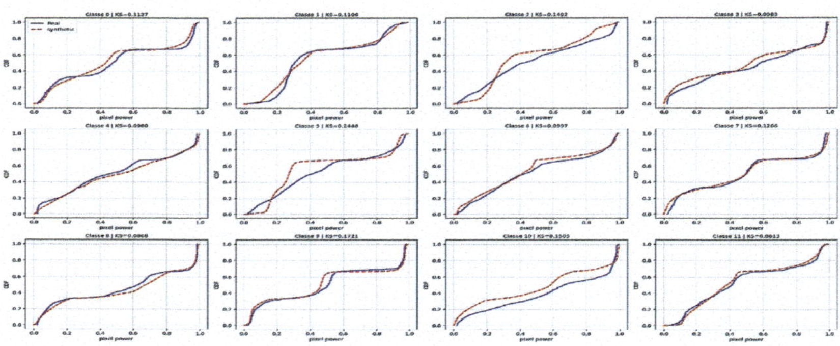

FIGURE 5.18 Cumulative distribution functions of pixel intensities for real and synthetic images in DCGAN–VGG16 generated IR thermal images. The figure shows the CDFs of pixel intensities for real and synthetic IR images generated by the DCGAN-VGG16 model. It illustrates how closely the synthetic images reproduce the statistical intensity patterns of real IR data, indicating the model's effectiveness in generating realistic thermal defect images (*CDF*, Cumulative distribution function; *DCGANs*, deep convolutional GANs; *IR*, infrared).

contributes to reducing cumulative discrepancies and enhances the statistical fidelity of synthetic images compared to the baseline adversarial approach.

Conditional variational autoencoder – qualitative analysis on electroluminescence images

For the EL modality, the CVAE-based architecture exhibits a significant improvement in the perceptual quality of the generated samples compared to

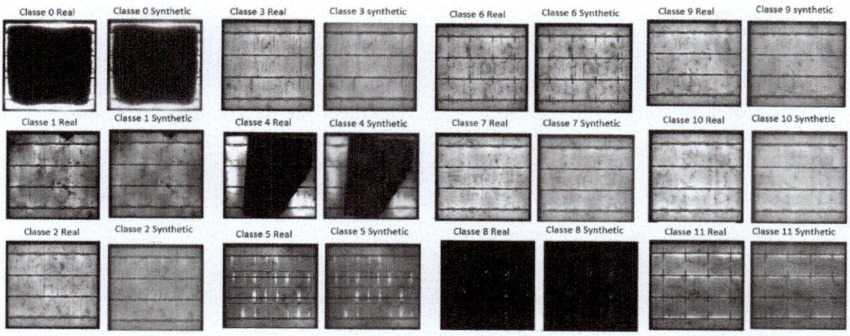

FIGURE 5.19 Qualitative comparison between real EL defect images and synthetic counterparts generated by VAE. The figure presents a qualitative comparison between real EL defect images and synthetic counterparts generated by a VAE. It highlights the model's ability to reproduce realistic defect patterns, demonstrating how VAE can generate diverse and representative EL images for data augmentation and analysis (*EL*, Electroluminescence; *VAE*, variational autoencoder).

adversarial approaches (See Fig. 5.19). The synthetic EL images present a much clearer visual rendering, with well-defined defect regions and markedly reduced background noise. In several classes, the model succeeds in reproducing fine structural details of the defects, including crack propagation patterns and localized cell degradation, which were previously smoothed or partially lost. The spatial distribution of luminance is highly consistent with that of the real samples, indicating that the model has effectively captured the underlying distribution of the EL data. Even though minor residual noise persists in some classes, the overall visual fidelity of the CVAE-generated EL images is considerably higher, making them suitable for visual inspection and additional downstream analyses.

Fig. 5.20 presents the histogram analysis of the CVAE model, showing the strongest correspondence observed thus far between real and synthetic EL images. In particular, Classes 2 and 10 exhibit nearly identical distributions, with synthetic data closely overlapping the real histograms and appearing almost indistinguishable. This outcome confirms the high generative fidelity of the CVAE framework for these cases. Similarly, Classes 6 and 9 demonstrate strong alignment between real and synthetic intensity profiles, reflecting consistent improvements in capturing structurally complex and texturally diverse patterns. Some residual deviations remain, most notably in Classes 0 and 8, where synthetic images still slightly exaggerate the low-intensity peaks compared to real samples.

Despite these minor inconsistencies, the overall results indicate substantial progress, with the majority of classes exhibiting close statistical similarity between real and generated data. This comparative histogram analysis underscores the progressive enhancement of the generative process, illustrating how the CVAE model advances from limited structural fidelity to robust alignment

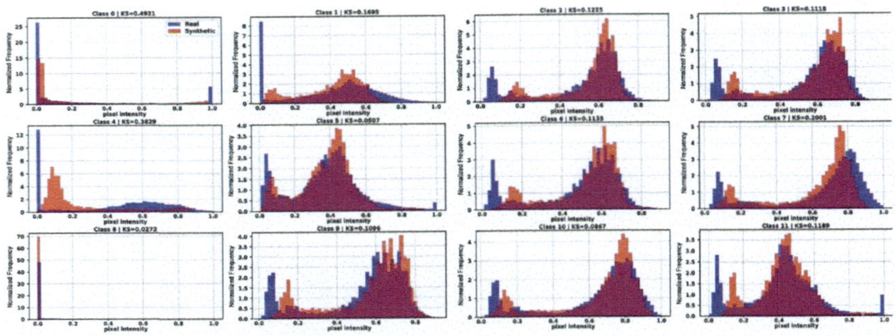

FIGURE 5.20 Histogram-based analysis of pixel intensity distributions in CVAE-generated EL images. The figure presents a histogram-based analysis of pixel intensity distributions in EL images generated by a CVAE. It compares the statistical properties of the synthetic images with real EL data to assess the model's ability to replicate realistic brightness and contrast patterns (*EL*, Electroluminescence; *CVAE*, conditional variational autoencoder).

with real distributions. The refined model achieves reliable statistical consistency across most defect categories, validating its suitability for advanced PV fault detection and diagnostic applications.

The CVAE framework achieves the most stable and consistent alignment between real and synthetic intensity distributions. The CDFs across nearly all classes exhibit tight overlap, with significantly lower KS values compared to both DCGAN and DCGAN–VGG16. This is particularly evident in classes 4, 6, and 11, where the synthetic distributions nearly replicate the real pixel power curves, demonstrating the CVAE's strength in learning smooth latent representations conditioned on class information. The improved consistency stems from the probabilistic nature of the CVAE, which enforces latent regularization and prevents mode collapse, ensuring balanced generation across all classes. While minor deviations persist in class 2 and class 10, their magnitude is substantially smaller than in adversarial frameworks. Consequently, the CVAE demonstrates the highest capacity for reproducing realistic electroluminescence intensity distributions, making it particularly well-suited for applications requiring balanced synthetic augmentation across diverse defect classes (See Fig. 5.21).

Conditional variational autoencoder – qualitative analysis on infrared images

The qualitative results obtained on IR images are even more satisfactory. The CVAE generates highly realistic thermal maps with sharp and coherent representations of the defect-related hot-spot regions. In all classes, the synthetic images exhibit smooth and physically consistent temperature gradients, and the location as well as the spatial extent of the thermal anomalies are very close to those observed in the real data (See Fig. 5.22).

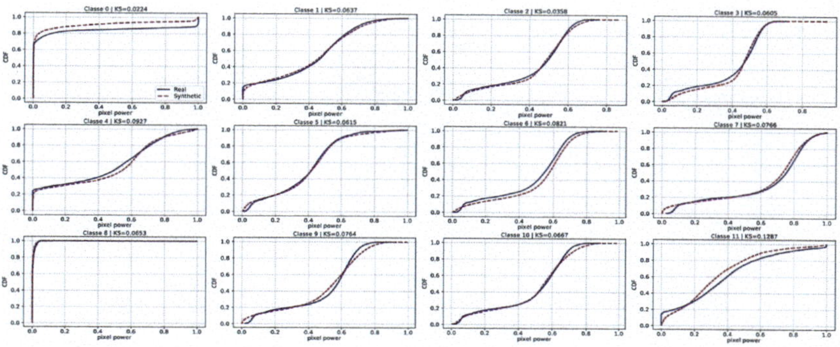

FIGURE 5.21 Cumulative distribution functions of pixel intensities for real and synthetic images in CVAE-generated EL images. The figure shows the cumulative CDFs of pixel intensities for real and synthetic EL images generated by a. It illustrates how closely the synthetic images replicate the statistical intensity patterns of real EL data, indicating the effectiveness of the CVAE model in generating realistic defect images (*CVAE*, Conditional variational autoencoder; *EL*, electroluminescence).

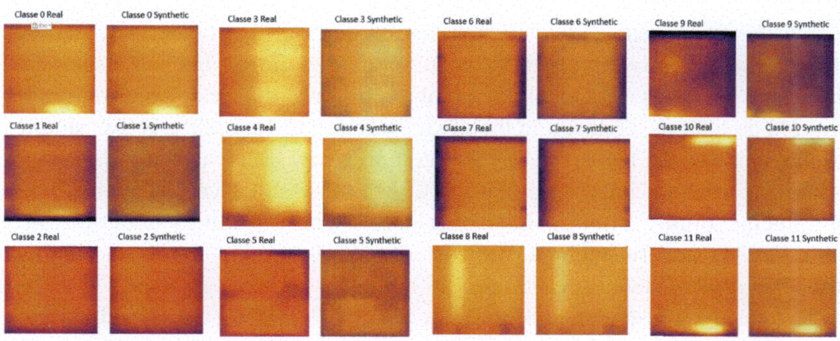

FIGURE 5.22 Qualitative comparison between real IR defect images and synthetic counterparts generated by DCGAN-VGG16. The figure presents a qualitative comparison between real IR defect images and synthetic counterparts generated by the DCGAN-VGG16 model. It highlights the model's ability to produce realistic thermal defect patterns, closely resembling those observed in actual IR data (*DCGANs*, Deep convolutional GANs; *IR*, infrared).

Fine-scale thermal variations and subtle defect signatures are clearly visible and largely preserved, demonstrating the model's strong capacity to encode and reconstruct complex IF patterns. In addition, the level of background noise is substantially reduced compared to the GAN-based outputs, resulting in high-quality synthetic IR images that are visually close to real samples and highly relevant for data augmentation and anomaly detection purposes.

Fig. 5.23 presents the result produced by the CVAE model, which exhibits the closest overall alignment between real and synthetic data, as reflected by

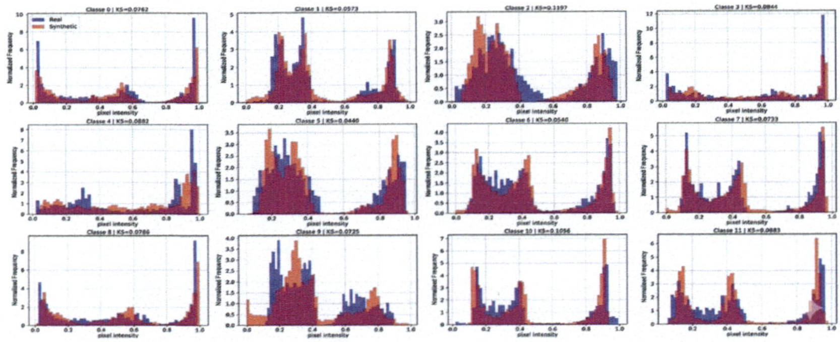

FIGURE 5.23 Histogram-based analysis of pixel intensity distributions in VAE-generated IR images. The figure presents a histogram-based analysis of pixel intensity distributions in IR images generated by a VAE. It compares the statistical characteristics of the synthetic images with real IR data to evaluate the model's ability to replicate realistic thermal intensity patterns (*IR*, Infrared; *VAE*, conditional variational autoencoder).

the generally lower distributional differences. For instance, in Class 0, the overlap between real and synthetic distributions is nearly perfect, demonstrating the stability of the CVAE in modeling global intensity statistics. Similarly, in Class 8, the synthetic data follows the real histogram with minimal deviations, confirming the model's ability to generalize across simpler intensity structures. However, in more complex cases such as Class 5, residual differences remain, indicating that the CVAE may still face challenges when reproducing highly multimodal distributions. These findings underline that while CVAEs ensure stable and consistent reconstructions, they may compromise fine-grained detail preservation in complex classes.

The results presented in Fig. 5.24, corresponding to the CVAE model, show the closest overall agreement between real and synthetic CDFs among all evaluated frameworks. In Class 0, the two curves are virtually indistinguishable, indicating that the CVAE accurately captures the global distribution of pixel intensities and preserves overall luminance balance. Similarly, in Class 8, the synthetic CDF closely follows the real distribution across the entire intensity range, exhibiting only minimal divergence. However, in more structurally complex cases such as Class 5, slight discrepancies remain visible, particularly within the mid-intensity region. These deviations suggest that while the CVAE effectively models unimodal and moderately complex distributions, it still faces challenges in fully representing highly multimodal intensity behaviors. Overall, these findings confirm that the CVAE generates stable and globally consistent intensity distributions, while residual fine-scale mismatches highlight the need for further refinement when modeling defects characterized by intricate textural and luminance variations.

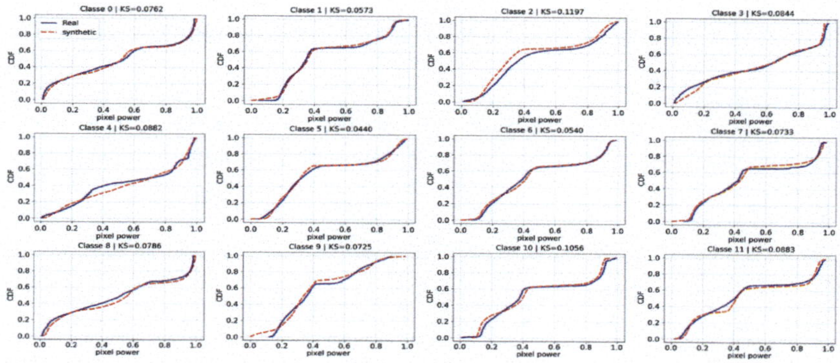

FIGURE 5.24 Cumulative distribution functions of pixel intensities for real and synthetic images in CVAE-generated IR images. The figure shows the cumulative CDFs of pixel intensities for real and synthetic IR images generated by a CVAE. It illustrates how effectively the synthetic images replicate the statistical intensity patterns of real IR data, indicating the model's capability to generate realistic thermal defect images (*CVAE*, Conditional variational autoencoder; *IR*, infrared).

TABLE 5.9 Overall averages of the results for the synthetic IR images.

Model	*MSE*	*PSNR*	*FID*	*SSIM*
CVAE	0.173	8.21	396.90	0.244
DCGAN with VGG16	0.222	6.94	457.54	0.183
DCGAN	0.209	7.08	458.86	0.175

CVAE, Conditional variational autoencoder; *DCGANs*, deep convolutional GANs; *IR*, infrared.

Quantitative results

The results obtained on synthetic IR images reveal a clear performance gap between the different generative models (See Table 5.9). The Conditional Variational Autoencoder (CVAE) achieves the best overall balance with the lowest MSE (0.173), indicating reduced reconstruction errors, and the highest Peak Signal-to-Noise Ratio (PSNR) (8.21), reflecting superior fidelity in generated image quality. Additionally, the CVAE yields the most favorable Structural Similarity Index Measure (SSIM) (0.244), highlighting improved structural similarity with real images. Moreover, its FID (396.90) is significantly lower than that of the GAN-based models, suggesting that the CVAE generates images that are perceptually closer to real IF samples.

TABLE 5.10 Overall averages of the results for the synthetic EL images.

Model	*MSE*	*PSNR*	*FID*	*SSIM*
CVAE	0.0386	27.27	404.98	0.654
DCGAN with VGG16	0.0195	20.77	263.87	0.581
DCGAN	0.0383	18.31	274.63	0.476

CVAE, Conditional variational autoencoder; *DCGANs*, deep convolutional GANs; *EL*, electroluminescence.

Between the two GAN-based approaches, the DCGAN with VGG16 integration clearly improves structural similarity (SSIM=0.183 vs 0.175 for DCGAN) and reduces perceptual divergence, achieving a lower FID (457.54) compared to the baseline DCGAN (458.86). This indicates that the incorporation of VGG16 features enhances the generative process, leading to images that are closer to real IF data. Conversely, the baseline DCGAN exhibits weaker performance, with relatively poor SSIM and slightly higher FID, confirming the added value of the VGG16 backbone. The CVAE remains the most effective model for IR images; the DCGAN–VGG16 demonstrates a clear advantage over the standard DCGAN, highlighting the benefit of feature transfer from a pretrained deep network.

For synthetic EL images (See Table 5.10), the DCGAN with VGG16 provides a significant improvement compared to the baseline DCGAN. It achieves the lowest MSE (0.0195 vs 0.0383), the lowest FID (263.87 vs 274.63), and a much higher SSIM (0.581 vs 0.476), confirming its ability to produce more structurally coherent and perceptually realistic images. These results demonstrate the importance of leveraging pretrained VGG16 features to enhance image quality and distribution-level similarity. The CVAE exhibits a different advantage: it produces the highest PSNR (27.27) and SSIM (0.654), reflecting excellent pixel-wise fidelity and structural consistency. However, its relatively higher FID (404.98) suggests that, although accurate at the pixel level, its generated distribution deviates more from real EL samples compared to DCGAN-VGG16.

Available dataset

Table 5.11 provides a comparative overview of publicly available EL and IR photovoltaic image datasets, summarizing their key characteristics in terms of size, resolution, defect types, and accessibility. This comparison facilitates the selection of suitable datasets for defect diagnosis, anomaly detection, and benchmarking of machine learning models.

TABLE 5.11 Comparative analysis of publicly available EL and IR photovoltaic image datasets.

Ref.	Type of image	Size (images)	Collection method/ Quality/Resolution	Labels/ Annotations	Access
PVEL-AD (2024)	EL	2000+ images (12 defect classes)	High-resolution EL images were collected in a laboratory with controlled acquisition. Covers a broad spectrum of degradation modes.	12 clearly defined defect categories (class-level labels). Imbalanced class distribution.	Public (Hebei Univ. of Technology & BUAA).
PV-IV-EL (2017)	EL	613 module-level EL sets (IV traces + EL images)	EL images acquired in laboratory with flash testers on modules removed from field; paired IV measurements → high quality, calibrated, useful for performance correlation.	EL images, IV metadata, and timestamps are good experimental metadata.	Public (Data.gov) ZIP downloads. (catalog.data.gov)
Deitsch et al. (2019)	EL	2624 cell images (300×300, 8-bit grayscale)	High-resolution EL images processed to cell crops (extracted from module EL photos). Good SNR for cell defects; consistent cropping/size.	Binary labels (functional/ defective) + defect type metadata in repo.	Public (GitHub)

(Continued)

TABLE 5.11 (Continued)

Ref.	Type of image	Size (images)	Collection method/ Quality/Resolution	Labels/ Annotations	Access
Benchmark EL images (2024)	EL	~500–1000 (varies selon source; e.g., 593 in one benchmark paper)	Aggregated from multiple public & private sources; preprocessed to 512×512 in some repos. Good for segmentation / semantic labels.	Pixel-level masks (in some subsets), class labels for defect types in the benchmark.	Public (GitHub/paper supplementary). (ResearchGate)
Infrared Solar Modules (2024)	IR	Varies (Kaggle listing) dataset collated from field inspections; some public releases report hundreds of thousands; Raptor Maps aggregated large corpora.	Aerial and ground thermographic surveys (radiometric TIFFs in some releases). Radiometric images enable temperature readings, varying resolution (drone/panel imager dependent).	Often labeled by anomaly type (hot-spot, soiling, string/ diode faults) in some releases; others are unlabeled.	Public/community (Kaggle), sometimes with usage restrictions. (Kaggle)
Photovoltaic System Thermography (2024)	IR	Several hundred (Kaggle dataset page)	Aggregated from literature/ paper datasets; images supplied as TIFF/JPEG; mixed quality & resolutions.	Some annotations (module bounding boxes/anomaly labels) depend on the upload.	Public (Kaggle)

Solar-panel-infrared-images project collection (2024)	IR	Small/variable (e.g., 29–196 images depending on collection)	Often small academic collections for object detection tasks, mixed resolutions, useful for prototyping but limited generalization.	Bounding boxes, object classes (depending on the Roboflow project).	Public (Roboflow)
PVEL-AD (2024)	IR	~20,000 images	Large aerial thermography corpus collated for anomaly classification (very low-res versions exist e.g., 24×40 pixels in some public releases); labeled across ~12 classes (11 faults + healthy).	Class labels (anomaly types) may include geo/metadata.	Public summary/paper; raw access may require contact or specific repo. (ai4earthscience.github.io)

EL, Electroluminescence; *IR*, infrared.

Conclusion

This chapter presented a comprehensive framework for improving PV defect detection through advanced data augmentation and synthetic data generation applied to both EL and IR imagery. The results demonstrate that data augmentation techniques combining geometric and photometric transformations significantly enhance model generalization by simulating real-world variability in PV module orientation, lighting, and temperature, thereby addressing dataset imbalance in both modalities. Building on this foundation, three generative architectures, DCGAN, DCGAN–VGG16, and CVAE, were evaluated for their capacity to synthesize high-fidelity defect images. The baseline DCGAN produced visually realistic but distributionally inconsistent samples, especially in complex or multimodal classes. The integration of VGG16 into the discriminator improved statistical alignment, as evidenced by reduced KS distances and closer correspondence between real and synthetic histograms and CDFs.

The CVAE achieved the most consistent results, generating stable and globally coherent distributions across nearly all defect categories while maintaining interpretability through its structured latent space. Collectively, these findings confirm that combining augmentation with modern generative models can mitigate class imbalance, improve data diversity, and enhance the robustness of deep learning frameworks for PV fault detection. Future work should extend these models to larger multimodal datasets, enable real-time integration into UAV-based inspection systems, and link synthetic data generation with predictive maintenance algorithms to support intelligent, data-driven PV monitoring and decision-making.

AI Disclosure

During the preparation of this work, the author(s) used ChatGPT in order to Proofreading. After using this tool/service, the author(s) reviewed and edited the content as needed and take(s) the full responsibility for the content of the publication.

References

Arjovsky, M., Chintala, S., Bottou, L., 2017. Wasserstein generative adversarial networks. In: Proceedings of the International Conference on Machine Learning.

Benchmark EL images 2024 https://github.com/TheMakiran/BenchmarkELimages

Bian, J., Hui, X., Sun, S., Zhao, X., Tan, M., 2019. A novel and efficient CVAE-GAN-based approach with informative manifold for semi-supervised anomaly detection. IEEE Access 7, 88903–88916. Available at https://doi.org/10.1109/access.2019.2920251.

Deitsch, S., Christlein, V., Berger, S., Buerhop-Lutz, C., Maier, A., Gallwitz, F., et al., 2019. Automatic classification of defective photovoltaic module cells in electroluminescence images. Solar Energy 185, 455–468. Available at https://doi.org/10.1016/j.solener.2019.02.067.

Drir, N., Chekired, F., Mellit, A., Blasuttigh, N., 2025a. Hybrid CNN-EML model for fault diagnosis in electroluminescence images of photovoltaic cells. Renewable Energy 250, 123343. Available at https://doi.org/10.1016/j.renene.2025.123343.

Drir, N., Mellit, A., Bettayeb, M., Dhimish, M., 2025b. Enhanced photovoltaic defect detection using perceptual loss in DCGAN and VGG16-integrated models on electroluminescence images. IEEE Journal of Photovoltaics 15 (6), 759–769. Available at https://doi.org/10.1109/jphotov.2025.3542829.

Ebied, M.A., Munshi, A., Alhuzali, S.A., El-sotouhy, M.M., Shehta, A.I., Elborlsy, M.S., 2025. Advanced deep learning modeling to enhance detection of defective photovoltaic cells in electroluminescence images. Scientific Reports 15 (1). Available at https://doi.org/10.1038/s41598-025-14478-y.

Goodfellow, I., Pouget-Abadie, J., Mirza, M., Xu, B., Warde-Farley, D., Ozair, S., et al., 2020. Generative adversarial networks. Communications of the ACM 63 (11), 139–144. Available at https://doi.org/10.1145/3422622.

Gulrajani, I., Ahmed, F., Arjovsky, M., Dumoulin, V., Courville, A., 2017. Neural information processing systems foundation Canada Improved training of wasserstein GANs 2017. Advances in Neural Information Processing Systems 5768–5778.

Harrou, F., Dairi, A., Taghezouit, B., Khaldi, B., Sun, Y., 2024. Automatic fault detection in grid-connected photovoltaic systems via variational autoencoder-based monitoring. Energy Conversion and Management 314, 118665. Available at https://doi.org/10.1016/j.enconman.2024.118665.

Hijjawi, U., Lakshminarayana, S., Xu, T., Piero Malfense Fierro, G., Rahman, M., 2023. A review of automated solar photovoltaic defect detection systems: approaches, challenges, and future orientations. Solar Energy 266, 112186. Available at https://doi.org/10.1016/j.solener.2023.112186.

Infrared Solar Modules 2024 https://www.kaggle.com/datasets/marcosgabriel/infrared-solar-modules

Karimi, A.M., Fada, J.S., Parrilla, N.A., Pierce, B.G., Koyuturk, M., French, R.H., et al., 2020. Generalized and mechanistic PV module performance prediction from computer vision and machine learning on electroluminescence images. IEEE Journal of Photovoltaics 10 (3), 878–887. Available at https://doi.org/10.1109/JPHOTOV.2020.2973448, https://ieeexplore.ieee.org/xpl/RecentIssue.jsp?punumber=5503869.

Kasi, G., Abirami, S., Lakshmi, R.D., 2023. A deep learning based cross model text to image generation using DC-GAN. In: Proceedings of the 12th International Conference on Advanced Computing (ICoAC).

Li, D., Zhang, Y., Yang, Z., Jin, Y., Xu, Y., 2024. Sensing anomaly of photovoltaic systems with sequential conditional variational autoencoder. Applied Energy 353, 122124. Available at https://doi.org/10.1016/j.apenergy.2023.122124.

Mellit, A., Kalogirou, S., 2021. Artificial intelligence and internet of things to improve efficacy of diagnosis and remote sensing of solar photovoltaic systems: challenges, recommendations and future directions. Renewable and Sustainable Energy Reviews 143, 110889. Available at https://doi.org/10.1016/j.rser.2021.110889.

Millendorf, M., Obropta, E., Vadhavkar, N., 2020. Infrared solar module dataset for anomaly detection. In: Proceedings of the International Conference on Learning Representations (ICLR).

Neloy, A.A., Turgeon, M., 2024. Disentangled conditional variational autoencoder for unsupervised anomaly detection. In: Proceedings of the IEEE International Conference on Big Data (BigData).

Photovoltaic System Thermography. 2024 https://www.kaggle.com/datasets/marcosgabriel/photovoltaic-system-thermography

PVEL-AD 2024 https://github.com/binyisu/PVEL-AD

PV-IV-EL. 2017 https://catalog.data.gov/dataset/photovoltaic-module-current-voltage-and-electroluminescence-image-data-pv-iv-el

Radford, A., Metz, L., Chintala, S., 2016. Unsupervised representation learning with deep convolutional generative adversarial networks 1 2016/01 4th International Conference on Learning Representations, ICLR 2016 - Conference Track Proceedings International Conference on Learning Representations, ICLR United States https://dblp.org/db/conf/iclr/iclr2016.html

Rocha, D., Alves, J., Lopes, V., Teixeira, J.P., Fernandes, P.A., Costa, M., et al., 2023. Multidefect detection tool for large-acale PV plants: segmentation and classification. IEEE Journal of Photovoltaics 13 (2), 291–295. Available at https://doi.org/10.1109/JPHOTOV.2023.3236188, https://ieeexplore.ieee.org/xpl/RecentIssue.jsp?punumber=5503869.

Shorten, C., Khoshgoftaar, T.M., 2019. A survey on image data augmentation for deep learning. Journal of Big Data 6 (1). Available at https://doi.org/10.1186/s40537-019-0197-0, https://journalofbigdata.springeropen.com.

da Silveira Junior, C.R., Sousa, C.E.R., Fonseca Alves, R.H., 2025. Automatic fault classification in photovoltaic modules using denoising diffusion probabilistic model, generative adversarial networks, and convolutional neural networks. Energies 18 (4) Available from. Available at https://doi.org/10.3390/en18040776, http://www.mdpi.com/journal/energies/.

Solar-panel-infrared-images project collection, Available at https://universe.roboflow.com/solarpanelimages/solar-panel-infrared-images.

Su, B., Chen, H., Zhu, Y., Liu, W., Liu, K., 2019. Classification of manufacturing defects in multicrystalline solar cells With novel feature descriptor. IEEE Transactions on Instrumentation and Measurement 68 (12), 4675–4688. Available at https://doi.org/10.1109/tim.2019.2900961.

Tan, H., Guo, Z., Lin, Z., Chen, Y., Huang, D., Yuan, W., et al., 2024. General generative AI-based image augmentation method for robust rooftop PV segmentation. Applied Energy 368, 123554. Available at https://doi.org/10.1016/j.apenergy.2024.123554.

Tang, C., Ren, H., Xia, J., Wang, F., Lu, J., 2023. Automatic defect identification of PV panels with IR images through unmanned aircraft. IET Renewable Power Generation 17 (12), 3108–3119. Available at https://doi.org/10.1049/rpg2.12831.

Tang, W., Yang, Q., Dai, Z., Yan, W., 2024. Module defect detection and diagnosis for intelligent maintenance of solar photovoltaic plants: techniques, systems and perspectives. Energy 297, 131222. Available at https://doi.org/10.1016/j.energy.2024.131222.

Wang, J., Cheng, Z., 2025. Photovoltaic panel defect detection algorithm based on infrared imaging and improved YOLOv8. PeerJ Computer Science 11, e2776. Available at https://doi.org/10.7717/peerj-cs.2776.

Zhang, Y., Zhang, X., Tu, D., 2024. Solar photovoltaic module defect detection based on deep learning. Measurement Science and Technology 35 (12), 125404. Available at https://doi.org/10.1088/1361-6501/ad7d28.

Chapter 6

Smart monitoring systems for photovoltaic power plants

Adel Mellit[1], Nassiha Boutana[2], and Soteris Kalogirou[3]
[1]Faculty of Sciences and Technology, University of Jijel, Jijel, Algeria, [2]Department of Electronics, USTHB University, Algiers, Algeria, [3]Department of Mechanical Engineering and Materials Science and Engineering, Cyprus University of Technology, Cyprus

Introduction

A large number of photovoltaic (PV) plants have recently been installed, and with reference to the International Energy Agency, the global cumulative installed capacity of solar PV reached approximately 2.2 TW by the end of 2024 (Snapshot of Global PV Markets, 2025). This capacity highlights the rapid growth and widespread adoption of solar PV technologies in recent years. To keep these PV plants safe and reliable during their expected lifetime, the systems have to be monitored continually (Mellit and Kalogirou, 2021). Recent technological advances have transformed PV monitoring from passive data acquisition to smart, autonomous system supervision. The integration of Internet of Things (IoT) technologies enables real-time data collection through distributed sensors, while standardized communication protocols ensure seamless data transmission to cloud platforms. In parallel, artificial intelligence (AI) and artificial intelligence of things (AIoT) enhance monitoring by enabling advanced analytics, including anomaly detection, root cause analysis, and estimation of the remaining useful life (RUL) of PV components. These capabilities support predictive maintenance strategies, allowing operators to anticipate failures before they occur and improve both energy yield and system reliability.

This chapter introduces the evolution of PV monitoring systems, starting with traditional solutions and progressing toward modern IoT and AI-enabled architectures. It provides a detailed review of IoT sensors, data transmission protocols, cloud-based analytics, and the role of autonomous monitoring functions. A literature review captures the state-of-the-art of smart PV monitoring techniques, positioning emerging AIoT solutions within the broader research context. To demonstrate practical implementation, two

Embedded Artificial Intelligence and the Internet of Things for Photovoltaic Systems.
DOI: https://doi.org/10.1016/B978-0-443-44835-5.00001-1

case studies will be presented: an IoT-based PV monitoring prototype and an AIoT-powered anomaly detection framework for PV systems. The chapter also examines the challenges associated with communication reliability, data interoperability, cybersecurity, and the scalability of smart monitoring solutions, while proposing potential pathways for improvement. It highlights also how PV monitoring is transitioning from manual, periodic inspection toward autonomous, intelligent systems capable of improving performance, reducing operational costs, and extending system lifetime. The integration of IoT and AI technologies marks a significant step toward next-generation PV digitalization, enabling smarter, more resilient, and highly efficient solar power plants.

Photovoltaic monitoring systems

Most utility-scale PV plants are equipped with a supervisory control and data acquisition (SCADA) system to monitor their performance. The SCADA system provides continuous real-time monitoring of PV system parameters such as solar irradiance, temperature, DC/AC voltages, currents, power output, and other meteorological parameters. This helps operators detect anomalies early and take corrective actions to minimize downtime, ensure continuity of power supply, and optimize maintenance cost (Woyte et al., 2013; Rahman et al., 2018). While these systems enable fundamental monitoring and operational control, they often lack the ability to perform granular data analysis, detect subtle anomalies, or support predictive maintenance. As PV systems expand in size and complexity, particularly in utility-scale and distributed deployments, continuous monitoring and data-driven decision making have become essential to maximize system performance and reduce downtime. A configuration of a SCADA system in PV plants is shown in Fig. 6.1. A SCADA system generally comprises an on-site control cabinet and specialized software with advanced analytical capabilities, allowing for remote control and real-time monitoring of the generation portfolio. Its functionality is based on several core components, including controllers that gather data from sensors distributed throughout the PV plant.

A PV SCADA system is typically structured around three key components:

Hardware at the inverter stations and the master control room

- *Data acquisition devices*: These include input/output modules, terminal servers, and Ethernet switches that collect data from inverters, combiner boxes, transformers, and weather stations across the PV plant.
- *Control and monitoring servers*: Redundant servers oversee plant-wide data acquisition, monitoring, and control. They are designed according to industry standards, featuring open architecture and compatibility with standard communication protocols to ensure reliable performance.

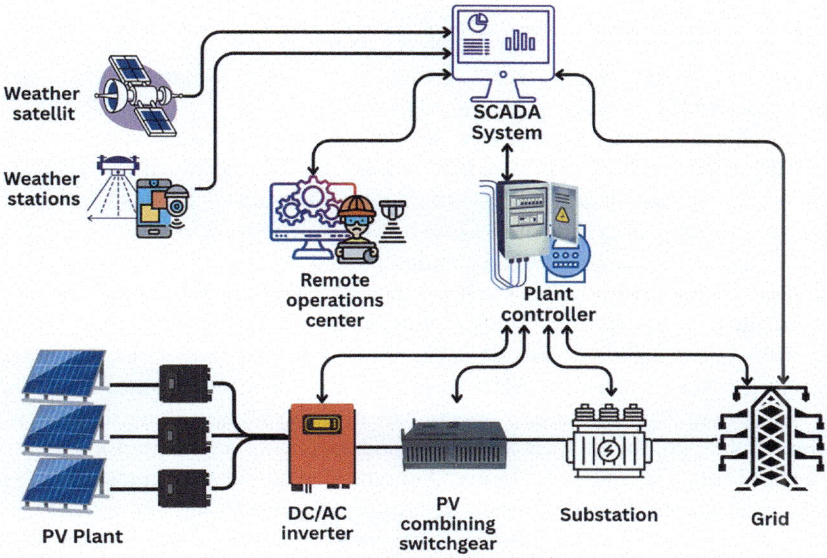

FIGURE 6.1 A utility-scale PV power plant with a SCADA system. The figure shows a PV SCADA system with three main parts: field devices, communication links, and a central control station. It highlights how data flows from the PV site to the monitoring center (*PV*, Photovoltaic; *SCADA*, supervisory control and data acquisition).

Operator consoles

- *Human–machine interface systems*: These provide intuitive, user-friendly interfaces that allow operators to supervise and control both the PV plant and the substation.

Network infrastructure

- *Communication network*: A fiber-optic ring links all devices in the system, ensuring resilience and uninterrupted operation even in the event of component failures.
- *Substation integration*: The SCADA system interfaces with the substation gateway, enabling seamless data exchange with Load Dispatch Centers.

A variety of PV monitoring systems have been developed and commercialized, with the primary objective of collecting operational data and verifying system performance based on key parameters such as power, current, voltage, and meteorological parameters. These monitoring systems play a critical role in supervising PV installations (Rahman et al., 2018). They are generally classified into centralized and decentralized configurations, where centralized systems rely on a single control unit for the entire plant, while decentralized systems distribute monitoring and control equipment across multiple local units to enhance reliability and scalability (Mohammed and Al-Naib, 2023; Madeti and Singh, 2017). Table 6.1 shows a brief comparison between centralized and decentralized PV SCADA systems.

The choice of either a centralized SCADA or decentralized SCADA system for PV power plants mainly depends on the scale, geographical distribution, and operating conditions of the installation. Centralized SCADA systems represent a simpler architecture since all data is processed at one control centre; such systems are easier to maintain, and are the most cost-effective solution for small to medium-sized plants. However, these systems have the disadvantage of relying on control from a central unit, which may affect reliability and scalability, and can delay on-the-spot responses to local events. In contrast, decentralized SCADA represents a growing trend, since it allows the control and data processing on any part of the system, making the system more reliable, improving local responses, and enhancing scalability for large or geographically dispersed PV farms. Although the initial investment may be higher, decentralized systems are increasingly adopted in modern large-scale installations and micro-grids due to their robustness and flexibility. Commercially, centralized SCADA solutions are often a good fit for utility-scale plants up to 100 MW, while decentralized architectures support multi-site and high-capacity projects exceeding 50 MW, reflecting a growing trend toward distributed control in the PV industry.

Generally, a PV SCADA system offers significant advantages for monitoring and controlling solar PV power plants. It ensures compliance with national and international grid standards, facilitates flexibility by adapting to various PV technologies and configurations, and offers broad compatibility with diverse communication protocols and devices (Moreno-Garcia et al., 2016). In addition, a SCADA system improves cost-effectiveness by automating and facilitating predictive diagnostics that reduce monthly operation, maintenance, and repair efforts while ensuring transparency and end-to-end visibility from the planning stage to immediate operation. Nevertheless, there are limitations associated with a PV SCADA system, including high initial infrastructure and software investment costs, difficulties with integrating the SCADA system with heterogeneous devices and legacy systems, potential cybersecurity risks from remote connectivity, and the requirement for qualified personnel to operate the SCADA system in terms of day-to-day operations, maintenance, repairs, and analysing data. All these challenges need to be managed effectively to fully maximize SCADA's potential in PV applications.

Modern photovoltaic monitoring systems based on the Internet of Things and artificial intelligence

The smart PV monitoring system is an advanced solution that combines sensors, IoT technologies, and smart analytics to improve the performance and

TABLE 6.1 Comparative analysis of centralized versus decentralized SCADA systems in PV plants.

Aspect	Centralized SCADA	Decentralized SCADA
Architecture	A single central server/ master control room manages all monitoring and control.	Multiple local controllers or subsystems manage monitoring and control independently.
Data Processing	Data is collected and processed at a central location.	Data processed locally at the subsystem level, with summarized data sent to higher-level SCADA.
Reliability	More vulnerable to a single point of failure.	Higher fault tolerance; local units continue operating even if one fails.
Scalability	Limited scalability; adding new units increases central load.	Highly scalable; new units can be added with local controllers.
Communication requirements	Requires robust and high-capacity communication lines to the central unit.	Lower bandwidth per unit; local control reduces communication load.
Complexity	Simpler to design, operate, and maintain.	More complex due to distributed architecture and integration requirements.
Cost	Lower initial cost (fewer controllers and infrastructure).	Higher initial investment (more local units and communication links).
Response time	May experience latency if the central system is overloaded.	Faster local response due to distributed processing.
Maintenance	Easier maintenance at a single control point.	Requires coordination across multiple control units, but fault isolation is easier.
Typical use case	Utility-scale PV plant with a single control room.	Large-scale PV farms or geographically dispersed PV systems.
Commercial Adoption	Widely adopted in small to medium utility-scale PV plants; popular with Siemens, Schneider Electric, and ABB SCADA solutions.	Increasing adoption in large PV farms, microgrids, and remote installations; solutions by ABB, GE, Schneider, and Hitachi.
Typical capacity range	1 –100 MW	Small-scale or distributed PV systems (rooftop, microgrids, mini-PV farms, typically < 50 MW each)

PV, Photovoltaic; *SCADA*, supervisory control and data acquisition.

reliability of solar power plants (Ando et al., 2015). Unlike traditional monitoring, which is limited to tracking basic electrical parameters, smart monitoring provides real-time data on environmental conditions, energy production, and equipment status, and transmits this information to cloud platforms for advanced processing (Rahman et al., 2018). By integrating AI and predictive analytics, the system can detect faults early, predict failures, and suggest corrective actions, reducing downtime and maintenance costs (Mellit and Kalogirou, 2021). Furthermore, the system offers remote control capabilities, user-friendly dashboards, and seamless integration with smart grids and storage systems, transforming raw data into actionable insights that improve efficiency and extend system life. Fig. 6.2 depicts the basic structure of a smart PV monitoring system. The IoT ensures universal sensing and connectivity, while AI transforms raw data into intelligent insights and predictive actions, making photovoltaic systems more reliable, efficient, and cost-effective.

The key elements of a smart PV monitoring system are:

- Sensors and data collection: Collect electrical data (voltage, current, power), environmental data (radiation, temperature, wind), and unit-level data in real-time.

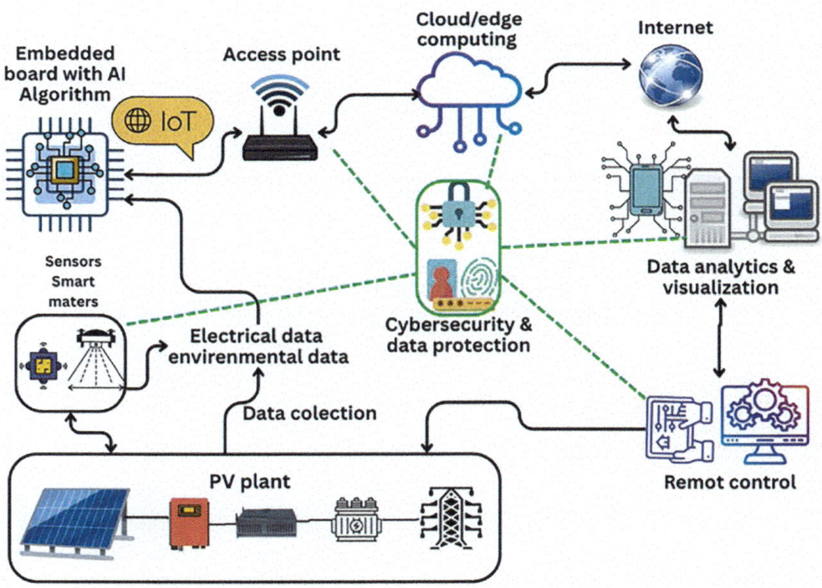

FIGURE 6.2 Basic structure of a smart PV monitoring system. The figure shows a smart PV monitoring system composed of sensors/devices in the field, a communication layer, and a cloud or central platform. It illustrates how data is collected, transmitted, and visualized for intelligent monitoring and decision-making (*PV*, Photovoltaic).

- IoT connectivity: Use wireless protocols (Wi-Fi, LoRa, NB-IoT, ZigBee, 5G, etc.) and edge devices to securely and reliably transmit data to central/cloud platforms.
- Cloud/Edge computing platforms: Store, consolidate, and preprocess data; edge computing supports low-latency analysis near the source.
- AI-based algorithms: Algorithms for fault detection, performance prediction, anomaly detection, predictive maintenance, and energy production forecasting.
- Data analytics and visualization: Dashboards and mobile/web applications with real-time visualization, key performance indicators, and alerts to support decision-making.
- Remote control and automation: Remote configuration of inverters, trackers, and storage systems, enabling automatic response to faults or grid signals.
- Cybersecurity and data protection: Secure data transmission, encryption, and authentication to protect against cyber threats in IoT systems.
- Integration with smart grids and energy management: Enable demand response, load balancing, and coordination with storage systems, electric vehicle charging, or microgrids.

Internet of Things sensors for photovoltaic monitoring

In PV systems, a variety of IoT-enabled sensors play a vital role in ensuring accurate and complete monitoring. Temperature sensors capture both ambient conditions and module surface temperatures, helping to identify heat-related efficiency losses. Solar irradiance sensors (pyranometers or reference solar cells) measure solar energy intensity, providing the data needed to calculate performance ratios and optimize the system (Gopal et al., 2020). To assess the electrical behaviour, current and voltage sensors track power output across various points in the system, enabling real-time monitoring and early detection of any performance degradation. Weather stations, equipped with instruments such as humidity sensors and anemometers, provide broader environmental data that directly impacts PV performance (Asnil et al., 2023). For surface-related issues, dust and dirt sensors determine the amount of particle buildup on panels, allowing operators to optimize cleaning schedules and maintain efficiency (Lay-Ekuakille et al., 2018).

In more advanced configurations, thermal imaging sensors detect hotspots, while string monitors help to accurately identify underperforming modules (Mellit and Kalogirou, 2025). Combined, these sensors feed data to acquisition and IoT platforms, creating a comprehensive picture of the system's condition. By integrating diverse sensor inputs, operators obtain accurate performance assessments and enable predictive maintenance scheduling. Modern IoT systems significantly enhance the monitoring and control of PV plants by leveraging advanced data acquisition, cloud connectivity, and real-time analytics. By collecting data from multiple sensors, including irradiance, temperature, voltage, current, and inverter performance, these systems provide a holistic view of the plant's operational status. Advanced algorithms, often incorporating machine learning or AI techniques, process this multi-dimensional data to detect anomalies, predict faults, optimize energy output, and improve maintenance scheduling. Beyond fault detection, IoT-enabled PV monitoring platforms can generate actionable insights, such as adjusting inverter settings for maximum efficiency, predicting energy yield based on weather forecasts, or identifying underperforming modules. Some systems even provide personalized recommendations to operators and maintenance teams, enabling proactive decision-making and minimizing downtime. Overall, IoT transforms traditional PV monitoring from a reactive process into a predictive, data-driven operation that maximizes plant performance, reduces operational costs, and supports smarter grid integration.

Data gathering, transmission protocol, and data analysis

In monitoring systems with IoT-enabled PV, sensors in the array measure electrical (voltage, current, power) and environmental parameters (irradiance, temperature, humidity, wind). Data are collected by Data Acquisition (DAQS) or IoT nodes, converted to digital data through Analog-to-Digital Converter (ADCs), and are often preprocessed locally with steps such as filtering, compression, and feature extraction. Additional postprocessing may occur at edge units, which enables lightweight AI to detect anomalies. The preprocessed data are transferred through wired (CAN bus, RS485) or wireless (Zigbee, LoRaWAN, Wi-Fi, 5G) communication protocols via gateways to cloud servers or SCADA systems. Analytical and dashboard applications can perform analysis and generate actionable decision-making. Fig. 6.3 displays a block diagram of data acquisition, communication, and analytics within the smart PV systems.

Data gathering

The process of collecting data is a vital part of IoT-based PV monitoring systems. Multiple sensors and smart devices are installed across the PV array to continuously collect and measure key operational parameters. Examples of common sensors include (1) pyranometers for solar irradiance, (2) thermocouples, or resistance temperature detectors for module and ambient temperature, (3) voltage and current sensors for electrical performance, (4) inverter-capable monitoring units for device status, efficiency, or fault events, etc. Generally, these types of sensors record high-resolution time-series data at varying intervals (from a few seconds to minutes), depending on the particular system's applications.

Edge devices can perform a preprocessing function on the data locally to filter out noise, detect anomalies, and reduce latency before transmitting the data to a gateway or cloud. Once the data are collected, they are stored in structured databases or time-series database platforms and used for advanced analytics,

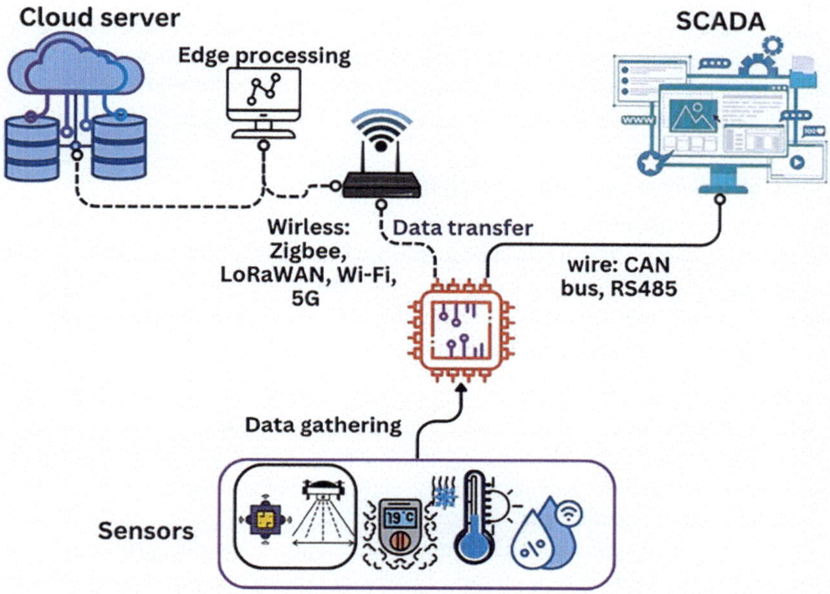

FIGURE 6.3 Data gathering, communication protocols, and data analysis. The figure shows the process of collecting data from sensors, sending it through communication protocols, and then analyzing it on a platform or system. It highlights the flow from data acquisition to transmission and final analysis.

predictive maintenance, or AI-based decision-making and similar purposes. This comprehensive data acquisition capability enables PV operators to detect underperforming modules, gain economic benefit via the maximization of energy yield, and improve overall PV system reliability (Adiguzel et al., 2023).

Communication protocols

Efficient and reliable communication is critical for IoT-based PV monitoring, ensuring seamless transmission of sensor data to edge devices, gateways, or cloud platforms. Common communication protocols include (Ansari et al., 2021):

- Wi-Fi: Suitable for small-scale PV systems with high data rates and local coverage.
- Zigbee: Low-power protocol ideal for short-range sensor networks.
- LoRaWAN: Long-range, low-power protocol for geographically dispersed PV arrays.
- Narrowband Internet of Things (NB-IoT)/Long Term Evolution for Machines (LTE-M): Cellular-based protocols offering wide-area coverage for remote PV installations.
- Modbus/Modbus TCP: Standard industrial protocol used for inverter and energy meter integration.

These protocols ensure secure and efficient data transmission, supporting near real-time monitoring and control. By combining the appropriate protocol with edge computing, IoT systems reduce bandwidth usage, enhance response times, and support large-scale, distributed PV installations with minimal infrastructure cost.

Cloud platforms and data analytics

Once data are collected and transmitted, cloud platforms provide the storage, computation, and visualization capabilities necessary for intelligent PV monitoring. Popular platforms include Amazon Web Services (AWS) IoT, Microsoft Azure IoT Hub, Google Cloud IoT, Siemens MindSphere, and IBM Watson IoT. These platforms facilitate:

- *Data aggregation and storage:* Large volumes of multi-sensor time-series data are organized and retained for long-term analysis.
- *Advanced analytics:* AI and machine learning algorithms process the data to detect anomalies, forecast energy yield, optimize inverter settings, and predict maintenance needs.
- *Visualization dashboards:* Operators can monitor system performance in real time through web or mobile interfaces, receiving alerts for faults or abnormal behavior.
- *Decision support:* Actionable insights and recommendations are generated automatically, enabling proactive control and minimizing downtime.

By integrating cloud analytics with IoT data, PV operators can shift from reactive to predictive maintenance strategies, improve plant efficiency, and make data-driven decisions that enhance both energy yield and operational reliability (Rezk et al., 2017).

Autonomous monitoring

Autonomous monitoring in PV systems refers to the use of advanced sensing, communication, and intelligent processing technologies to manage and oversee PV system operation without human intervention. The ultimate goal is to automate the entire monitoring lifecycle, reducing reliance on manual inspections and interventions, while increasing reliability, efficiency, and predictive capabilities (Aghaei et al., 2025). Key aspects include:

Automated fault and anomaly detection

Autonomous monitoring systems continuously collect data from PV modules, inverters, strings, and environmental sensors. Using ML algorithms, statistical models, or rule-based systems, they can automatically detect deviations from expected performance, such as partial shading, module degradation, inverter malfunctions, or connection failures. This allows for real-time identification of faults and anomalies before they escalate into major system issues.

Root cause analysis

Beyond identifying a fault, advanced autonomous systems can analyze the collected data to determine its underlying cause. For example, a sudden drop in current in a string could be traced back to a module hotspot, dirt accumulation, or an inverter fault. This capability significantly reduces troubleshooting time and allows for targeted maintenance, preventing unnecessary inspection of healthy components.

Remaining useful life estimation

By leveraging predictive models based on historical performance, environmental conditions, and operational stress factors, autonomous monitoring systems can estimate the remaining useful life of critical components, such as PV modules, inverters, and batteries. RUL prediction enables proactive maintenance scheduling, minimizing downtime and extending system lifespan.

Integration with Internet of Things and artificial intelligence technologies

Autonomous monitoring relies heavily on the IoT to gather high-resolution data from multiple sources in real time. Coupled with AI and machine learning, the system can learn normal operating patterns, detect anomalies, predict failures, and even recommend corrective actions automatically.

Benefits of autonomous photovoltaic monitoring

- Reduced human labor and operational costs: Minimal need for manual inspections or data analysis.
- Improved reliability and efficiency: Faults and degradations are detected and addressed promptly.
- Predictive maintenance: Maintenance actions are scheduled based on real data and predictions rather than fixed intervals.
- Scalability: Suitable for large-scale PV plants where manual monitoring would be time-consuming and costly.
- Data-driven decision-making: Operators can receive actionable insights and recommendations automatically.

Artificial intelligence in analysing collected photovoltaic data

Data collected from solar PV plants can be efficiently analysed, including electrical parameters, atmospheric inputs, and even visual information (thermal and Red-Green-Blue images). AI techniques enable multiple advanced applications. For example, AI performance prediction models allow for estimating energy production or performance under a variety of operational and weather conditions to help optimise systems and improve their grid integration. Anomaly, or fault detection algorithms can notify stakeholders of anomaly conditions and issues (such as hotspots, shading effects, inverter failures), or unexpected output drops, often at an early stage before they escalate into major problems. By leveraging predictive maintenance strategies, AI models can anticipate component deterioration (such as dirty or aged panels or inverter wear) and recommend timely interventions, reducing downtime and extending system life. Furthermore, energy production forecasting models integrate weather forecasts with historical PV performance data to provide short- and long-term energy generation estimates, improving scheduling, energy trading, and demand management. In more advanced applications, computer vision techniques can be used on infrared or RGB color images from PV plants to automate the recognition of defects at the module level, such as cracks, peeling, and others. Overall, AI can transform multisource/raw PV data into valuable actionable insights that enhance the efficiency, reliability, and economic performance of solar power plants (Mellit and Kalogirou, 2025; Costa et al., 2023).

AI used for analysing big data for PV plants represents a key advancement in modern solar plant management (Pingulkar and Joshi, 2021). The purpose of AI techniques is to analyse and interpret the large volumes of data produced by PV systems, reducing the need for human intervention. Advanced sensing networks consistently collect electrical, environmental, and operational data, while AI algorithms detect faults, anomalies, or performance deviations automatically in real time. In addition to detection, these algorithms can execute root cause analysis, which determines the underlying causes of a performance drop, for example, inverter malfunctions, module degradation, and partial shading, to enable targeted and timely maintenance. Furthermore, predictive AI models can predict the RUL of key components, allowing for proactive maintenance planning and reduced downtime while extending the overall life of the system. AI's effectiveness is based on the big data produced by large-scale PV plants. Through the generation of massive amounts of heterogeneous data at high velocity, PV plants produce various measurements, including voltage, current, power production, irradiance, temperature, and other environmental parameters. By leveraging big data, AI can analyse additional subtle trends and patterns to identify and predict possible differences in degradation or emerging faults to optimize system-level performance and energy yield. Predictive analytics (combined with historical and real-time data) enables accurate RUL estimates, proactive maintenance or service plans, and decisions regarding system expansion or retrofitting. Additionally, the advancement of big data enables accurate solar generation forecasting, grid integration, and smart energy management (Preda et al., 2018). However, using AI to process big data also presents challenges.

Big data necessitates reliable computing, processing, and storage infrastructure, and model dependability may be lowered by noisy, inconsistent, or missing sensor data. Keeping sensitive operational data creates cybersecurity risks, and integrating diverse datasets from several sensors and inverters increases complexity (Sharabati et al., 2023). In order to maximize the benefits of AI in big data analysis, it is suggested to adopt efficient storage solutions (cloud or edge computing), implement data cleaning and preprocessing procedures, utilize advanced AI and machine learning, and establish secure data transmissions and access. Standards-based communication protocols, for example, MQTT, Modbus, or OPC-UA, can also provide better operational conditions and scalability.

Literature review of smart photovoltaic monitoring systems

To provide a clearer comparative overview of the available literature on smart PV monitoring systems, the selected works are summarized in Table 6.2. The table highlights key aspects such as the type of Microcontrollers (MCUs) or IoT platforms used, the nature of collected data, the type of web interface (free platforms or developed dashboards), as well as indicators of simplicity, scalability, implementation cost, and the scale of the PV system under consideration. This structured comparison facilitates the identification of technological trends, trade-offs, and research gaps across different generations of monitoring solutions.

The evolution of smart PV monitoring systems from the year 2016 to 2026 reveals a clear transition from low-cost, small-scale prototypes to sophisticated AIoT-integrated platforms designed for large-scale and utility deployments. In the early stage (2016–21), most systems relied on inexpensive microcontrollers such as Arduino and ESP modules, often connected to free IoT platforms like ThingSpeak or Blynk, or to simple custom dashboards. These works primarily measured basic electrical and environmental parameters such as voltage, current, temperature, and irradiance, with the main objective of demonstrating proof-of-concept prototypes. Their strengths were simplicity and affordability, making them well-suited for small stand-alone PV installations, rooftop systems, or educational laboratories.

However, their scalability and robustness were limited, restricting practical use in commercial or grid-connected plants. During the intermediate phase (2022–23), research began to incorporate cloud computing, machine learning, and more advanced hardware such as Raspberry Pi, Boltuino, and IoT gateways. These systems expanded beyond simple data logging, introducing intelligent monitoring, predictive analytics, and fault detection. For instance, Emamian et al. (2022) applied cloud-based ML for PV fault diagnostics, while Utama et al. (2023) presented an IoT-enabled platform capable of supervising multiple PV plants. Similarly, Rani et al. (2023), Mondal et al. (2022), Nuhin et al. (2022), Mellit et al. (2023) demonstrated cost-effective solutions integrating cloud dashboards and AI features for improved decision-making. This phase reflects a shift toward reliability and scalability, with applications extending to grid-connected systems and larger PV plants.

TABLE 6.2 Literature review of smart PV monitoring systems.

Ref.	Year	MCU/Platform	Collected data	Web interface (Free/ Developed)	Application	Simplicity	Scalability	Cost	PV scale
Adhya et al. (2016)	2016	PIC18F46K22 + GPRS	PV Voltage, PV Current, Temperature, Irradiance, Battery voltage	Custom dashboard	PV performance monitoring	High	Low	Very Low	Stand-alone/Lab
Shrihariprasath and Rathinasabapathy (2016)	2016	Arduino Uno R3	Voltage, Current	Custom dashboard	Monitoring the power conditioning unit	High	Low	Very Low	Small-scale PV
Kekre and Gawre (2018)	2017	Arduino Uno/GPRS	Voltage, Current, and Temperature	-	Remote monitoring	High	Low-Medium	Low	Residential PV
López-Vargas et al. (2018)	2018	Arduino Uno	Voltage, Current, Temperature, Irradiance, and Battery current	Developed webpage	Small stand-alone PV monitoring	Medium	Medium	Moderate	Small PV systems
Hamied et al. (2018)	2018	Arduino + IoT shield	Voltage, Current, Temperature	Developed platform	PV array monitoring	Medium	Medium	Low	Experimental PV
Khan et al. (2019)	2019	Raspberry Pi	Voltage, Current, Power, Energy	Cloud dashboard	Real-time energy monitoring	Medium	High	Moderate	Small-medium PV
Deriche et al. (2019)	2019	Arduino + Sensors	Voltage, Current, Temperature	Cloud dashboard	Panel condition monitoring	Medium	Medium	Low	PV panels
Aghenta and Iqbal (2019)	2019	Arduino Mega + SCADA	Voltage, Current, Power	Open-source SCADA	PV SCADA monitoring	Medium	High	Moderate	Multi-plant PV

Victor et al. (2019)	2019	Various IoT platforms	Voltage, Current, Irradiance	Cloud dashboard	Review on PV monitoring relevance	High	High	Varies	Decentralized PV
Kavitha and Malathi (2019)	2019	Arduino Uno + ESP8266	Voltage, Current	Custom webpage	Smart PV monitoring	High	Low-Medium	Low	Small-scale PV
Hamied et al. (2020)	2020	Arduino + ESP32	Voltage, Current, Temperature, and Fault Detection	ThingSpeak (Free)	Fault detection in stand-alone PV	Medium	Medium	Low	Stand-alone PV
Sharma et al. (2020)	2020	Arduino + IoT Gateway	Voltage, Current, Power	Cloud dashboard	General IoT remote monitoring	Medium	Medium	Moderate	Small PV
Shapsough et al. (2021)	2021	IoT + Smart devices	Voltage, Current, Power, Temperature	Cloud-based	Efficiency optimization of solar farms	Low	High	Moderate	Large-scale PV farms
Ramu et al. (2021)	2021	Arduino Mega	Voltage, Current, Temperature	Developed webpage	Smart PV monitoring	High	Medium	Low	Stand-alone PV
Rasal et al. (2021)	2021	ESP32 + IoT	Voltage, Current, Power	Custom dashboard	PV monitoring	High	Medium	Low	Rooftop PV
Shi et al. (2021)	2021	AIoT MCU + Sensors	Vibration, Temperature (Smart home)	Cloud dashboard	Smart home (conceptual extension to PV)	Medium	High	Moderate	Prototype/ Smart home
Malik et al. (2022)	2022	Raspberry Pi + Cloud	Voltage, Current, Power	Cloud dashboard	Monitoring & management	Medium	High	Moderate	Grid-connected PV
Ali (2023)	2022	Arduino Mega	Voltage, Current, Temperature	Developed webpage	Cost-effective PV monitoring	High	Medium	Low	Small-scale PV

(Continued)

TABLE 6.2 (Continued)

Ref.	Year	MCU/Platform	Collected data	Web interface (Free/ Developed)	Application	Simplicity	Scalability	Cost	PV scale
Emamian et al. (2022)	2022	Raspberry Pi + Cloud ML	Voltage, Current, Temperature, Irradiance	Cloud with ML integration	Intelligent PV monitoring with ML	Low	High	Moderate	PV plants
V and G (2022)	2022	Arduino Uno	Voltage, Current	ThingSpeak (Free)	Remote monitoring	High	Low	Very Low	Small PV
Mondal et al. (2022)	2022	Boltuino IoT board	Voltage, Current, Irradiance	Custom dashboard	Smart solar panel monitoring	Medium	High	Moderate	Residential PV
Nuhin et al. (2022)	2022	Arduino + IoT board	Power, Grid data	Custom webpage	Grid-connected PV monitoring	Medium	Medium	Moderate	Grid PV
Rani et al. (2023)	2023	ESP8266	Voltage, Current	Cloud dashboard	Smart solar monitoring	Medium	Medium	Low	Rooftop PV
Utama et al. (2023)	2023	IoT Cloud + AI	Voltage, Current, Multi-plant data	Cloud IoT platform	Multi-PV plant monitoring	Medium	Very High	Moderate	Utility PV plants
Mellit et al. (2023)	2023	Embedded system + IoT	Voltage, Current, and Fault Detection	Cloud + ML	Fault diagnosis with ML	Low	High	Moderate	Large-scale PV
Krishna Rao et al. (2024)	2024	IoT MCU + Sensors	Voltage, Current, Power	Cloud dashboard	Intelligent PV monitoring	Medium	High	Moderate	PV power generation
Mamun et al. (2024)	2024	ESP32 + Cloud	Voltage, Current, Temperature	Developed webpage	Smart solar monitoring for sustainability	High	Medium	Low	Rooftop PV
Rao et al. (2024)	2024	IoT Gateway + Cloud	Voltage, Current, Grid data	AI-enabled platform	Energy optimization of PV	Medium	High	Moderate	Grid PV

Lodhi et al. (2025a)	2025	AIoT (Edge + Cloud)	Voltage, Current, Faults	AIoT-integrated platform	Fault diagnosis, predictive maintenance	Low	Very High	High	Utility-scale PV
Boucif et al. (2025)	2025	AIoT (Processor/ Field Programmable Gate Array)	Voltage, Current, Control signals	Cloud + AI	Monitoring & control	Low	Very High	High	Large-scale PV
Lodhi et al. (2025b)	2025	AIoT (Edge devices)	Voltage, Current, Faults	AIoT dashboard	Remote monitoring & fault diagnostics	Low	Very High	High	Utility PV
Alshawabkeh et al. (2026)	2026	AIoT + Image Sensors	Image (dust, shading), Color recognition	Cloud AI dashboard	Dust/soiling detection on panels	Low	High	High	Desert PV farms

Implementation costs were moderate but remained significantly lower compared to traditional SCADA solutions. In the recent phase (2024–26), the literature shows a paradigm shift toward AIoT and edge intelligence. Works such as (Krishna Rao et al., 2024; Mamun et al., 2024; Rao et al., 2024; Lodhi et al., 2025a,b; Boucif et al., 2025) propose frameworks that integrate AI-driven fault diagnosis, predictive maintenance, and real-time decision-making, supported by edge devices, FPGA/ARM platforms, and cloud-based dashboards. These systems offer high scalability, robustness, and autonomy, making them suitable for utility-scale PV plants. Although more complex and cost-intensive, they represent a significant step toward fully automated PV management. A more specialized approach is highlighted in Alshawabkeh et al. (2026), which applied AIoT and image processing for dust and soiling detection, addressing environmental and shading-related challenges in desert climates. Fig. 6.4 illustrates the timeline of the evolution of smart monitoring systems (2016–26).

Case studies

This section presents two case studies that will be particularly valuable for researchers and engineers working on PV monitoring and predictive maintenance, as well as for industry practitioners aiming to implement smart monitoring solutions in large-scale solar plants. The description and methodology have been structured so that graduate students, academics, and professionals in renewable energy will be able to follow the implementation and adapt it to their own contexts. The case studies were carried out with the involvement of a multidisciplinary team of experts, including PV engineers, IoT developers, data scientists, and AI researchers, ensuring both technical accuracy and practical relevance. These case studies, therefore, serve as both a practical demonstration and a reference framework for integrating IoT and AIoT into PV monitoring systems. By bridging theoretical concepts with real-world applications, it highlights how emerging technologies can enhance system reliability, optimize performance, and support the wider adoption of smart energy solutions.

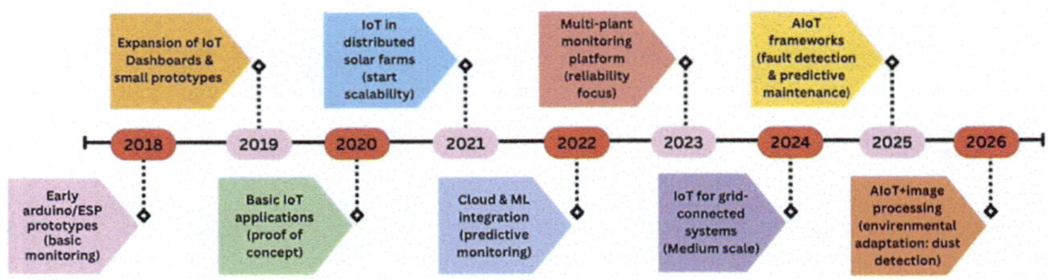

FIGURE 6.4 Timeline of the evolution of smart monitoring systems (2018–26). The figure presents a timeline showing how smart monitoring systems have evolved from 2018 to 2026. It highlights key technological advancements over the years, illustrating the transition toward more intelligent, connected, and automated monitoring solutions.

Case study #1

Internet of Things-based photovoltaic monitoring system

The first case study presents a comprehensive design for a low-cost prototype for monitoring electrical and meteorological data from a PV array in real-time, including array current, voltage, ambient temperature, and solar radiation. IoT technology is used to transmit the monitored data to the cloud via a Wi-Fi module (ESP-01). The collected data are then stored and displayed remotely on a dedicated web page. Two approaches are used to visualize the data: either using an available free IoT platform, such as ThingSpeak (MathWorks), or developing a custom web interface from scratch using software and programming languages such as MySQL, PHP, and JavaScript.

Fig. 6.5 shows the basic block diagram of the IoT-based PV monitoring prototype.

The system is divided into three main parts:

- *Sensing part:* This includes sensors for measuring PV voltage, PV current, module temperature, solar irradiance, and ambient air temperature. These parameters are collected in real-time from the PV module/array.
- *Processing and transmission part:* A MCU processes the acquired data and transmits it via IoT protocols (e.g., Wi-Fi) to a remote server for storage and further analysis.
- *Data visualization part:* A web-based dashboard and mobile application retrieve the stored data from the server and display it to the user in real-time. This interface allows system operators to monitor PV performance.

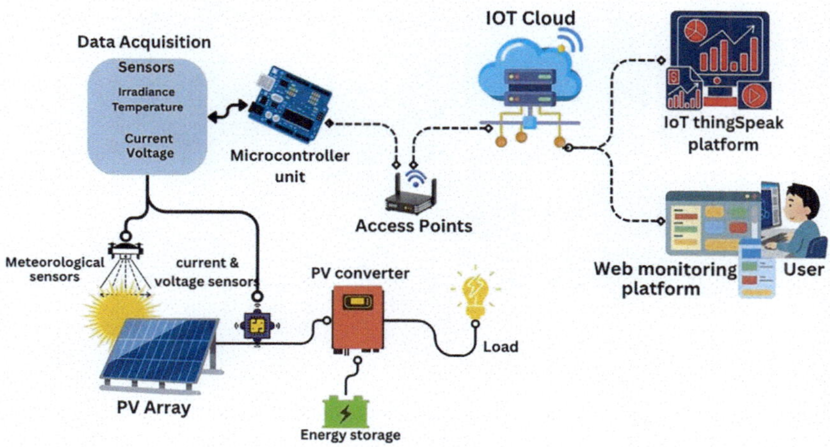

FIGURE 6.5 Block diagram of the IoT-based PV monitoring prototype. The figure shows how data flows from PV sensors through communication modules to a monitoring platform. It illustrates the main components of the IoT-based PV monitoring prototype and their connections (*IoT*, Internet of Things; *PV*, photovoltaic).

Experimental design of an Internet of Things-based photovoltaic monitoring system

This subsection presents the experimental setup designed for the development and validation of the IoT-based PV monitoring system, with particular emphasis on the hardware components. The architecture of the proposed system is illustrated in Fig. 6.6. At its core, the system employs an Arduino Mega MCU to process the acquired signals and coordinate the operation of the different sensors. A light dependent resistor (LDR) is included to measure light intensity, while an air temperature sensor (LM35CZ) provides environmental temperature readings. In addition, current and voltage sensors are integrated to continuously monitor the electrical performance of the PV system. The measured parameters are displayed locally using a Liquid Crystal Display (LCD) module, and the collected data are further transmitted to the IoT platform for remote monitoring and analysis. Table 6.3 summarises the hardware components used in the prototype, including their specifications and associated costs, as well as the overall implementation cost of the system.

Solar irradiance can be approximately estimated using an LDR, although its response is nonlinear and wavelength dependent. The resistance of the LDR decreases with increasing light intensity and is often expressed as:

$$R_{LDR} = A.\ E^{-k} \tag{6.1}$$

FIGURE 6.6 The designed data-acquisition board is based on the IoT technique. The figure shows the data-acquisition board designed using IoT technology. It highlights the board's main components used for collecting and transmitting PV system data (*IoT*, Internet of Things; *PV*, photovoltaic).

TABLE 6.3 Components used in the IoT-based PV monitoring prototype.

Component	Specification	Price (€)
Arduino Mega	ATmega2560, 54 digital I/O pins, 16 analog inputs, 16 MHz clock	30
ESP-8266-01	Wi-Fi module, 2.4 GHz, TCP/IP stack, UART interface	5
Current sensor (ACS712)	Hall-effect sensor, ±30 A measurement range	15
Voltage sensor	0–25 V measurement range, resistive divider module	10
LDR (light dependent resistor)	Photoconductive sensor, spectral response ~400–700 nm	3
LM35CZ	Temperature sensor, -55°C to +150°C, ±0.5°C accuracy	2
Analog-to-Digital Converter (ADC)	8-channel, 10–12 bit resolution (e.g., ADS1115-8ch)	15
Total cost		80

I/O, Input/output; *IoT*, Internet of Things; *PV*, photovoltaic.

where E is illuminance in lux and A and k are calibration constants. When used in a voltage divider, the resistance can be calculated from the output voltage as:

$$R_{LDR} = R\frac{V_{in}}{V_{out}} - 1 \tag{6.2}$$

Illuminance is then converted to solar irradiance G using the approximate relation G=E/120, since 1 W/m^2 of sunlight corresponds to about 120 lux under typical daylight conditions.

The LM35CZ is a widely used, low-cost analog temperature sensor designed for measuring ambient temperature with good accuracy. If powered at 5 V and connected to a microcontroller ADC, the temperature can be derived from the output voltage $T(°C)=V_{out}(mV)/10$.

To get cell temperature, we need, for example, to convert the measured ambient temperature into module temperature using an empirical relation that accounts for solar irradiance and the nominal operating cell temperature (NOCT). The most common model is expressed as:

$$T_C = T_a + \frac{G}{800}(NOCT - 20) \tag{6.3}$$

where Tc is the PV cell temperature (°C), Ta is the ambient temperature (°C), and NOCT is a module-specific constant (typically around 45°C)

This relation shows that under high irradiance, cell temperature can rise significantly above the ambient value, which directly impacts PV performance.

The ACS712 current sensor is a low-cost Hall-effect device widely used for measuring both AC and DC currents in PV monitoring applications. It provides an analog output voltage proportional to the current, with a zero-current reference at about 2.5 V (for a 5 V supply). When connected to the Arduino's analog input, the sensor output can be read and converted into current values using the sensor's sensitivity (e.g., 185 mV/A, 100 mV/A, or 66 mV/A, depending on the model). This makes it suitable for real-time monitoring of PV array output current, offering electrical isolation and safe integration with microcontrollers.

The 25 V voltage sensor module is a simple resistive voltage divider designed to work with microcontrollers such as Arduino for monitoring DC voltages. It typically uses two precision resistors to scale down an input voltage (up to 25 V) into a safe 0–5 V range that the Arduino's ADC can read.

Data collection and visualization

The simplest way to visualize the collected data online is by employing widely available IoT platforms such as ThingSpeak (MathWorks), Blynk, etc. In this study, ThingSpeak was selected due to its user-friendly interface, MATLAB® integration for advanced analysis, and suitability for rapid prototyping of IoT-based monitoring applications. As an example, Fig. 6.7 illustrates the collected data (air temperature, solar irradiance, current, and voltage) over a short observation period. The ability to store, visualize, and analyze these parameters remotely demonstrates the practicality and scalability of the proposed IoT-based PV monitoring system for both research and industrial applications.

ThingSpeak offers a simple, cloud-based solution for IoT data visualization with advantages, such as real-time monitoring, ease of use, built-in MATLAB analytics for advanced processing, and application programming interface support that facilitates integration with IoT devices. Its free tier and user-friendly interface make it particularly attractive for academic research, prototyping, and small-scale PV monitoring applications. However, it also has limitations, including restricted data upload frequency and storage in the free version, limited scalability for large industrial projects, reliance on stable internet connectivity, fewer customization options compared to platforms like Grafana, and potential concerns about data privacy. Therefore while ThingSpeak is highly suitable for research and pilot applications, more robust platforms may be required for large-scale or commercial PV system monitoring.

To overcome the inherent limitations of existing IoT platforms, such as restricted customization, limited storage capacity, or dependence on third-party services, developing a dedicated webpage from scratch provides a more

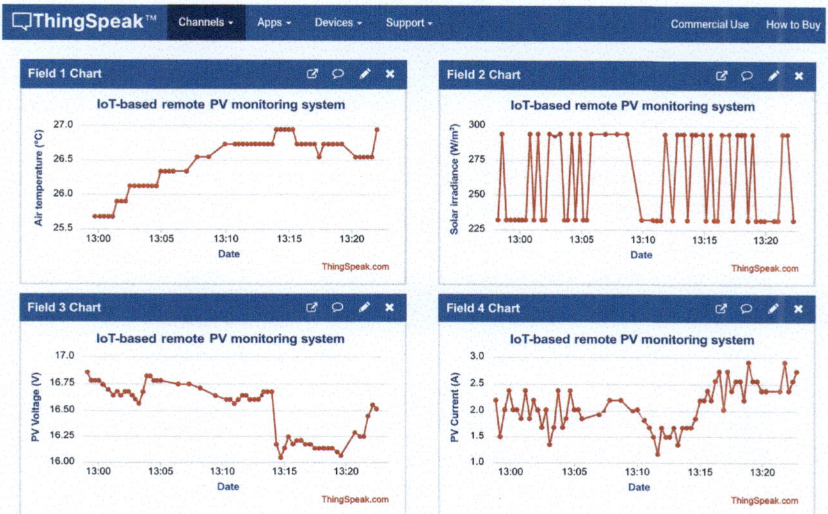

FIGURE 6.7 Collected data using the IoT ThingSpeak platform. The figure displays data collected from the PV monitoring system on the IoT ThingSpeak platform. It shows how measurements are visualized in real time through graphs or dashboards.

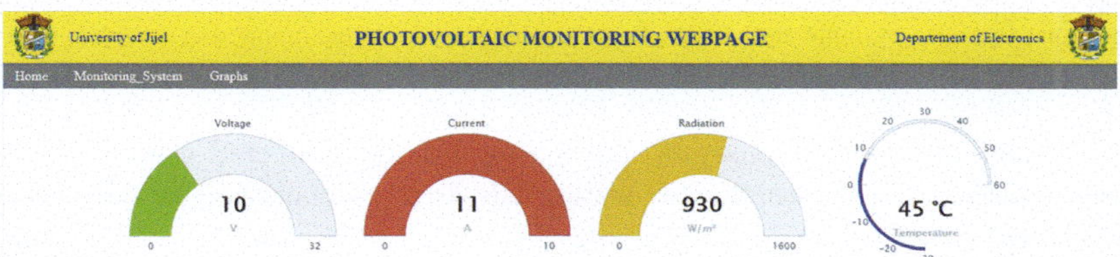

FIGURE 6.8 Developed a webpage for visualizing the collected data. The figure shows the developed webpage used to visualize the collected monitoring data. It presents the data in an organized and user-friendly layout for easy interpretation.

flexible and robust solution. This approach, however, requires proficiency in web programming languages (e.g., HTML, CSS, JavaScript) as well as expertise in database management using MySQL.

Fig. 6.8 illustrates the developed webpage, which enables dynamic visualization of the monitored parameters. The interface allows users to view the collected data in multiple formats, including interactive graphs, tabular displays, and exportable Excel files for offline analysis. Compared to off-the-shelf IoT platforms, a custom-designed webpage offers significant advantages in terms of scalability, data security, and adaptability to specific research or industrial requirements, thereby making it a suitable option for large-scale and long-term PV monitoring applications.

	A	B	C	D	E
1	THE GRAPH OF MONITORED DATA				
2	Date	Voltage	Current	Radiation	Temperature
3	2018-06-18 13:02:20.868653	10,47	2,1	1073.00	18,07
4	2018-06-18 13:02:49.945982	12,08	2.00	1076.00	21.00
5	2018-06-18 13:03:51.938748	12,16	2.00	1072.00	21.00
6	2018-06-18 13:04:53.931602	12,13	2.00	1072.00	21,48
7	2018-06-18 13:05:55.927298	12,13	2.00	1071.00	21,48
8	2018-06-18 13:06:57.925859	12,18	2.00	1016.00	21.00
9	2018-06-18 13:07:59.910509	12,18	2.00	1086.00	21,48
10	2018-06-18 13:08:04.997076	12,18	2.00	1086.00	21,48
11	2018-06-18 13:09:06.997226	12,23	2.00	1088.00	21,48
12	2018-06-18 13:10:08.996869	12,3	2.00	1088.00	21,48
13	2018-06-18 13:11:10.987182	12,33	2.00	1090.00	21,48
14	2018-06-18 13:12:12.983227	12,38	2,05	1090.00	21,48
15	2018-06-18 13:13:14.980817	12,28	2,05	1094.00	21,48
16	2018-06-18 13:14:16.970283	12,21	2,05	1097.00	21,48
17	2018-06-18 13:15:18.955525	12,23	2,05	1100.00	21,48

FIGURE 6.9 Example of the downloaded Excel sheet used to store and analyse the collected data. The figure shows an example of an Excel sheet containing the collected monitoring data. It is used to store the measurements and facilitate further analysis.

To facilitate offline analysis and reporting, the collected data can be downloaded in Excel format. This feature allows researchers and practitioners to perform statistical analysis, generate customized plots, and easily share the results with collaborators (See Fig. 6.9).

Case study #2

Smart monitoring and anomaly detection in photovoltaic modules based on the concept of artificial intelligence of things

Fig. 6.10 shows the basic configuration of the application of the concept of AIoT in monitoring and anomaly detection in PV systems.

The process consists of three main functional modules:

- **Data acquisition system**

This system is responsible for measuring the main environmental and electrical parameters of the PV system. It includes several types of sensors, such as:

- *Air and cell temperature sensors*: Monitors ambient conditions that affect PV efficiency.

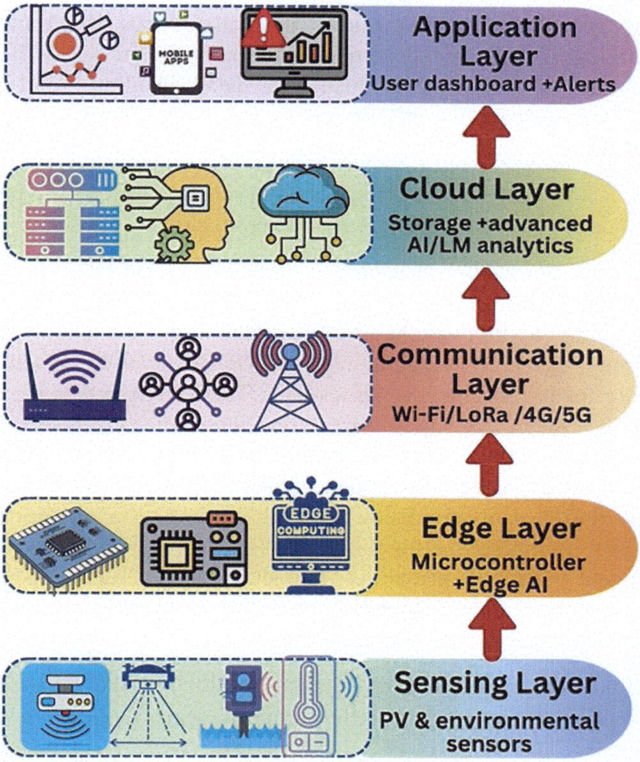

FIGURE 6.10 Basic configuration of the AIoT-based PV monitoring system. The figure shows the basic configuration of an AIoT-based PV monitoring system, where data is collected from PV sensors and processed using AI techniques. It illustrates how IoT connectivity enables real-time data transmission and intelligent analysis (*AI*, Artificial intelligence; *AIoT*, artificial intelligence of things; *PV*, photovoltaic).

- *Reference cell sensor*: Provides a benchmark for the solar irradiance, which directly affects the output power.
- *Voltage and current sensors*: Measure the electrical characteristics of the PV array in real-time.

The accuracy and reliability of these sensors are critical, as the quality of the input data directly affects the performance of subsequent analysis and anomaly detection.

- **Data storage and visualization**
  Once data is collected, the sensor data is compiled, processed, and stored for monitoring and analysis purposes. A custom interface has been developed to display the data on a web-based dashboard, enabling real-time visualization of the PV system's performance. Additionally, a mobile

app has been launched that allows users to access data remotely via smartphones, making the system portable and easy to use. The storage module ensures archiving of historical data, enabling trend analysis, performance evaluation, and long-term system diagnostics.

- **Central unit**

The MCU forms the core of the system and serves as the central processing and communication centre. Its main functions include:

- Transferring data via IoT protocols (such as MQTT/HTTP) to a web server for display and storage.
- Implementing a lightweight AI-based anomaly detection algorithm. Specifically, a recurrent neural network (RNN) is used, given its high suitability for processing time series data. The RNN predicts the expected PV output power using input characteristics collected from sensors.
- The predicted output value is continuously compared with the actual measured values. Large deviations between the two values indicate the potential presence of anomalies (such as shading, contamination, partial failures, or equipment malfunction).
- In the event of any abnormalities, the MCU triggers alerts by displaying warning messages on a web page, ensuring that operators are immediately informed and able to take corrective action.

This architecture enables the system to monitor PV performance in real-time, in addition to integrating predictive intelligence, enhancing the reliability, efficiency, and safety of PV power plants.

Experimental design of an artificial intelligence of things-based photovoltaic monitoring system

This subsection presents the experimental setup and methodology used to design and implement the IoT-based PV monitoring system, highlighting the hardware components, data flow, and integration of AI for anomaly detection. Fig. 6.11 shows the designed prototype, which consists of an MCU type NodeMCU ESP8266, a reference solar cell, an air temperature and humidity sensor, a type K module temperature, current, and voltage sensors, and an LCD displayer. Table 6.4 lists the used components to design this prototype, as well as the total cost.

Data collection

Fig. 6.12 illustrates the remote monitoring dashboard developed for the IoT-based PV system, presenting real-time measurements obtained from the prototype. The monitored PV module corresponds to the specifications summarized in Table 6.5. The dashboard provides a comprehensive visualization of key performance indicators, including PV current, PV voltage, output

FIGURE 6.11 The designed prototype is based on the AIoT concept. The figure illustrates the designed prototype based on the AIoT concept, integrating sensing, embedded intelligence, and wireless communication within a compact hardware platform. It demonstrates how data acquisition, on-device AI processing, and IoT connectivity are combined to enable real-time monitoring and intelligent decision-making at the edge (*AI*, Artificial intelligence; *AIoT*, artificial intelligence of things; *PV*, photovoltaic).

power, air temperature, solar irradiance, and PV cell temperature. These parameters are critical for evaluating the operational status and energy yield of the PV module, as well as for identifying potential performance anomalies. The remote accessibility of the dashboard through a web interface ensures continuous supervision and facilitates timely decision-making by users and operators.

TABLE 6.4 Components used in the design of the prototype.

Item	Specification	Price (€)
NodeMCU ESP8266	32-bit, 80 Mhz, IEEE 805.11b/g/n, 4MB, 64 KB	15
Current sensor (ACS712)	±30 A, Hall-effect based	15
Voltage sensor	0–25 V, resistive divider type	10
Solar cell reference	Monocrystalline, ~0.5 V @ STC	95
Module temperature sensor (Type K)	-50°C to 700°C, thermocouple with amplifier	10
DHT sensor (DHT22 recommended)	Temp: -40°C–80°C ±0.5°C, Humidity: 0%–100% ±2%5% RH	8
Analog-to-Digital Converter (ADC)	8-channel, 10–12 bit resolution (e.g., ADS1115-8ch)	15
DC–DC converter	12 V to 3.3 V, 5 V	12
Total cost	—	180

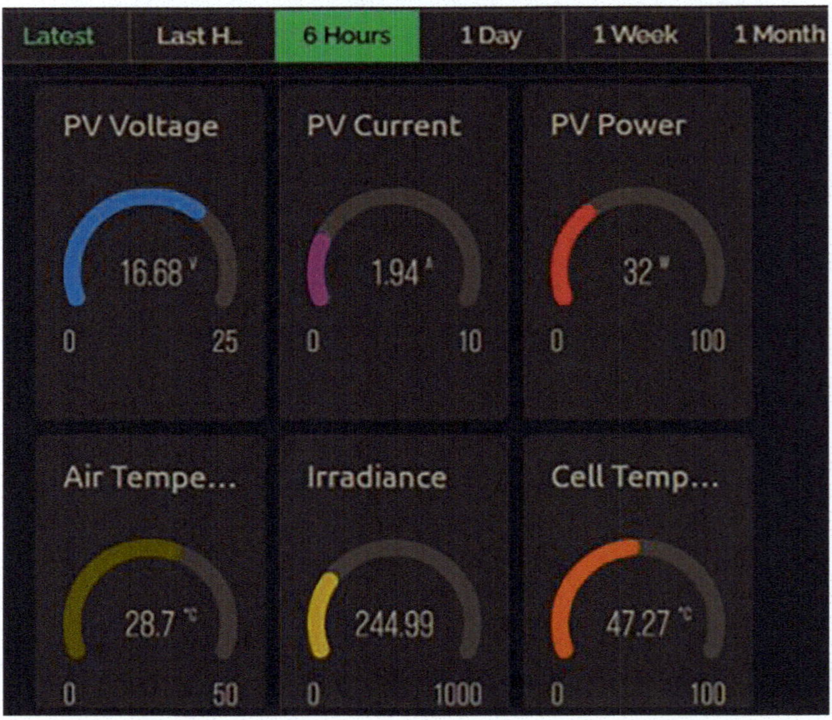

FIGURE 6.12 Dashboard of the developed IoT-based PV monitoring system. The figure shows the dashboard of the developed IoT-based PV monitoring system, displaying real-time data and key performance indicators for easy monitoring and analysis (*IoT*, Internet of Things; *PV*, photovoltaic).

TABLE 6.5 PV module characteristics.

Module type	100P (36)
Maximal power	100 W
Tolerance	±3 %
Voltage at Pmax (Vmp)	17.45 V
Current at Pmax (Imp)	5.73 A
Open-circuit voltage (Voc)	21.87 V
Short-circuit current (Isc)	5.98 A

PV, Photovoltaic.

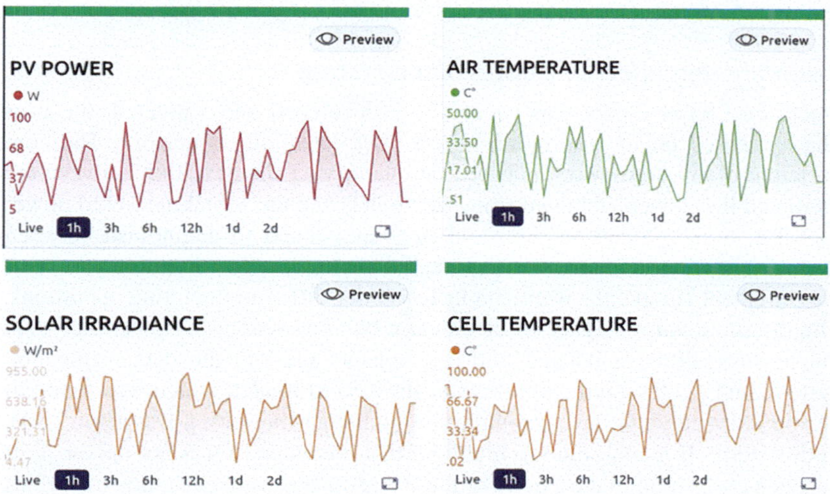

FIGURE 6.13 The evolution of the measured data using a phone App for a period of one hour. The figure shows the evolution of the measured PV data over one hour using a phone app, illustrating real-time monitoring and trends (*PV*, Photovoltaic).

Fig. 6.13 illustrates an example of the real-time data collected over a short observation period. The displayed parameters include PV output power, solar irradiance, PV cell temperature, and ambient air temperature. These measurements highlight the dynamic behavior of the PV system under varying environmental conditions. In particular, the figure demonstrates how fluctuations in solar irradiance and temperature directly influence the PV output power. Such real-time monitoring not only provides valuable insights into the instantaneous performance of the PV module but also forms the basis for subsequent analysis, including anomaly detection, fault diagnosis, and performance optimization. A dedicated dataset was collected to support the development and training of an RNN for potential PV power prediction.

Application of an artificial intelligence-based algorithm to detect anomalies in the photovoltaic system

Photovoltaic power model development

The RNN architecture employed in this study is the well-known long short-term memory (LSTM) network (Mellit et al., 2021), which is designed to capture temporal dependencies in sequential data. The LSTM model was trained to predict PV output power using historical measured data as input features. For model development, 80% of the dataset was allocated for training, while the remaining 20% was reserved for testing and performance evaluation.

Anomaly detection in photovoltaic system

Once the LSTM model was successfully developed and validated, the next phase focused on its integration into a fault detection framework. The core principle of this framework is based on quantifying the deviation between the predicted PV power (obtained from the model) and the actual measured power (See Eq. 6.4). A decision threshold (μ) was defined to distinguish between normal operating fluctuations and potential anomalies. To determine μ, several measurement campaigns were conducted under different operating conditions. The threshold was designed to incorporate two key sources of uncertainty: (1) the intrinsic error associated with the sensors used in the data acquisition system, and (2) the prediction error of the LSTM model, which was found to be approximately 3%. By accounting for both contributions, μ was set to ±5 W in this study. It is important to highlight that the value of μ is not universal; it depends on the nominal capacity of the PV array under investigation as well as the degradation rate of PV modules. For larger PV systems, the threshold must be scaled accordingly to ensure reliable fault detection without generating excessive false alarms.

$$\Delta P_{pv} = |Ppv - \hat{P}pv| \leq \mu \tag{6.4}$$

where $\hat{P}pv$ is the forecasted power, and Ppv is the measured power.

The total experimental run spanned 10,000 minutes, during which both normal and faulty operating conditions were recorded. Each faulty scenario was defined by its corresponding start and end times, enabling the creation of labelled datasets that facilitate model training, validation, and fault detection analysis. Three operational scenarios under abnormal conditions were investigated, as illustrated in Fig. 6.14.

(A) (B) (C)

FIGURE 6.14 The investigated anomalies: (A) scenario #1, (B) scenario #2, and (C) scenario #3. The figure shows three investigated anomalies in the PV system, labeled as scenario #1, scenario #2, and scenario #3. It highlights different types of irregularities detected during monitoring (*PV*, Photovoltaic).

The anomalies were introduced by applying different levels of sand deposition on the PV module surface:

- *Scenario #1:* Heavy sand deposition covering approximately 90% of the PV module.
- *Scenario #* 2: Partial sand deposition covering about 25% of the PV module.
- *Scenario #* 3: Complete coverage of the PV module surface.

The obtained results are summarized in Table 6.6. It can be clearly observed that, for all three investigated cases, the power deviation (ΔP) exceeds 5W, indicating the occurrence of an anomaly in the PV array. It should be noted that the proposed fault detection procedure is not designed to differentiate between specific types of anomalies (e.g., sand deposition, complete coverage, or shading). Instead, it reliably identifies the presence of anomalies by comparing the predicted power from the developed RNN model with the actual measured values. The primary objective is to design a low-cost, online, and automatic monitoring system capable of detecting anomalies in PV arrays based on forecasted PV power. This goal has been successfully achieved with satisfactory accuracy. Furthermore, by integrating additional experiments and incorporating AI-based classification techniques, the system can be further extended to identify and classify various fault types, thereby enhancing its diagnostic capabilities.

Challenges and solutions

The designed prototypes were initially validated through experimental testing on a single PV module. Leveraging the same foundational configuration, it is feasible to develop a scalable model for PV strings or arrays by accounting for

TABLE 6.6 Experimental results.

Scenario	Gi (W/m^2)	Tc (°C)	Measured power (W)	forecasted power (W)	ΔP (W)
Scenario #1	897	48	8	142	134
Scenario #2	989	50	85	180	95
Scenario #3	785	45	2	128	126

cumulative current and voltage measurements. Accordingly, the selected sensors would need to be reconfigured to accommodate higher current and voltage capacities. While scaling will inevitably increase the cost, we aim to mitigate this by prioritising cost-effective and efficient component selection. Notably, we are actively pursuing this line of research to make the prototype adaptable across diverse PV technologies and system sizes. However, several challenges were encountered during the prototype design process, including:

- Developing the LSTM forecaster involved extensive hyperparameter optimization and selection, a process that was quite time-consuming, as well as the quantization process.
- After finalizing the model, another challenge was estimating the threshold interval, which varied depending on the type of PV technology and environmental conditions.
- Data collection was conducted in a severe environment in terms of temperature, where the module was tested.
- The prototype design could be useful for similar research applications, but its effectiveness depends on the data collected, as this shapes the LSTM structure.

As a recommendation, the procedure itself should be extended to handle multiple faults. In this context, the trade-off between computational efficiency and model performance must be carefully assessed, particularly for real-time or resource-constrained environments. The main limitation of the current model lies in its dependence on a location-specific database and a single PV system configuration, which may restrict its generalizability to other regions or PV technologies. This drawback can be addressed by expanding the dataset to include measurements from diverse geographical locations and PV module types, thereby enhancing the robustness and adaptability of the model under different environmental conditions.

The trajectory of smart PV monitoring research points toward the convergence of AIoT, blockchain, and self-healing. While current AIoT frameworks already enable predictive diagnostics and edge intelligence, future systems will need to address interoperability, cybersecurity, and regulatory compliance to ensure robust deployment in large-scale PV plants. Integration of blockchain can enhance data integrity and energy transaction transparency, particularly in decentralized energy markets, while self-healing and adaptive monitoring systems will provide resilience against faults and cyberattacks. Moreover, lightweight TinyML models and specialized hardware accelerators are expected to make real-time AI feasible on low-power edge devices, ensuring both efficiency and scalability. Ultimately, the evolution from simple MCU-based prototypes to complex autonomous AIoT platforms highlights a clear shift toward fully digitalized, intelligent, and secure PV monitoring ecosystems, capable of supporting the next generation of sustainable smart grids.

Conclusion

The evolution of smart PV monitoring systems has significantly broadened their scope of application, supported by continuous technological advancements in IoT, AI, and data-driven platforms. While these developments have demonstrated remarkable potential in enhancing accuracy, reliability, and automation, they also introduce new challenges related to scalability, interoperability, privacy, and security of data exchange between nodes. Addressing these issues will be crucial for enabling seamless integration across different vendors, devices, and large-scale PV infrastructures. Furthermore, the lack of standardization and regulatory frameworks remains a barrier to widespread adoption. Establishing clear guidelines for communication protocols, cybersecurity, and grid compliance will ensure consistency, resilience, and trust in PV monitoring deployments, particularly for commercial and utility-scale applications. The combination of autonomous monitoring and big data analytics transforms PV system management from reactive maintenance to a proactive, predictive, and highly efficient operation. By continuously analysing real-time and historical data, operators can detect faults promptly, predict component degradation, optimize energy yield, and make informed decisions, significantly reducing operational costs and increasing the overall reliability and lifespan of PV plants. Even with data acquisition enhancements, there is still room for improving its implementation regarding data security.

Looking ahead, emerging trends such as AIoT and blockchain are shaping the next generation of PV monitoring. AIoT empowers real-time fault detection, predictive maintenance, and autonomous decision-making at the edge, while blockchain provides secure and transparent mechanisms for energy data management, trading, and warranty validation. Together, these advancements point toward the development of intelligent, decentralized, and resilient PV monitoring infrastructures that can adapt dynamically to environmental and operational uncertainties, ensuring long-term stability and high availability of solar PV power plants.

AI Disclosure

During the preparation of this work, the author(s) used ChatGPT for proofreading. After using this tool/service, the author(s) reviewed and edited the content as needed and take(s) full responsibility for the content of the publication.

References

Adhya, S., Saha, D., Das, A., Jana, J., Saha, H., 2016. An IoT based smart solar photovoltaic remote monitoring and control unit. In: 2016 2nd International Conference on Control, Instrumentation, Energy and Communication, CIEC 2016, pp. 432–436. Institute of Electrical and Electronics Engineers Inc. India. Available at https://doi.org/10.1109/CIEC.2016.7513793.

Adiguzel, E., Gurkan, K., Ersoy, A., 2023. Design and development of data acquisition system (DAS) for panel characterization in PV energy systems. Measurement 221, 113425.

Aghaei, M., Kolahi, M., Nedaei, A., Venkatesh, N.S., Esmailifar, S.M., Moradi Sizkouhi, A.M., et al., 2025. Autonomous intelligent monitoring of photovoltaic systems: an in-depth multidisciplinary review. Progress in Photovoltaics: Research and Applications 33 (3), 381–409. Available at https://doi.org/10.1002/pip.3859.

Aghenta, L.O., Iqbal, M.T., 2019. Development of an IoT Based Open Source SCADA System for PV System Monitoring. In: 2019 IEEE Canadian Conference of Electrical and Computer Engineering, CCECE 2019. Institute of Electrical and Electronics Engineers Inc. Canada. Available at http://ieeexplore.ieee.org/xpl/mostRecentIssue.jsp?punumber=8852749, https://doi.org/10.1109/CCECE.2019.8861827.

Ali, M.I., 2023. IoT based smart solar PV monitoring system; A cost effective and reliable solution. Sukkur IBA Journal of Computing and Mathematical Sciences 6 (2), 8–14. Available at https://doi.org/10.30537/sjcms.v6i2.1160.

Alshawabkeh, O.J., Nisirat, M.A., Qawasmeh, B.R., 2026. The use of AIoT-integrated approaches to detect accumulated dust on solar panels utilizing color recognition-based techniques. Renewable Energy 56.

Ando, B., Baglio, S., Pistorio, A., Tina, G.M., Ventura, C., 2015. Sentinella: smart monitoring of photovoltaic systems at panel level. IEEE Transactions on Instrumentation and Measurement 64 (8), 2188–2199. Available at https://doi.org/10.1109/TIM.2014.2386931.

Ansari, S., Ayob, A., Lipu, M.S.H., Saad, M.H.M., Hussain, A., 2021. A review of monitoring technologies for solar PV systems using data processing modules and transmission protocols: progress, challenges and prospects. Sustainability 13 (15), 8120. Available at https://doi.org/10.3390/su13158120.

Asnil, A., Krismadinata, K., Husnaini, I., Hazman, H., Astrid, E., 2023. Real-time monitoring system using IoT for photovoltaic parameters. TEM Journal 12 (3), 1316–1322. Available at https://doi.org/10.18421/TEM123-11, http://www.temjournal.com/.

Boucif, O.H., Lahouaou, A.M., Boubiche, D.E., Toral-Cruz, H., 2025. Artificial intelligence of things for solar energy monitoring and control. Applied Sciences 15 (11), 6019. Available at https://doi.org/10.3390/app15116019.

Costa, V., Scianni, L., Miranda, R.C., Bonatto, B., 2023. Assessment of the status and trends of photovoltaic distributed generation in Brazil: an in-depth approach based on big data processing. Solar Energy 249, 694–711.

Deriche, M., Raad, M.W., Suliman, W., 2019. An IOT based sensing system for remote monitoring of PV panels. In: 16th International Multi-Conference on Systems, Signals and Devices, SSD 2019. 393–397. Institute of Electrical and Electronics Engineers Inc. Saudi Arabia. Available at http://ieeexplore.ieee.org/xpl/mostRecentIssue.jsp?punumber=8885406, https://doi.org/10.1109/SSD.2019.8893161.

Emamian, M., Eskandari, A., Aghaei, M., Nedaei, A., Sizkouhi, A.M., Milimonfared, J., 2022. Cloud computing and IoT based intelligent monitoring system for photovoltaic plants using machine learning techniques. Energies 15 (9), 3014. Available at https://doi.org/10.3390/en15093014.

Gopal, M., Prakash, T.C., Ramakrishna, N.V., Yadav, B.P., 2020. IoT based solar power monitoring system. In IOP Conference Series: Materials Science and Engineering 981 (3).

Hamied, O., Mellit, A., Zoulid, M.A., Birouk, R., 2018. IoT-based experimental prototype for monitoring of photovoltaic arrays. In 2018 International conference on applied smart systems (ICASS) 1–5.

Hamied, O., Mellit, A., Zoulid, M.A., Birouk, R., 2020. A low-cost monitoring and fault detection system for stand-alone photovoltaic systems using IoT technique. ELECTRIMACS 2019 1, 349–358.

Kavitha, V., Malathi, V., 2019. A smart solar PV monitoring system using IoT. Journal of Physics: Conference Series 1367 (1).

Kekre, A., Gawre, S.K., 2018. Solar photovoltaic remote monitoring system using IOT. 2018 International Conference on Recent Innovations in Signal Processing and Embedded Systems, RISE 2017. 2018-, pp. 619–623. Institute of Electrical and Electronics Engineers Inc. India. Available at http://ieeexplore.ieee.org/xpl/mostRecentIssue.jsp?punumber=8371184, https://doi.org/10.1109/RISE.2017.8378227.

Khan, M.S., Sharma, H., Haque, A. 2019 IoT Enabled Real-Time Energy Monitoring for Photovoltaic Systems. Proceedings of the International Conference on Machine Learning, Big Data, Cloud and Parallel Computing: Trends, Prespectives and Prospects, COMITCon 2019, pp. 323–327. Institute of Electrical and Electronics Engineers Inc. India. Available at http://ieeexplore.ieee.org/xpl/mostRecentIssue.jsp?punumber=8851231, https://doi.org/10.1109/COMITCon.2019.8862246.

Krishna Rao, C., Sahoo, S.K., Yanine, F.F., 2024. An IoT-based intelligent smart energy monitoring system for solar PV power generation. Energy Harvesting and Systems 11 (1).

Lay-Ekuakille, A., Ciaccioli, A., Griffo, G., Visconti, P., Andria, G., 2018. Effects of dust on photovoltaic measurements: a comparative study. Measurement 113, 181–188. Available at https://doi.org/10.1016/j.measurement.2017.06.025.

Lodhi, E., Liu, X., Xiong, G., Khan, M.A., Lodhi, Z., Nawaz, T., et al., 2025a. SmartPV-AIoT: an AIoT-integrated framework for fault diagnosis and remote monitoring in photovoltaic systems. Energy Conversion and Management: X 27, 101117.

Lodhi, X., Liu, G., Xiong, M.A., Khan, Z., Lodhi, T., Nawaz, et al., 2025b. An Efficient Aiot-Based Framework for Remote Monitoring and Fault Diagnostics in Photovoltaic Systems. SSRN.

López-Vargas, A., Fuentes, M., Vivar, M., 2018. On the application of IoT for real-time monitoring of small stand-alone PV systems: results from a new smart datalogger. In: 2018 IEEE 7th World Conference on Photovoltaic Energy Conversion, WCPEC 2018 - A Joint Conference of 45th IEEE PVSC, 28th PVSEC and 34th EU PVSEC, pp. 605–607. Institute of Electrical and Electronics Engineers Inc. Spain. Available at http://ieeexplore.ieee.org/xpl/mostRecentIssue.jsp?punumber=8510986, https://doi.org/10.1109/PVSC.2018.8547612.

Madeti, S.R., Singh, S.N., 2017. Monitoring system for photovoltaic plants: a review. Renewable and Sustainable Energy Reviews 67, 1180–1207. Available at https://doi.org/10.1016/j.rser.2016.09.088, https://www.journals.elsevier.com/renewable-and-sustainable-energy-reviews.

Malik, A., Haque, A., Kurukuru, V.S.B., 2022. IoT-based monitoring and management for photovoltaic system. Fault Analysis and its Impact on Grid-connected Photovoltaic Systems Performance. Wiley Blackwell, India, pp. 291–318. Available at https://www.wiley.com/en-us/Fault+Analysis+and+its+Impact+on+Grid+connected+Photovoltaic+Systems+Performance-p-9781119873778, https://doi.org/10.1002/9781119873785.ch9.

Mamun, A.A., Shuvo, M.H., Islam, T., Islam, D., Islam, M.J., Tanvir, F.A., 2024. Developed an IoT-based smart solar energy monitoring system for environmental sustainability. In: 2024 3rd International Conference on Advancement in Electrical and Electronic Engineering, ICAEEE 2024. Institute of Electrical and Electronics Engineers Inc. Bangladesh. Available at http://ieeexplore.ieee.org/xpl/mostRecentIssue.jsp?punumber=10561631, https://doi.org/10.1109/ICAEEE62219.2024.10561689.

Mellit, A., Kalogirou, S., 2021. Artificial intelligence and internet of things to improve efficacy of diagnosis and remote sensing of solar photovoltaic systems: challenges, recommendations and future directions. Renewable and Sustainable Energy Reviews 143, 110889. Available at https://doi.org/10.1016/j.rser.2021.110889.

Mellit, A., Kalogirou, S., 2025. Recent advances in the application of infrared thermographic imaging and embedded artificial intelligence for fault diagnosis and predictive maintenance of photovoltaic plants: challenges and future directions. Renewable and Sustainable Energy Reviews 223, 116057. Available at https://doi.org/10.1016/j.rser.2025.116057.

Mellit, A., Pavan, A.M., Lughi, V., 2021. Deep learning neural networks for short-term photovoltaic power forecasting. Renewable Energy 172, 276–288. Available at https://doi.org/10.1016/j.renene.2021.02.166, http://www.journals.elsevier.com/renewable-and-sustainable-energy-reviews/.

Mellit, A., Benghanem, M., Kalogirou, S., Massi Pavan, A., 2023). An embedded system for remote monitoring and fault diagnosis of photovoltaic arrays using machine learning and the internet of things. Renewable Energy 208, 399–408. Available at https://doi.org/10.1016/j.renene.2023.03.096.

Mohammed, M.I., Al-Naib, A.M.I., 2023. Design of a SCADA system for a solar photovoltaic power plant. NTU Journal of Engineering and Technology 2 (2), 55–62.

Mondal, A., Ali, M.J., Dutta, P., 2022. IoT Enabled Smart Solar Panel Monitoring System Based on Boltuino Platform. In: 2022 IEEE International IOT, Electronics and Mechatronics Conference, IEMTRONICS 2022. Institute of Electrical and Electronics Engineers Inc. India. Available at http://ieeexplore.ieee.org/xpl/mostRecentIssue.jsp?punumber=9795657, https://doi.org/10.1109/IEMTRONICS55184.2022.9795836.

Moreno-Garcia, I.M., Palacios-Garcia, E.J., Pallares-Lopez, V., Santiago, I., Gonzalez-Redondo, M.J., Varo-Martinez, M., et al., 2016. Real-time monitoring system for a utility-scale photovoltaic power plant. Sensors (Switzerland) 16 (6). Available at https://doi.org/10.3390/s16060770, http://www.mdpi.com/1424-8220/16/6/770/pdf.

Nuhin, M.I.I., Shariar, M.A., Khan, J.M., Kongkon, T.N., Imam, M.H., 2022). Design of an IoT based power monitoring system model for a grid connected solar PV. International Journal of Advanced Technology and Engineering Exploration 9 (92), 923–940. Available at https://doi.org/10.19101/IJATEE.2021.875084, https://www.accentsjournals.org/PaperDirectory/Journal/IJATEE/2022/7/6.pdf.

Pingulkar, C., Joshi, A., 2021. AI and data analytics for predictive maintenance in solar power plants. International Journal of Progressive Research in Engineering Management and Science 1, 52–69.

Preda, S., Oprea, S.V., Bâra, A., Belciu, A., 2018. PV forecasting using support vector machine learning in a big data analytics context. Symmetry 10 (12). Available at https://doi.org/10.3390/sym10120748, https://res.mdpi.com/symmetry/symmetry-10-00748/article_deploy/symmetry-10-00748.pdf?filename=&attachment=1.

Rahman, M.M., Selvaraj, J., Rahim, N.A., Hasanuzzaman, M., 2018. Global modern monitoring systems for PV based power generation: a review. Renewable and Sustainable Energy Reviews 82, 4142–4158. Available at https://doi.org/10.1016/j.rser.2017.10.111, https://www.sciencedirect.com/science/journal/13640321.

Ramu, S.K., Irudayaraj, G.C.R., Elango, R., 2021. An IoT-based smart monitoring scheme for solar PV applications. In Electrical and Electronic Devices, Circuits, and Materials 211–233.

Rani, D.P., Suresh, D., Kapula, P.R., Akram, C.M., Hemalatha, N., Soni, P.K., 2023. IoT based smart solar energy monitoring systems. Materials Today: Proceedings 80.

Rao, C.K., Sahoo, S.K., Yanine, F.F., 2024. Designing an intelligent smart energy monitoring system for optimizing the utilization of PV energy. Energy Systems. Available at https://doi.org/10.1007/s12667-024-00703-6, https://www.springer.com/journal/12667.

Rasal, V., Raulkar, V., Reddy, S., Sanap, J., Sarap, A., Mhetre, M., 2021. IoT-based smart solar monitoring system. Intelligent Sustainable Systems 701–708.

Rezk, H., Tyukhov, I., Al-Dhaifallah, M., Tikhonov, A., 2017. Performance of data acquisition system for monitoring PV system parameters. Measurement 104, 204–211. Available at https://doi.org/10.1016/j.measurement.2017.02.050.

Shapsough, S., Takrouri, M., Dhaouadi, R., Zualkernan, I.A., 2021. Using IoT and smart monitoring devices to optimize the efficiency of large-scale distributed solar farms. Wireless Networks 27 (6), 4313–4329. Available at https://doi.org/10.1007/s11276-018-01918-z, https://www.springer.com/journal/11276.

Sharabati, A.A.A., Allahham, M., AbuSaimeh, H., Ahmad, A.Y.B., Sabra, S., Daoud, M.K., 2023. Effects of artificial integration and big data analysis on economic viability of solar microgrids: mediating role of cost benefit analysis. Operational Research in Engineering Sciences: Theory and Applications. 6 (3), 370–379. Available at https://doi.org/10.31181/oresta/060318, https://oresta.org/article-view/?id=649.

Sharma, M., Singla, M.K., Nijhawan, P., Ganguli, S., Rajest, S.S., 2020. An application of IoT to develop concept of smart remote monitoring system. Business Intelligence for Enterprise Internet of Things.

Shi, Q., Zhang, Z., Yang, Y., Shan, X., Salam, B., Lee, C., 2021. Artificial Intelligence of Things (AIoT) enabled floor monitoring system for smart home applications. ACS Nano 15 (11), 18312–18326. Available at https://doi.org/10.1021/acsnano.1c07579, http://pubs.acs.org/journal/ancac3.

Shrihariprasath, B., Rathinasabapathy, V., 2016. A smart IoT system for monitoring solar PV power conditioning unit. 2016 World Conference on Futuristic Trends in Research and Innovation for Social Welfare. IEEE, pp. 1–5.

Snapshot of Global PV Markets. 2025 Available at https://iea-pvps.org/wp-content/uploads/2025/04/Snapshot-of-Global-PV-Markets_2025.pdf.

Utama, I.B.K.Y., Pamungkas, R.F., Faridh, M.M., Jang, Y.M., 2023. Intelligent IoT platform for multiple PV plant monitoring. Sensors 23 (15). Available at https://doi.org/10.3390/s23156674, http://www.mdpi.com/journal/sensors.

V, V.K., G, S., 2022. Arduino based smart solar photovoltaic remote monitoring system. Malaysian Journal of Science 41 (3), 58–62.

Victor, J.L.F., Jucá, S.C.S., Pereira, R.I.S., Carvalho, P.C.M., Fernández-Ramírez, L.M., 2019. IoT monitoring systems applied to photovoltaic generation: the relevance for increasing decentralized plants. Renewable Energy and Power Quality Journal 17, 536–545. Available at https://doi.org/10.24084/repqj17.368, www.icrepq.com/RE&PQJ-papers.html.

Woyte, A., Richter, M., Moser, D., Mau, S., Reich, N., Jahn, U., 2013. Monitoring of photovoltaic systems: good practices and systematic analysis. In Proc. 28th European Photovoltaic Solar Energy Conference. 3686–3694.

Chapter 7

Advances in anomaly detection and fault diagnosis for photovoltaic power plants

Adel Mellit and Chourouk Moussaoui
Faculty of Sciences and Technology, University of Jijel, Jijel, Algeria

Introduction

Due to the demand for sustainable fossil fuel substitutes, photovoltaic (PV) power plants have emerged as a crucial component of the world's energy mix in recent years. Long-lasting but expensive investments, solar energy systems are susceptible to malfunctions that shorten their service life and result in revenue losses during the recovery process. For this reason, it is crucial to both prevent malfunctions from occurring and quickly fix them; however, no early warning system of this kind has been created as of yet (Mellit et al., 2018). Despite their increasing significance, a variety of issues, from partial shade and module soiling to inverter failures and wiring flaws, frequently impair the performance and dependability of PV systems. Significant energy losses, increased operating expenses, and even safety risks can result from such problems.

Recently, there has been a surge in interest in assessing the performance, quality, and lifespan of solar modules. As a result, maintaining an ideal output efficiency requires routine PV module inspection. PV module inspection techniques include infrared (IR) thermography, satellite surveillance, laser detection, manual examination, and electroluminescence (EL) imaging (Kellil et al., 2023). Fault detection and diagnostic systems have to be installed alongside solar PV plants to guarantee their safe and dependable operation. This will help identify issues early and address them. Numerous fault detection and diagnostic techniques have been examined in the literature to solve these problems. They vary in their demands for speed, complexity, sensors, and the capacity to detect numerous defects (Hijjawi et al., 2023).

For example, Table 7.1 summarizes the commonly used approaches.

Embedded Artificial Intelligence and the Internet of Things for Photovoltaic Systems.
DOI: https://doi.org/10.1016/B978-0-443-44835-5.00011-4

TABLE 7.1 Summary of the commonly employed fault detection and diagnosis (*FDD*) methods in PV systems.

Fault detection methods				
Ref.	**PV system level**	**Dataset type**	**Complexity level**	**Used approach**
Takashima et al. (2008)	String of modules (module-string)	Electrical measurement (earth-capacitance measurement) – not I–V or images specifically	medium	Earth capacitance method to detect a disconnected module in a string
Ma et al. (2021)	Module/panel level (automated defect detection)	I–V curves	Low	This method identifies the step and inflection point of the I–V curve by detecting changes in its concavity.
S et al. (2024)	Module level (PV modules)	Images (visual faults) – stacking ensemble (viewed via visual faults context)	Medium-High	Ensemble machine-learning model for visual fault detection from module images
Parvin et al. (2025)	System/array level	Electrical/operational PV signals	High	DNN feature engineering + ensemble classifier
Fault diagnosis methods				
Spataru et al. (2015)	Module/array level	I–V curves	Medium	Analytical comparison of measured versus modeled I–V characteristics
Tsanakas et al. (2015)	Module level	IR images	Medium	Image processing + Canny edge detection
(Garoudja et al., 2017)	System level	I–V curves	High	Enhanced machine learning model combining statistical and physical indicators
(Jaffery et al., 2017)	Panel level	IR images	Medium	Predictive fault diagnosis using IR imaging

(Huang et al., 2019)	Module/string level	I–V curves	Medium–High	I–V curve measurement + diagnostic algorithm based on deviation features
Fadhel et al. (2019)	Panel/isolated PV system	I–V curves	Medium	Principal component analysis (PCA) for fault detection and classification
Zbib and Al Sheikh (2020)	System/array level	I–V curves	Medium	I–V curve analysis with fault signature extraction
Alves Dos Reis Benatto et al. (2020)	Module level	EL images	High	Drone-based daylight EL imaging for defect identification
Bommes et al. (2021)	Module level	Aerial IR videos	High	Computer vision tool for automatic fault detection and mapping
Kim et al. (2021)	Module level	IR images	Medium	Image enhancement + cumulative distribution function (CDF) analysis
Mellit and Kalogirou (2022)	Module level	I–V curves	Medium	ML and ensemble learning methods
Moussaoui et al. (2022)	Module level	IR images	Medium	Low-cost edge device + CNN model for PV fault diagnosis
Benghanem et al. (2023)	Module level	IR images	High	Embedded hybrid CNN–ML model for fault diagnosis
Tang et al. (2023)	Module level	Aerial IR images	High	Automatic defect identification using UAV and deep learning
Boubaker et al. (2023)	System level	IR images	High	Machine and deep learning comparison for thermography-based fault detection

(*Continued*)

TABLE 7.1 (Continued)

Fault detection methods				
Ref.	**PV system level**	**Dataset type**	**Complexity level**	**Used approach**
El-Otum (2023)	Module level	EL images	High	Deep learning-based defect classification (CNN)
Mellit et al. (2024)	Module level	IR images	High	Hybrid Deep CNNs (DCNNs) for fault classification
Nassreddine et al. (2025)	Module level	I–V curves	High	Machine learning (Support vector machine, Decision Trees) for fault detection and classification
Lin et al. (2024)	Array level	I–V curves	High	Machine learning (Support vector machine , Decision Trees) for fault detection and classification
Liu et al. (2025)	Module level	Visual images + multiangle features	High	Similarity Learning
(Ren et al., 2025)	Module or string level	Global I–V curve features	Medium	1D-CNN
Ali and Ding (2025)	Array level	1D PV array signals (time series)	High	2D-CNN + XAI (Explanaible AI)
Shin et al. (2025)	Module or string level	Normalized I–V curves	High	Digital Twin + Physics Modeling
Chen et al. (2025)	Module level	Real-time PV module data + digital twin simulation	High	Similarity learning

Based on the above-reviewed papers in Table 7.1, the following points can be highlighted:

- Fault detection methods vary depending on the system level, from individual modules to complete PV power systems.
- Electrical techniques (e.g., ground capacitance, current-voltage diagnostics) are simple and effective for basic fault conditions.
- Image-based methods (IR and EL imaging) combined with AI provide automated and scalable fault identification.
- Review articles offer comprehensive overviews but lack depth in practical application.
- Hybrid approaches utilizing both electrical and visual data improve detection accuracy.
- Diagnostic methods heavily rely on current–voltage (I–V) curve analysis, often enhanced by machine learning (ML) algorithms (e.g., suport vector machine, random forest and neural networks).
- Drones and aerial thermal imaging are emerging in large-scale automated inspections.
- Edge computing and embedded AI models enable real-time and low-cost fault classification.
- DL dominates highly complex methods, particularly in IR and EL image-based diagnostics.
- Ensemble and hybrid models show promising potential for improving the robustness and accuracy of fault classification

A visual network of frequently occurring keywords from examined publications is shown in Fig. 7.1.

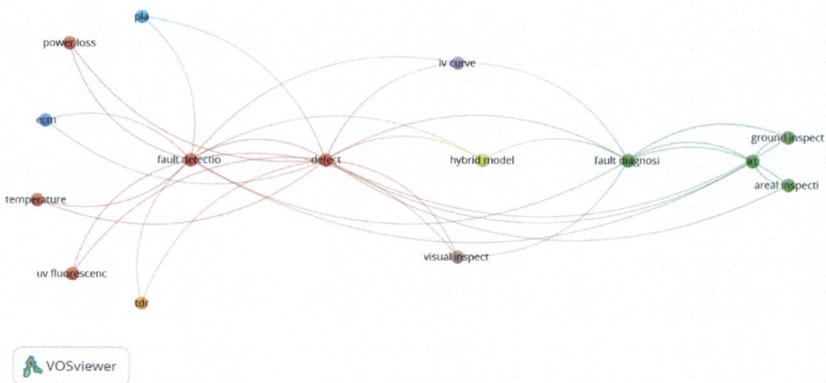

FIGURE 7.1 Keywords map in the presented papers using the visualization of similarities (VOS) viewer. This figure illustrates a keywords map using VOS viewer to regroup the keywords used in this chapter and form links between them.

This chapter presents advanced anomaly detection methods and their real-time implementation for PV power plants. It covers both traditional electrical techniques and modern image-based approaches, highlighting their potential in identifying and classifying different types of faults. Electrical methods based on I–V curve analysis are examined alongside image processing techniques, including visual, IR, and EL imaging. Hybrid paradigms that combine I–V analysis with image-based methods are also discussed, offering a multi-perspective approach for robust fault detection. A major emphasis of the chapter is placed on artificial intelligence (AI)-driven solutions. ML and deep learning (DL) models, such as convolutional neural networks (CNNs), Visual Geometry Group (VGG), MobileNet, and You Only Look Once (YOLO), are introduced as powerful tools for automated fault classification.

These models are trained and tested on real-world datasets comprising visual, IR, and EL images of PV modules and arrays. Their practical application in real-time scenarios is demonstrated through illustrative examples and end-to-end solutions. In addition, the chapter explores embedded AI frameworks for deploying lightweight models on edge devices, enabling real-time, on-site monitoring without relying heavily on cloud infrastructure. The benefits of embedded AI for reducing latency, improving scalability, and enhancing reliability are highlighted. A real-world case study is included to illustrate practical deployments, showcasing the transition from algorithm design to field implementation. The chapter concludes with an outlook on future trends and a roadmap for integrating AI and embedded AI into PV fault diagnosis. This includes potential improvements in model efficiency, data acquisition strategies, and the fusion of multimodal monitoring techniques.

Types of defects

PV system defects can be classified into two main classes: permanent and temporary failures. In this section, a review of common defects in the PV modules is presented (Ahmad et al., 2019). Fig. 7.2 illustrates a diagram summarizing the types of defects in PV plants, categorized 1 according to El-Banby et al. (2023) and Hijjawi et al. (2023).

Permanent defects

Ethylene vinyl acetate failures

Ethylene vinyl acetate encapsulating materials can yellow and brown due to UV exposure, high temperatures, humidity, or poor production. This deterioration can lead to delamination, corrosion, and discoloration, affecting efficiency and short-circuit current. Thorough testing is recommended to reduce these failures, especially in dump heat tests.

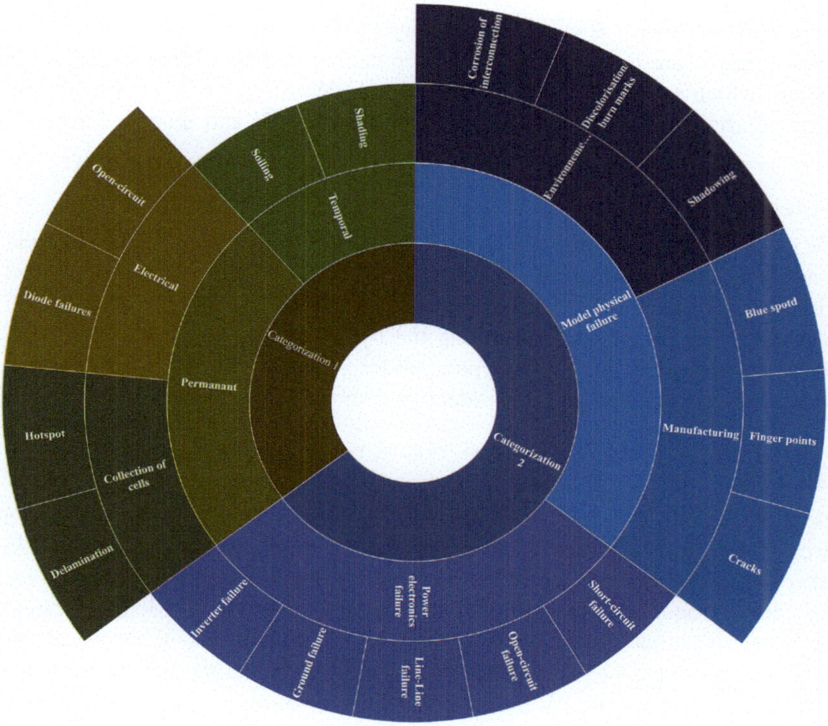

FIGURE 7.2 Types of defects in PV power plants. The figure illustrates the distribution of various defects commonly observed in photovoltaic installations. It categorizes defect types according to two different papers.

Micro cracks

Tiny fractures in PV cells can cause physical harm and current imbalances, impacting performance. Damaged cells may decrease the submodule's characteristic curve. Models simulate healthy cell behavior.

Disconnected cells and faulty string-ribbon interconnection

Ribbons are used in the tabber-stringer manufacturing process to create a series connection between individual solar cells. High mechanical stress caused by this technique may result in latent cracks from shoddy soldering, which might cause future cell cracks or ribbon disconnections. The series resistance of a PV module can be increased by a single detached ribbon, leading to up to 35% power losses. In the past, PV modules employed two busbars that, when disconnected, would cut off the current and turn on a bypass diode to safeguard the cells. Since there are now more than three busbars in operation, any disconnections jeopardize the cells' ability to generate electricity.

Potential induced degradation

PV specialists are increasingly aware of potential induced deterioration in PV strings, which can reach up to 1500 V in grid-connected facilities. This occurs when a large voltage gradient forms between PV cells and module frames.

Defective bypass diode

PV modules are equipped with bypass diodes to shield PV cells from internal flaws and current mismatches caused by shadowing. A diode is typically fitted for a group of around 20–24 cells since the breakdown voltage of a single cell is taken into account when installing the bypass diodes in the modules. The short circuit issue is one of the bypass diodes' primary flaws. The PV modules' power decrease makes it easy to identify a short-circuited diode.

Mismatch

PV modules in a PV array are connected in either series or parallel. When a string, which refers to several PV modules connected in series, is connected to other strings in series and parallel, multiple dysfunctions can take place. One possible dysfunction is that the current produced from one of the strings could be lower than that of the other strings. This situation can occur when PV modules connected in strings are not capable of generating electricity in the normal way because the PV module has been somehow polluted or damaged.

Junction box

The junction box, often known as the J-box, is an essential part of a PV plant. It holds bypass diodes for current protection and links the output cables of the solar modules. The J-box may, however, experience a number of issues, including weak or corroded connections, inadequate sealing that permits moisture intrusion, overheating brought on by reverse current or inadequate contact, and bypass diode failure. Hot spots, electrical arcing, power outages, and even fire dangers might result from these problems. Because bad design, subpar materials, or incorrect installation are the main causes of failures, J-box reliability is crucial to the PV system's long-term performance and safety.

Temporal defects

Soiling

The buildup of dust, sand, birds, tree leaves, or any other item on the surface of PV modules is referred to as PV panel soiling. The effectiveness of producing power is decreased by this dirt, which prevents sunlight from reaching the solar cells.

Shading

Because of accumulated debris, poor system design, and external obstacles such as buildings and trees, shading on solar surfaces lowers energy production. Because PV cells are connected in series, even slight shadowing has a substantial influence on performance, which is why bypass diodes are used to lessen these effects.

Fault detection methods

In solar PV systems, PV panel fault detection is an essential technique for locating and diagnosing malfunctions, including power loss, cracks, and hot spots that can endanger public safety and shorten component lifespan and performance (Aghaei et al., 2022).

Electrical-based models

Detection techniques using current–voltage curves

The I–V curve provides the values of the main parameters, including fill factor, maximum power output, open-circuit voltage, and short-circuit current, by displaying the output combinations between voltages and currents (Lin et al., 2024). I–V curve analysis is a conventional fault detection and diagnosis (FDD) technique that uses a module's electrical readings to determine its short-circuit current, open-circuit voltage, and other parameters that may indicate a system failure. When an external electronic load or power source is applied, the voltage or current across the module changes, and the I–V curve is tracked and recorded (Hijjawi et al., 2023).

Statistical and signal processing techniques

Waveform signal analysis is the foundation of signal processing techniques, including time domain reflectometry (TDR), power losses analysis, speared spectrum, and earth capacitance measurement (ECM). To identify and locate effective PV module arrays, the TDR approach is employed. Regretfully, it could rely on the installation circumstances, such as the wiring and materials of the PV components (Hijjawi et al., 2023).

Earth capacitance measurements

ECM is typically used to identify where electrical cables are disconnected. The idea behind this technique for PV modules is to measure the capacitance between the module's terminal and the grounding terminal to identify capacitance created as a result of flaws. As a result, this method is not affected by variations in irradiance (Lin et al., 2024).

Time domain reflectometry

Instead of measuring capacitance, the TDR approach measures the change in impedance between an input signal to the strings of the PV module and the reflected signal that results from an anomaly or defect. Following that, the reflected signal is compared to the reference signal in terms of various electrical properties, such as (1) the shift in the signal from the input signal, which indicates the location of the fault in the string; (2) the change in the waveform, which indicates the type of mismatch (e.g., open circuit, short circuit, resistance increase, etc.); and (3) the change in the impedance, which indicates the fault's severity (Lin et al., 2024).

Power losses analysis

Using this method, the simulated power output of a healthy PV module is compared to the real power output of the PV module. To find ascribed flaws, the resultant error deviations are further examined. For example, a self-adaptive PV production model has been presented by Yang et al. (2024) to continually compare the generated PV power to the anticipated power. A warning signal is displayed if differences are found. However, neither the location nor the type of defect was identified.

Image-based methods

Image-based methods rely on four main techniques: (1) visual inspection, (2) IR thermography (IR), (3) EL imaging, and (4) ultraviolet fluorescence (UV-FL) imaging.

Visual inspection

For the image-based methods, the first step to detect PV failures is to view the PV modules from different angles (Stauffer et al., 2015; Moussaoui et al., 2024). The engineer can conduct periodic inspections of the panels using a visual camera to determine if the panel has any noticeable defects. Using this method, only some visual defects can be detected; Fig. 7.3 illustrates an example of a visual image of a PV module showing browning failure.

Infrared thermography

IR thermal (IRT) imaging is a detection method used to detect heat dissipation in solar cells, allowing for the prediction of failure before it occurs. There are two primary methods: active IR thermal, which uses an external heat source to raise the temperature, and passive IR thermal, which records thermal patterns directly using the PV module's natural infrared radiation. Passive IRT is popular for examining large PV facilities and is easier and less expensive to install. Fig. 7.4 represents an infrared thermal image of PV modules showing hotspot cell defects (Kellil et al., 2023).

FIGURE 7.3 Visual image of a PV module showing browning failure. This figure presents a photovoltaic module with visible discoloration, typically appearing as brown or darkened areas on the encapsulant. This browning highlights material degradation that can reduce light transmission and negatively impact the module's electrical performance (*PV*, Photovoltaic).

FIGURE 7.4 Infrared thermal image of PV modules showing hotspot cell defects. The figure displays PV modules captured with a thermal camera, revealing localized areas of elevated temperature. These hotspots indicate defective or stressed cells that can reduce module efficiency and may lead to further degradation if not addressed (*PV*, Photovoltaic).

Electroluminescence imaging

One of the well-known image-based techniques for testing PV modules or cells and identifying failure is the EL approach, which uses EL pictures as a data set. A DC current is applied to PV modules in order to encourage radiative recombination inside the solar cells. EL emission is measured using a charged silicon camera (Charge-Coupled Device) (Hijjawi et al., 2023). An example of an EL PV module image is indicated in Fig. 7.5 (Drir et al., 2026).

The ultraviolet fluorescence method

The discoloration of PV modules can be analyzed using the UV FL method (FL imaging) of EVA in PV cells, as illustrated in Fig. 7.6. Even in a dark outdoor environment, it can identify the quantity and location of cell cracks in PV modules, but it is unable to identify cracks on the cell's edge (Hijjawi et al., 2023). Fig. 7.6 presents a UV fluorescence PV module image.

Fault diagnosis methods

Electrical-based method

Current–voltage curve

I–V curve-based diagnostics is a prominent topic, as an I–V curve inherently contains rich information about the health state of PV modules. Conventional I–V tracers typically enable measurements at the level of a single module, a small-scale string, or an array. In recent years, hardware solutions integrated at the inverter level for the periodic acquisition of I–V curves at the power plant scale have become commercially available. Consequently, I–V curve-based diagnostic techniques can now be applied across most standard PV installations (Mellit and Kalogirou, 2022). Generally, I–V curve-based methods rely on comparing the measured I–V curves with a reference I–V curve

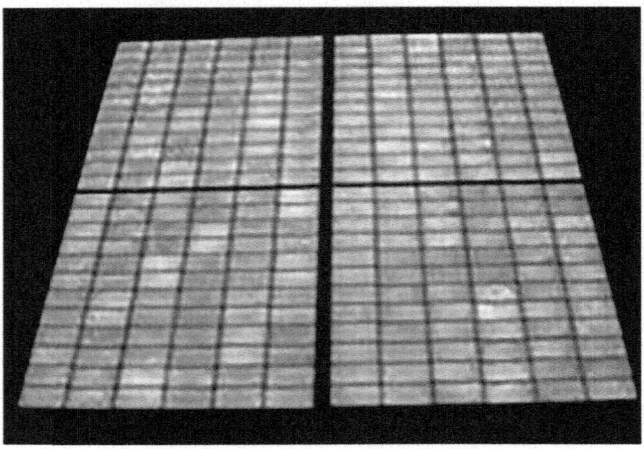

FIGURE 7.5 Electroluminescence image of PV modules. The figure shows PV modules captured under electroluminescence testing, revealing variations in brightness across the cells. These brightness differences highlight defects such as microcracks, inactive regions, or broken cells that may not be visible in standard imagery (*PV*, Photovoltaic).

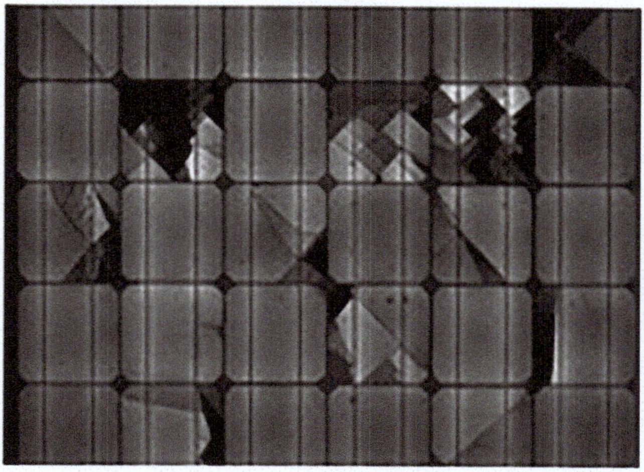

FIGURE 7.6 UV fluorescence image of a PV module. The figure presents the module under ultraviolet illumination, highlighting fluorescence patterns across the surface (*PV*, Photovoltaic; *UV*, ultraviolet).

corresponding to healthy PV modules. These methods can be implemented in various ways to extract diagnostic information, as reported in the literature. The most prevalent approaches can be classified into three categories: (1) Use the curve's important features (such as V_{OC}, I_{SC}, V_{MPP}, I_{MPP}, FF, R_s, and R_{sh}) for diagnosis using statistical methods, threshold analysis, ML techniques, etc. (2) Determine the steps on the I–V curves or compute the curve's first or second-order derivative. And (3) to provide residuals for examination, complete I–V curves are compared to simulated ones. From the block diagram of the I–V curve-based approach illustrated in Fig. 7.7, it can be observed that the method relies on the acquisition of I–V curves, followed by their analysis using various techniques, such as ML, DL, or signal processing (Li et al., 2021).

Power-loss analysis

Distributed maximum power point tracking systems, such as power optimizers and micro-inverters, enable precise monitoring of PV systems, allowing for the identification of faults and power losses. These distributed topologies, initially designed to mitigate losses from shading and mismatch in central maximum power point tracking systems, can achieve up to 30% instantaneous power improvements and 12% annual energy gains in shaded environments (see Fig. 7.8). The systems require module-level monitoring of voltage and current, which can be implemented cost-effectively through the addition of a communication module and the use of power line carrier signals to collect operational data from all modules (Solórzano and Egido, 2013).

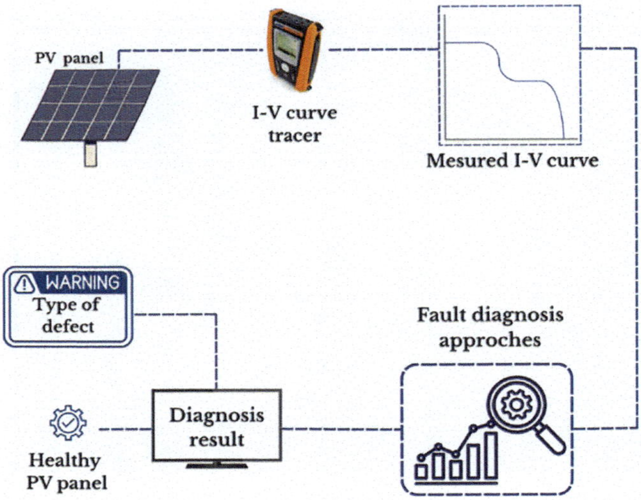

FIGURE 7.7 I–V curve-based method's diagram. The figure illustrates the overall workflow of a PV plant fault-detection process, starting from the collection of current–voltage (I–V) data from the PV arrays. It then shows how the data are analyzed and compared with expected performance indicators to determine whether the plant is operating normally or exhibiting defects (*PV*, Photovoltaic).

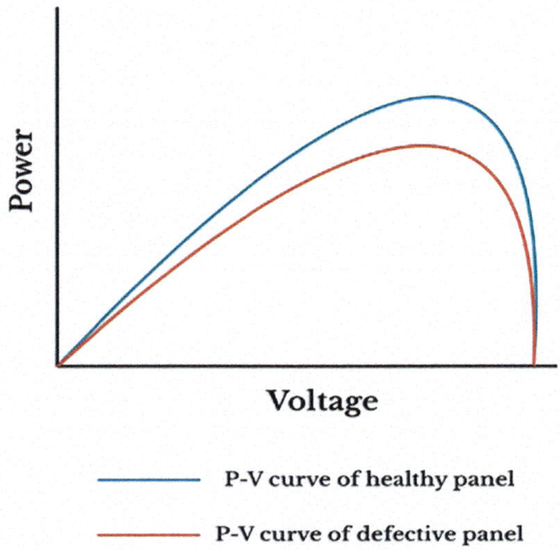

FIGURE 7.8 P-V curve for a healthy and defective PV module. The figure compares the power–voltage characteristics of a properly functioning panel with those of a faulty one. It shows how defects cause a noticeable reduction or distortion in the power curve, indicating decreased performance compared to the healthy panel (*PV*, Photovoltaic).

FIGURE 7.9 Visual inspection of PV plants. The figure shows on-site examination of photovoltaic installations to identify visible defects or abnormalities (*PV*, Photovoltaic).

Visual inspection

This type of inspection is conducted in the presence of an engineer who visually inspects the panels to identify faults (Fig. 7.9). The engineer can visually detect and classify some defects, such as dust, snow or bird drops, but can't detect deep problems as diode problems or hot spots. That's why the PV installation centers are moving toward the use of other types of images to clearly identify other problems, such as IR images and EL images.

Image-based methods

Infrared thermography imaging

IR thermography measures radiation from a body's surface and is used for PV inspections. Thermal cameras measure wavelengths between 7 and 14 meters, balancing price and product availability. Unconverted photons raise cell temperatures, causing a noticeable "temperature signature" in IRT pictures (de Oliveira et al., 2022). This method is based on differences in temperature profiles in the IR images, where healthy cells have a specific color depending on the type of camera used, and the cells that have a defect, causing a change in their temperature (usually an increase) have a different color, allowing for the detection of the fault. Additionally, the distribution of the color change helps in identifying the type of defect (Figgis et al., 2025).

Ground inspection using a thermal camera: This method is based on two approaches:

- In the first approach, IR images are acquired using a handheld thermal camera, with image capture performed manually. For diagnostic purposes, the engineer may either transmit the acquired images to a control room or laboratory for offline processing and defect identification, or perform on-site diagnosis using a complete prototype equipped with an embedded diagnostic system.
- In the second approach, integrated thermal cameras are mounted on unmanned ground vehicles (UGVs) to perform IR scanning of PV modules, followed by automated image analysis using various techniques, including conventional image processing and ML/DL methods.

Fig. 7.10(A) represents the diagram of manual ground inspection using a thermal camera, and Fig. 7.10(B) represents the diagram of the same method when the thermal camera is associated with a UGV.

Aerial inspection: Unmanned aerial vehicles (UAVs) have become increasingly popular for collecting images of solar farms due to their efficiency, cost-effectiveness, and increasing performance. The most advanced imaging tools are visual and thermal IR cameras, which can identify mechanical faults, operation and maintenance issues, and thermal signs. However, IR images are sensitive to environmental conditions, the operating state of PV strings, and the UAV and camera's operation. Obtaining high-quality aerial IR images requires precise and consistent methods (Kellil et al., 2024; Alves Dos Reis Benatto et al., 2020). The UAV, equipped with a thermal camera, scans the solar module field to capture images (see Fig. 7.11). These images are either processed in real time by an embedded system onboard the UAV, which then sends an alert to inform control-room engineers whether the generators are faulty, or transmitted to the

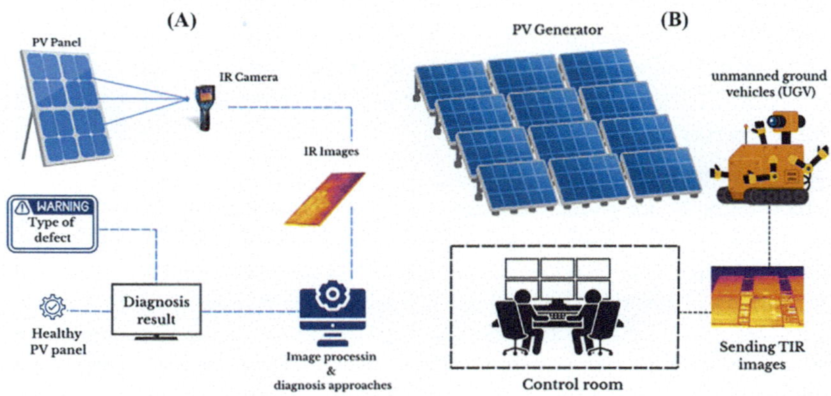

FIGURE 7.10 Ground inspection using thermal cameras. The figure illustrates how thermal imaging is conducted either manually by an operator or using a ground-based unmanned vehicle. It shows the workflow for capturing thermal data to detect hotspots and other temperature-related defects in PV modules (*PV*, Photovoltaic).

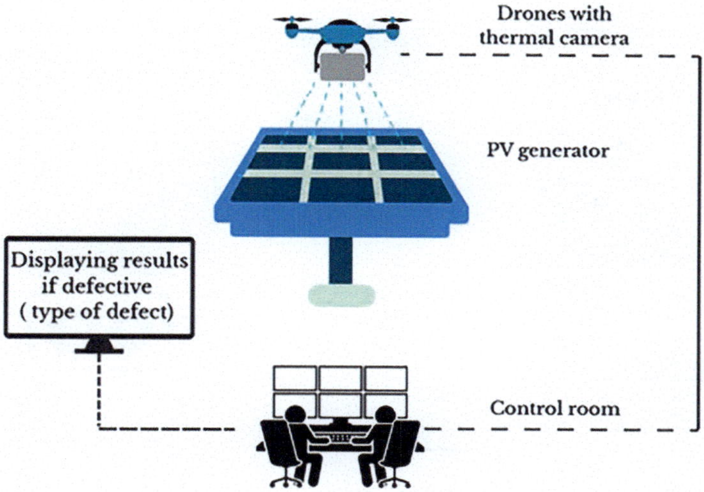

FIGURE 7.11 Aerial inspection with a UAV equipped with a thermal camera. The figure depicts the use of an unmanned aerial vehicle equipped with a thermal camera to survey PV installations from above. It illustrates the process of capturing thermal imagery to identify the defects (*PV*, Photovoltaic; *UAV*, unmanned aerial vehicle).

control room for offline processing using available methods to identify defects and support appropriate maintenance decisions.

Electroluminescence imaging

EL imaging is a PV module characterization method that offers high accuracy in identifying a wide range of flaws and defects, including cracks, broken cell interconnections, shunts, and other degradation mechanisms (Drir et al., 2025) (See Fig. 7.12).

The EL technique is also widely used due to its high level of detail and its direct correlation with the injected carrier density. However, because the luminescence signal of crystalline silicon is many orders of magnitude weaker than sunlight, this approach is typically limited to indoor measurements or outdoor use between twilight and early morning. This restriction limits the application of such a powerful technique for utility-scale inspections, which has motivated the development of electrical biasing instruments to enable fast and effective outdoor EL imaging (Drir et al., 2026). The processing of EL images is performed in the laboratory using specific techniques to predict failures.

Machine learning and deep learning for fault detection and diagnosis in photovoltaic power plants

Under certain conditions, where the database contains both healthy and defective panels, this technique is utilized to examine the data that has been gathered.

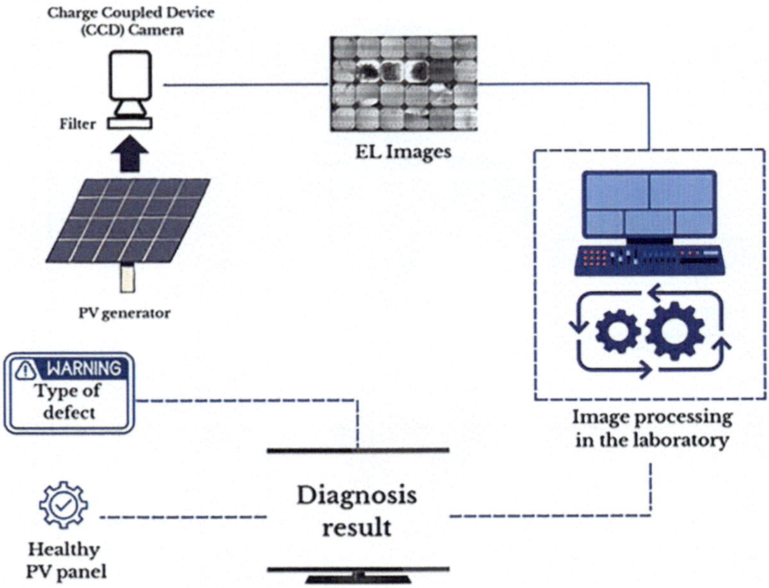

FIGURE 7.12 EL image-based process. The figure titled illustrates the workflow of using electroluminescence imaging for PV module inspection. It shows how EL images are captured and analyzed to detect the defects (*EL*, Electroluminescence; *PV*, photovoltaic).

Machine learning

Both techniques that depend on image processing and those that rely on electrical measurements involve ML. Once a database of the same kind has been gathered, first a binary classification is performed to determine whether the panel is healthy or not; then, if it is defective, a multiclass classification is performed to classify the types of defects using one of the ML algorithms (see Fig. 7.13). Two primary categories of ML methodologies exist:

- Supervised ML techniques require labeled data from the outset, so all classes are predefined. During training, the algorithms learn to extract relevant data features, which are later used to classify new, unseen data.
- In contrast, unsupervised ML techniques work with unlabeled data. The algorithms identify patterns and create classes based on extracted features during training, which are then used to analyze and categorize new data in the testing phase.

The following diagram, Fig. 7.13, summarizes the sequence of the PV module FDD process based on ML:

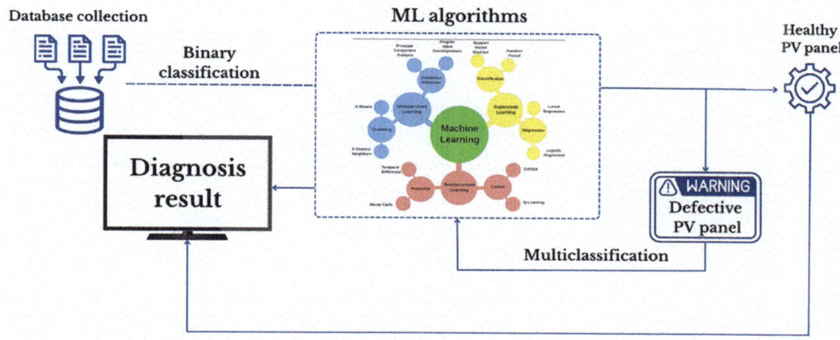

FIGURE 7.13 Block diagram of ML-based approaches for FDD of PV modules. The figure illustrates the workflow of using machine learning for PV plant fault detection or performance analysis. It shows the process from data collection and preprocessing, through training and testing ML models, to classify the defects (*FDD*, Fault detection and diagnosis; *PV*, photovoltaic).

Deep learning

DL offers an innovative approach for the automated inspection and condition monitoring of PV modules. Advanced architectures such as U-Net, ResNet, and other CNNs enable sophisticated image processing tasks, including defect classification, feature extraction, and image segmentation. Common faults, such as cracks, delamination, dust accumulation, hotspots, and shading effects, can be accurately detected and localized using thermal and visual data acquired from drones or fixed monitoring cameras. This AI-based approach reduces human intervention, lowers operation and maintenance costs, and enables real-time monitoring through preventive and predictive maintenance strategies. Consequently, incorporating DL techniques into solar plant management significantly improves defect detection accuracy, system reliability, and overall energy yield. The block diagram in Fig. 7.14 explains how to use DL approaches for FDD of PV modules.

Hybrid models

This approach combines DL and traditional ML within a single framework, particularly when feature extraction is required. In this setup, generic features are first extracted from the training database using ML techniques, and the classification is subsequently performed by DL models. Alternative strategies involve extracting features with deep convolutional neural networks, removing the final fully connected layer, and replacing it with a ML classifier, such as a support vector machine, to perform defect classification (see Fig. 7.15) (Benghanem et al., 2023).

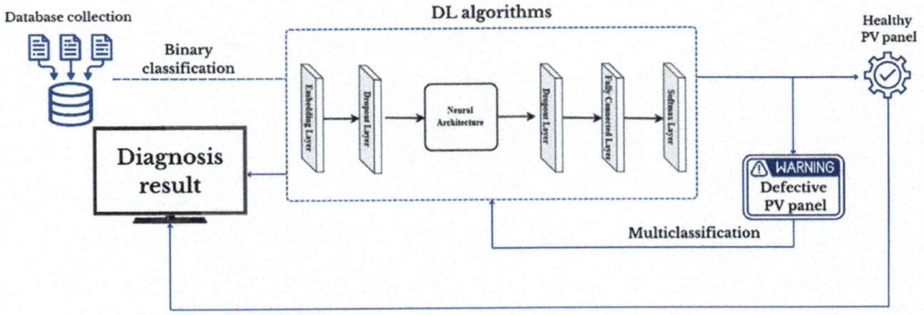

FIGURE 7.14 Block diagram of DL-based approaches for FDD of PV modules. The figure illustrates the workflow of using deep learning for PV module fault diagnosis. It shows the process from data acquisition and preprocessing, through training and validating deep neural networks, to automatically classifying defects (*FDD*, Fault detection and diagnosis; *PV*, photovoltaic).

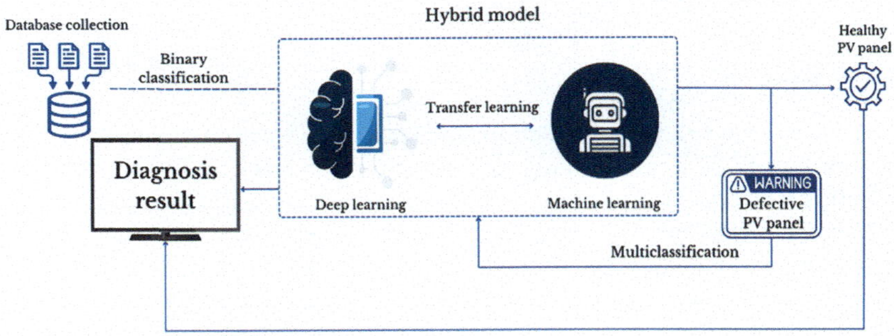

FIGURE 7.15 Block diagram of hybrid-based approaches. The figure illustrates a fault diagnosis or performance analysis workflow that combines multiple techniques, such as machine learning, deep learning, and traditional methods. It shows how data are collected and processed through different models or algorithms, with their results integrated to improve accuracy in PV plant diagnosis (*PV*, Photovoltaic).

You Only Look Once-based models

YOLO is one of the commonly used algorithms for object detection, and it can be used in our case to first verify the presence of solar panels within the studied area. Drones equipped with optical or thermal cameras are usually used to capture images, which are then examined to confirm the presence of the panels. After that, in order to identify faults and locate them, we typically use a thermal camera to collect images, and then process them using the YOLO algorithm to obtain the results. If no faults are detected, this means there are none (Kellil et al., 2024). But if a fault is present, the system will identify it as shown in the diagram presented in Fig. 7.16.

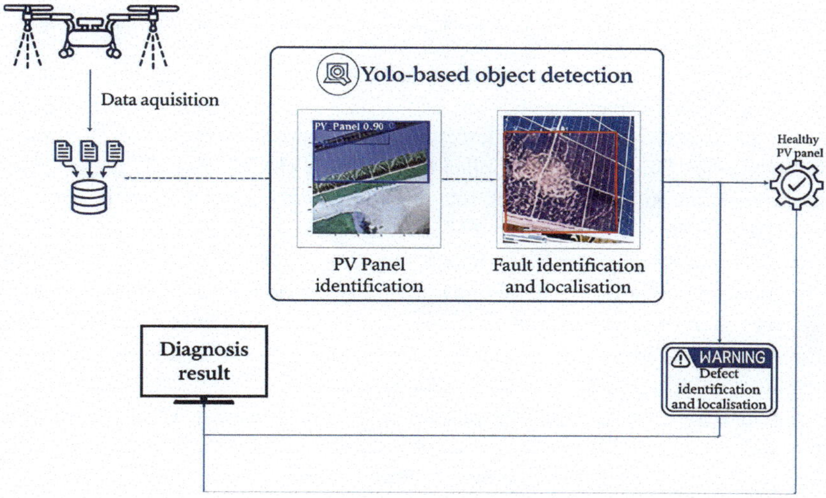

FIGURE 7.16 Diagram: YOLO-based diagnosis method. The figure illustrates the workflow of using the YOLO (You Only Look Once) object detection algorithm for PV module fault detection. It shows how images of PV panels are captured, processed by the YOLO model to detect defects in real time, and then classified to identify faulty cells or modules (*PV*, Photovoltaic).

Case study

This section presents a case study on anomaly detection and fault diagnosis in PV power plants using advanced techniques, including AI and DL. This approach relies on image processing, specifically the analysis of visible and infrared thermal images, to identify key hotspots and diagnose their underlying causes, such as dust accumulation, bird droppings, degradation, or localized cell damage. By integrating electrical and image-based diagnostics, the study provides a more comprehensive view of the state of PV systems. This case study is designed for a broad audience, including researchers, academics, and professionals in the field of PV fault detection and intelligent diagnostics, and demonstrates how image-based methods can enhance diagnostic accuracy, improve system reliability, and reduce downtime. PV operators and engineers may also benefit, as the findings highlight techniques that minimize performance losses and optimize maintenance schedules. The study was conducted through close collaboration between system operators, who provided field data and access, and research teams, who developed and validated the AI and deep learning models.

The motivation stems from the limitations of traditional monitoring, which often fails to detect subtle or early-stage problems such as microcracks, gradual degradation, or small-scale hotspots. These issues can escalate into major failures and lead to significant energy losses. By leveraging AI-driven methods, the study moves beyond basic detection towards precise fault identification, classification, and root cause analysis. Early detection not only prevents long-term damage but also enables predictive maintenance, reducing costs and extending asset lifespan. This case study highlights the potential of AI in transforming PV energy monitoring into a more accurate, automated, and intelligent diagnostic system, enhancing the performance, reliability, and sustainable integration of solar energy.

The method focuses on fault detection and diagnosis using IR thermography combined with DL. Although applicable to visible (RGB) and EL images, the emphasis is on infrared thermal imaging due to its proven effectiveness in identifying thermal anomalies in PV modules. By analyzing thermal profiles, infrared thermography reveals hotspots, localized defects, or excessively hot components, known to be critical and potentially dangerous faults. If undetected, hotspots can reduce energy production, lead to performance degradation, and pose safety risks, including fire hazards. Therefore, early detection and accurate classification are crucial for both efficiency and safety.

A real-world example demonstrates the use of embedded AI for automated fault detection in PV arrays. An intelligent prototype was developed, integrating low-cost infrared imaging with deep learning algorithms on an embedded system. This system is distinguished by its ability to detect, classify, and automatically report faults in real time, making it suitable for ground-mounted photovoltaic power plants and aerial inspections using drones. This integration highlights the potential for providing advanced and scalable self-monitoring solutions that enhance reliability, reduce maintenance costs, and improve safety in large-scale photovoltaic installations.

Dataset

The dataset employed in this case study was collected in Algeria (two different climatic regions, south and north of Algeria) and consists of IR thermography images representing three types of defects observed in a semiarid climate (the Sahara of Algeria). The most common and investigated defects in this region include: (1) dust accumulation, where the temperature profile varies depending on the distribution of dust particles; (2) sand deposits, which cause similar but more irregular temperature variations; and (3) short-circuits, in which a large number of solar cells become overheated. Fig. 7.17 illustrates an example from

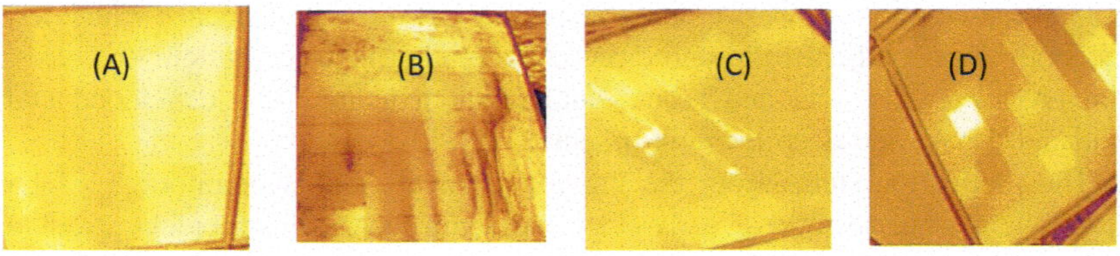

FIGURE 7.17 The collected IR images: (A) Normal (N), (B) sand deposit (D1), (C) dirt accumulation (D2), and (D) short-circuit (D3). The figure shows infrared thermography images of a photovoltaic module under normal, sand deposition, dirt accumulation, and short-circuit conditions. Each condition exhibits a distinct thermal pattern (*IR*, Infrared).

the collected dataset. As shown, in the case of short-circuit defects, a significant portion of the PV module cells exhibit overheating. By contrast, in cases of dust or sand accumulation, the temperature distribution depends strongly on the pattern and density of the deposited material. Under normal operating conditions, the module displays a homogeneous temperature profile across the surface.

Data augmentation is a widely used technique in computer vision to artificially expand the size and diversity of training datasets by applying random transformations to input images. Basic augmentations include operations such as flipping, rotation, translation, zooming, cropping, and brightness or color adjustments, which help the model learn to recognize objects under different orientations, lighting conditions, and scales. By generating new, slightly varied versions of the same image, augmentation reduces overfitting, improves generalization, and enables models to perform better on unseen data without the need for additional real-world samples. Basic data augmentation techniques were applied to increase the size of the dataset. As an example, Fig. 7.18 shows the augmented dataset. The total size after augmentation is 2000 images. This dataset will be used to train a CNN model for potential classification of various defects (Mellit and Kalogirou, 2025).

Lightweight convolutional neural network model for photovoltaic fault classification

A lightweight CNN based on the MobileNet architecture (Qin et al., 2024) was developed for classifying faults in PV IR images using the Edge Impulse platform. The model is specifically optimized for embedded applications, ensuring low latency and minimal power consumption while maintaining satisfactory accuracy. The overall procedure for designing the classifier within Edge Impulse can be summarized in the following key steps:

Data acquisition – Upload IR images and label them into four classes (D1, D2, D3, and N).

Impulse design – Configure the learning pipeline:

- Create impulse: Define image size, add preprocessing and learning blocks.
- Image processing: Choose image type (RGB or grayscale) and extract class features.
- Transfer learning: Set CNN hyperparameters (training cycles, learning rate, validation set size, number of neurons, dropout).

Model retraining – If overfitting occurs, retrain using more diverse data while keeping block configurations unchanged.

Live classification – Test the model on single unseen IR images for quick evaluation.

Model testing – Perform classification on the full test dataset; generate a confusion matrix and performance metrics.

Deployment – Deploy the trained lightweight model to embedded/edge devices for offline, low-latency, and energy-efficient operation.

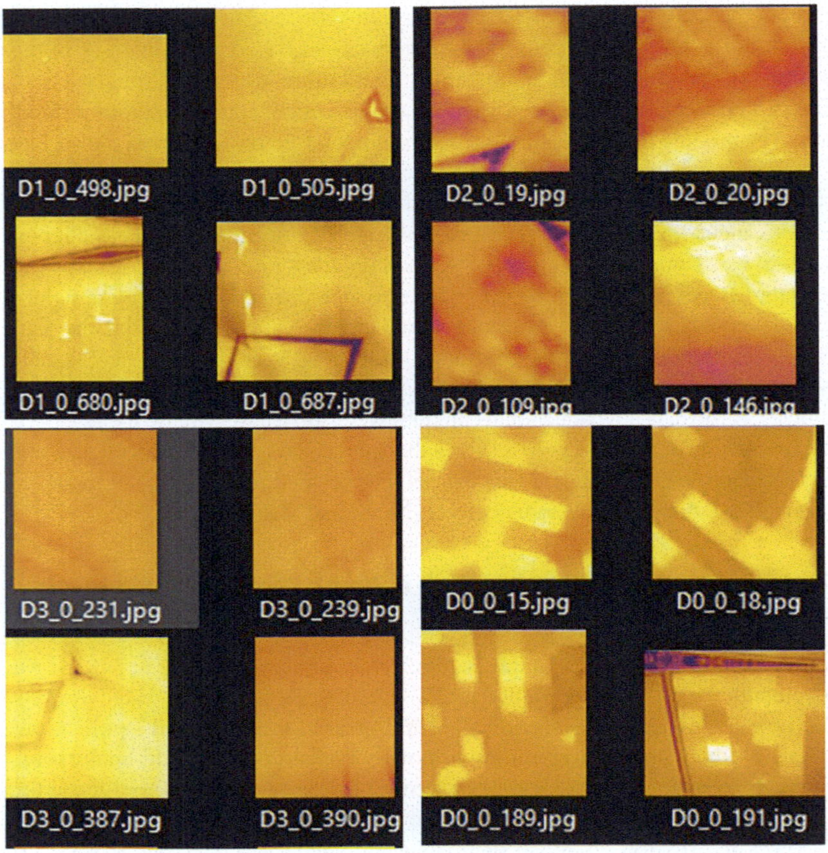

FIGURE 7.18 Augmented dataset using data augmentation techniques (e.g., rotation, flipping, translation/shifting, zooming, and brightness). The figure illustrates examples of the augmented dataset generated using data augmentation techniques, such as rotation, flipping, translation, zooming, and brightness adjustment. These transformations increase data diversity while preserving the original class characteristics.

Prototype design

The block diagram of the designed prototype is presented in Fig. 7.19. The system is composed of three main components: an IR thermography camera for data acquisition, an Arduino Portenta H7 microcontroller unit (MCU) for on-device processing and inference, and a display screen for real-time visualization of the results. Together, these components enable an efficient and lightweight embedded solution for PV fault detection and diagnosis (Mellit et al., 2025).

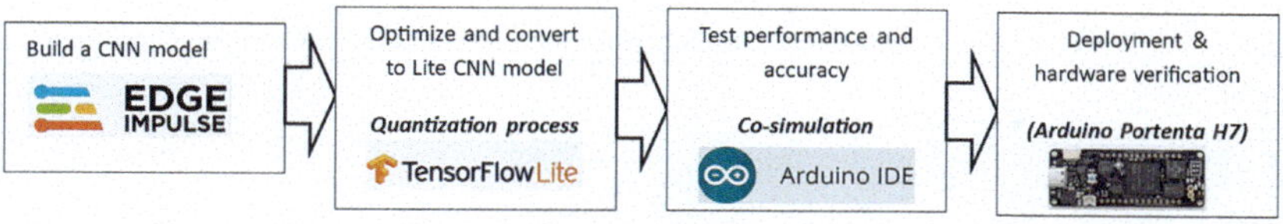

FIGURE 7.19 Block diagram of the prototype. The figure shows the block diagram of the proposed prototype, consisting of an infrared thermography camera for data acquisition, an Arduino Portenta H7 MCU for on-device inference, and a display for real-time visualization. These components together form a lightweight embedded system for photovoltaic fault detection and diagnosis (*MCU*, Microcontroller unit).

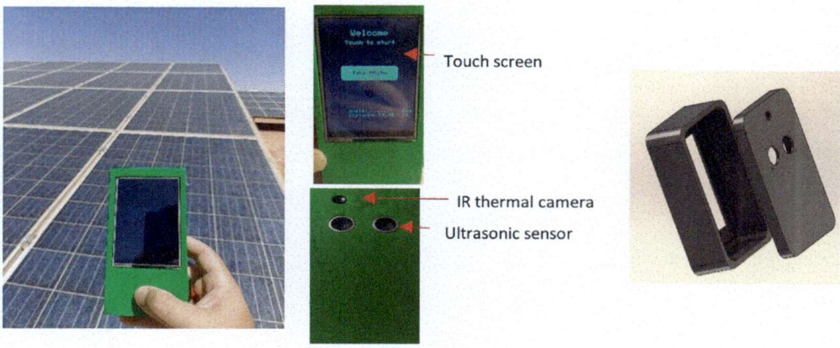

FIGURE 7.20 A real photograph of the designed prototype. The figure shows a real photograph of the designed prototype used for ground inspection of PV modules. The system integrates an IR camera and an embedded AI model to detect and analyze defects in real time (*AI*, Artificial intelligence; *IR*, infrared; *PV*, photovoltaic).

A real photograph of the developed prototype is presented in Fig. 7.20, illustrating the physical assembly of the system. The detailed hardware specifications, together with the total cost of the prototype, are summarized in Table 7.2. This information provides a clear overview of the components used, their technical characteristics, and the associated budget required to replicate the system. Moreover, the relatively low cost of the prototype highlights its potential for scalability and adaptation to larger photovoltaic installations, making it an attractive solution for both research and industrial applications.

Experimental verification

The designed prototype was experimentally evaluated and tested under real operating conditions. Fig. 7.21 presents examples of the experimental validation carried out on PV plants installed in two different regions of Algeria, one in the north and the other in the south, characterized by contrasting climatic conditions. The northern site is located in a Mediterranean climate, with moderate temperatures and seasonal rainfall, particularly during

TABLE 7.2 List of components used in the prototype, including specifications and costs.

Component	Specification	Price (€)
FLIR Lepton 2.5 IR Camera	Thermal resolution: 80 × 60 px, < 9 Hz; spectral range: 8–14 μm	200
TFT display (5 inch)	800 × 480 resolution, RGB interface, touch-enabled	40
Portenta H7 board	Dual-core STM32H747 (Cortex-M7 + M4), 480 MHz, Wi-Fi + Bluetooth	100
Ultrasonic sensor	Distance range= 2–400 cm, accuracy ±3 mm, sensing angle 15°	5
Battery	9 V with mico-c	10
SD card	16 GB, Class 10, microSD with adapter	10
PCB board	Custom design for signal routing and interfacing	10
3D-printed enclosure	PLA/ABS plastic, custom-designed box	20
Electronic components	Resistors, capacitors, small DC–DC inverter module	15
	Total estimated cost	410

FIGURE 7.21 An experimental test was conducted at two different regions in Algeria, with different climatic conditions and PV technologies. The figure illustrates the experimental tests conducted in two different regions of Algeria under distinct climatic conditions. This comparison highlights the system's performance across varying environmental settings (*PV*, Photovoltaic).

the winter. In contrast, the southern site is situated in an arid environment marked by high ambient temperatures, frequent sandstorms, and minimal precipitation throughout the year. These distinct climatic contexts provide an excellent opportunity to assess the robustness and adaptability of the prototype under varying operational and environmental stresses. Testing across both regions not only validates the system's performance in diverse weather conditions but also demonstrates its potential applicability for PV monitoring in geographically and climatically diverse settings, from temperate to harsh desert environments.

As an example, Fig. 7.22 shows the model detects and classifies the defect (short-circuit, class D3), achieving an accuracy of 87%. The key advantage of the designed prototype, compared to currently available industrial devices, lies in its ability not only to detect hotspots but also to identify and classify specific types of defects. Most commercial systems can accurately locate overheated regions but are generally limited to binary detection without providing insights into the underlying fault category. Our prototype integrates classification capability into a lightweight and low-cost design, making it both practical and accessible. Owing to its compactness and affordability, it is particularly well-suited for domestic and small-scale PV installations, where reducing maintenance costs and ensuring safe operation are essential. In addition, it offers portability, allowing operators or technicians to perform on-site inspections quickly and efficiently without relying on expensive industrial equipment.

FIGURE 7.22 Comparison between the designed smart prototype and a commercial device. The figure presents a comparison between the designed smart prototype and a commercial device. It highlights the main differences in functionality and performance.

Despite these advantages, the current version of the system has limitations in accuracy and image resolution, which should be addressed in future iterations through higher-resolution sensors, advanced algorithms, and larger, more diverse datasets. With these improvements, the prototype could evolve into a scalable solution that bridges the gap between household applications and industrial PV monitoring systems. In this case study, an end-to-end prototype for PV defect diagnosis has been designed and implemented, with a lightweight deep learning model developed and deployed on embedded hardware. This work underscores the importance of transitioning from laboratory-scale research to real-world applications. It also highlights the need for researchers to balance model complexity with practicality, especially for real-time implementations. Highly complex models often require significant computational resources and raise implementation costs, which can delay their adoption in operational PV plants. By contrast, lightweight yet efficient approaches, such as the one presented here, pave the way toward cost-effective, scalable, and field-ready PV monitoring solutions.

The prototype was originally designed for ground-based inspection of PV plants; however, its applicability can be extended to aerial inspections through integration with UAVs. Mounting the developed system on a drone would enable large-scale and rapid inspection of PV installations, significantly reducing the time and labor required compared to manual ground inspections. This approach would be particularly useful for utility-scale PV plants, where thousands of modules need to be monitored under varying environmental conditions. Nevertheless, some key recommendations must be considered when adapting the prototype for aerial applications. In particular, the choice of the IR camera is critical, as higher resolution and frame rate may be required to capture reliable thermal images during drone flights at different altitudes and speeds. Similarly, the selection of the MCU should ensure sufficient processing capacity to handle real-time image analysis while maintaining low power consumption to maximize flight duration. Careful attention to weight, energy efficiency, and communication protocols will also be necessary to ensure smooth integration into UAV platforms. With these modifications, the prototype has strong potential to evolve into a versatile, drone-based monitoring system capable of enabling fast, accurate, and cost-effective fault detection and diagnosis in large-scale PV plants.

Challenges and solutions

The development of smart prototypes for automatic defect detection and fault diagnosis in PV systems faces several challenges. Among these, the most critical are the scarcity of large and well-annotated datasets of PV defects, particularly for rare or complex multiclass faults, and the difficulty of achieving high classification accuracy under varying environmental and operational conditions. Furthermore, embedded hardware platforms often have limited computational capacity, memory, and power budgets, which constrain the deployment of DL models in real-time. Another major challenge lies in scalability, as prototypes designed and tested at the laboratory scale are not easily transferable to industrial PV plants with thousands of modules and diverse field conditions. To address these challenges, several strategies are recommended. Collaborative initiatives should be encouraged to establish standardized and shared PV fault databases collected across different climatic zones and technologies.

Lightweight and efficient DL models, such as MobileNet and TinyML, should be prioritized to ensure that embedded devices can perform real-time fault detection while maintaining low energy consumption. Furthermore, combining complementary diagnostic methods, such as infrared thermography, electroluminescence imaging, and I–V curve tracing, can significantly enhance the robustness and accuracy of fault detection. Finally, continuous model updates with new datasets are essential to ensure adaptability to evolving environmental factors and newly emerging fault types. Practical solutions to overcome current barriers include leveraging transfer learning and data augmentation techniques to compensate for limited datasets, as well as deploying hybrid edge–cloud architectures to balance real-time inference with large-scale, centralized analysis. Incorporating explainable AI can also improve interpretability, thereby increasing operator trust in automated fault diagnosis (Dwivedi et al., 2023). Moreover, the integration of ultra-low-power edge devices with optimized ML algorithms can reduce latency and energy consumption, facilitating sustainable operation in large-scale PV plants. These solutions collectively contribute to bridging the gap between experimental prototypes and field-ready applications.

Conclusion

In this chapter, the principal categories of defects affecting PV power plants have been reviewed, together with the state-of-the-art methods used for anomaly detection and fault diagnosis. Particular emphasis has been placed on the potential of artificial AI techniques applied to I–V curve analysis and infrared thermography for the automated detection and classification of various types of PV faults. The selected case study has demonstrated the effectiveness of AI-based approaches in advancing PV fault diagnosis research and has provided concrete evidence of their practical applicability. Furthermore, a smart monitoring prototype has been designed and experimentally validated under diverse climatic conditions in Algeria, encompassing both Mediterranean and arid environments. The experimental results have confirmed the feasibility of the proposed approach and have indicated that the developed prototypes have reached a sufficient level of technological maturity to enable further optimization toward industrial deployment.

To support the transition from laboratory-scale research to large-scale industrial implementation, a structured roadmap has been outlined. The initial phase has involved validation under controlled laboratory conditions using small-scale PV arrays to assess model accuracy, stability, and robustness. Subsequently, pilot-scale field trials have been conducted to evaluate system performance under real operating conditions. The next stage has focused on scaling both hardware and software architectures to utility-scale PV plants, while ensuring interoperability with existing supervisory control and data acquisition and monitoring infrastructures. In parallel, efforts have focused on establishing standardized practices for data acquisition, fault labeling, and benchmarking of AI-based monitoring systems. The final phase of this roadmap has targeted the commercialization of cost-effective, scalable, and reliable smart monitoring solutions capable of enabling predictive maintenance, reducing operational costs, enhancing overall system efficiency, and ensuring the long-term safety and reliability of PV installations. Overall, this roadmap has emphasized the necessity of moving beyond laboratory demonstrations toward industrial-scale solutions, thereby bridging the gap between research innovation and practical deployment in the PV industry.

AI Disclosure

During the preparation of this work, the author(s) used Quillbot in order to proofread and assist with paraphrasing. After using this tool/service, the author (s) reviewed and edited the content as needed and take(s) full responsibility for the content of the publication.

References

Aghaei, M., Fairbrother, A., Gok, A., Ahmad, S., Kazim, S., Lobato, K., et al., 2022. Review of degradation and failure phenomena in photovoltaic modules. Renewable and Sustainable Energy Reviews 159, 112160.

Ahmad, J., Ciocia, A., Fichera, S., Murtaza, A.F., Spertino, F., 2019. Detection of typical defects in silicon photovoltaic modules and application for plants with distributed MPPT configuration. Energies 12 (23).

Ali, Y.M., Ding, L., 2025. Fault diagnosis in photovoltaic arrays: a robust and efficient approach using feature engineering and 1D-CNN. International Journal of Electrical Power and Energy Systems 170.

Alves Dos Reis Benatto, G., Mantel, C., Spataru, S., Santamaria Lancia, A.A., Riedel, N., Thorsteinsson, S., et al., 2020. Drone-based daylight electroluminescence imaging of PV modules. IEEE Journal of Photovoltaics 10 (3), 872–877.

Benghanem, M., Mellit, A., Moussaoui, C., 2023. Embedded hybrid model (CNN–ML) for fault diagnosis of photovoltaic modules using thermographic images. Sustainability 15 (10), 7811.

Bommes, L., Pickel, T., Buerhop-Lutz, C., Hauch, J., Brabec, C., Peters, I.M., 2021. Computer vision tool for detection, mapping, and fault classification of photovoltaics modules in aerial IR videos. Progress in Photovoltaics: Research and Applications 29 (12), 1236–1251.

Boubaker, S., Kamel, S., Ghazouani, N., Mellit, A., 2023. Assessment of machine and deep learning approaches for fault diagnosis in photovoltaic systems using infrared thermography. Remote Sensing 15 (6), 1686.

Chen, Y., Ma, M., Ma, W., Zhou, X., Zhang, R., Fang, Z., 2025). An adaptive four layer digital twin with segmented diode model for real-time fault diagnosis and output characterization of PV modules. 300 (113755).

Drir, N., Chekired, F., Mellit, A., Blasuttigh, N., 2025. Hybrid CNN-EML model for fault diagnosis in Electroluminescence images of photovoltaic cells. Renewable Energy.

Drir, N., Mellit, A., Blasuttigh, N., Pavan, A.M., 2026. Enhancing defect detection in photovoltaic cells through advanced cascade CNN with attention mechanisms on electroluminescence images. Energy Conversion and Management 348, 120636.

Dwivedi, R., Dave, D., Naik, H., Singhal, S., Omer, R., Patel, P., et al., 2023. Explainable AI (XAI): core ideas, techniques, and solutions. ACM Computing Surveys 55 (9), 1–33.

El-Banby, G.M., Moawad, N.M., Abouzalm, B.A., Abouzaid, W.F., Ramadan, E.A., 2023. Photovoltaic system fault detection techniques: a review. Neural Computing and Applications 35 (35), 24829–24842.

El-Otum, H.M., 2023. Deep learning-based automated defect classification in electroluminescence images of solar panels. Advanced Engineering Informatics 58.

Fadhel, S., Delpha, C., Diallo, D., Bahri, I., Migan, A., Trabelsi, M., et al., 2019. PV shading fault detection and classification based on I-V curve using principal component analysis: application to isolated PV system. Solar Energy 179, 1–10.

Figgis, B., Azid, S.I., Parlevliet, D., 2025. Review of unmanned ground vehicles for PV plant inspection. Solar Energy 291.

Garoudja, E., Chouder, A., Kara, K., Silvestre, S., 2017. An enhanced machine learning based approach for failures detection and diagnosis of PV systems. Energy Conversion and Management 151, 496–513.

Hijjawi, U., Lakshminarayana, S., Xu, T., Piero Malfense Fierro, G., Rahman, M., 2023. A review of automated solar photovoltaic defect detection systems: approaches, challenges, and future orientations. Solar Energy 266, 112186.

Huang, J.M., Wai, R.J., Gao, W., 2019. Newly-designed fault diagnostic method for solar photovoltaic generation system based on IV-curve measurement. IEEE Access 7, 70919–70932.

Jaffery, Z.A., Dubey, A.K., Irshad, Haque, A., 2017. Scheme for predictive fault diagnosis in photo-voltaic modules using thermal imaging. Infrared Physics and Technology 83, 182–187.

Kellil, N., Aissat, A., Mellit, A., 2023. Fault diagnosis of photovoltaic modules using deep neural networks and infrared images under Algerian climatic conditions. Energy 263, 125902.

Kellil, N., Moussaoui, C., Mellit, A., Boudiaf, A., 2024. YOLOv8-based photovoltaic module detection using aerial imagery. International Conference on Artificial Intelligence in Renewable Energetic Systems 321–329.

Kim, B., Serfa Juan, R.O., Lee, D.-E., Chen, Z., 2021. Importance of image enhancement and CDF for fault assessment of photovoltaic module using IR thermal image. Applied Sciences 11 (18), 8388.

Li, B., Delpha, C., Migan-Dubois, A., Diallo, D., 2021. Fault diagnosis of photovoltaic panels using full I–V characteristics and machine learning techniques. Energy Conversion and Management 248, 114785.

Lin, W.T., Chang, C.M., Huang, Y.C., Wu, C.C., Kuo, C.C., 2024. Fault diagnosis in solar array I-V curves using characteristic simulation and multi-input models. Applied Sciences (Switzerland) 14 (13).

Liu, Q., Shi, H., Zhu, Y., Chen, L., Liu, M., Cao, W., et al., 2025. An efficient fault diagnosis method combining multi-angle feature expansion and visual image neural networks for solar photovoltaic modules. Energy 333, 137484.

Ma, M., Zhang, Z., Yun, P., Xie, Z., Wang, H., Ma, W., 2021. Photovoltaic module current mismatch fault diagnosis based on I-V data. IEEE Journal of Photovoltaics 11 (3), 779–788.

Mellit, A., Kalogirou, S., 2022. Assessment of machine learning and ensemble methods for fault diagnosis of photovoltaic systems. Renewable Energy 184, 1074–1090.

Mellit, A., Kalogirou, S., 2025. Recent advances in the application of infrared thermographic imaging and embedded artificial intelligence for fault diagnosis and predictive maintenance of photovoltaic plants: challenges and future directions. Renewable and Sustainable Energy Reviews 223, 116057.

Mellit, A., Tina, G.M., Kalogirou, S.A., 2018. Fault detection and diagnosis methods for photovoltaic systems: a review. Renewable and Sustainable Energy Reviews 91, 1–17.

Mellit, A., Moussaoui, C., Pastore, S., Pavan, A.M., 2024. Fault diagnosis of photovoltaic modules using infrared images and hybrid DCNNs: a comparative study. In: 2024 IEEE International Conference on Environment and Electrical Engineering and 2024 IEEE Industrial and Commercial Power Systems Europe, pp. 1–66.

Mellit, A., Blasuttigh, N., Pastore, S., Zennaro, M., Pavan, A.M., 2025. TinyML for fault diagnosis of photovoltaic modules using edge impulse platform and ir thermography images. IEEE Transactions on Industry Applications 61 (5), 8158–8169.

Moussaoui, C., Mellit, A., Bouzerdoum, M., Herrak, O., 2022. Fault diagnosis of photovoltaic panels using a low-cost edge device. In: International Conference on Advanced Renewable Energy Systems, pp. 651–658.

Moussaoui, C., Bouzerdoum, M., Kellil, N., Mellit, A., 2024. YOLO-based photovoltaic panel detection: a comparative study. In: International Conference on Artificial Intelligence in Renewable Energetic Systems, pp. 312–320.

Nassreddine, G., El Arid, A., Nassereddine, M., Al Khatib, O., 2025. Fault detection and classification for photovoltaic panel system using machine learning techniques. Applied AI Letters 6 (2).

de Oliveira, A.K.V., Aghaei, M., Rüther, R., 2022. Automatic inspection of photovoltaic power plants using aerial infrared thermography: a review. Energies 15 (6).

Parvin, M., Yousefi, H., Mohammadi-Ivatloo, B., 2025. Photovoltaic fault detection algorithm using ensemble learning enhanced with deep neural network feature engineering. Results in Engineering 27, 106491.

Qin, D., Leichner, C., Delakis, M., Fornoni, M., Luo, S., Yang, F., et al., MobileNetV4: universal models for the mobile ecosystem. In: European Conference on Computer Vision, pp. 78–96.

Ren, B., Wang, Q., Zhou, N., Lu, J., Mekhilef, S., 2025. Fault diagnosis of photovoltaic systems based on global I-V features and similarity learning. International Journal of Electrical Power & Energy Systems 171, 110987.

S, N.V., Sripada, D., Sugumaran, V., Aghaei, M., 2024. Detection of visual faults in photovoltaic modules using a stacking ensemble approach. Heliyon 10 (6), e27894.

Shin, W., Lee, J.S., Ju, Y.C., Hwang, H.M., Ko, S., 2025. CNN-based photovoltaic fault diagnosis using normalized I–V curves with explainability analysis. Solar Energy 301.

Solórzano, J., Egido, M.A., 2013. Automatic fault diagnosis in PV systems with distributed MPPT. Energy Conversion and Management 76, 925–934. Available at https://doi.org/10.1016/j.enconman.2013.08.055.

Spataru, S., Sera, D., Kerekes, T., Teodorescu, R., 2015. Diagnostic method for photovoltaic systems based on light I–V measurements. Solar Energy 119, 29–44. Available at https://doi.org/10.1016/j.solener.2015.06.020.

Stauffer, Y., Ferrario, D., Onillon, E., Hutter, A., 2015. Power monitoring based photovoltaic installation fault detection. In: 2015 International Conference on Renewable Energy Research and Applications, ICRERA, pp. 199-202. Available at https://doi.org/10.1109/ICRERA.2015.7418695

Takashima, T., Yamaguchi, J., Ishida, M., 2008. Disconnection detection using earth capacitance measurement in photovoltaic module string. Progress in Photovoltaics: Research and Applications 16 (8), 669–677.

Tang, C., Ren, H., Xia, J., Wang, F., Lu, J., 2023. Automatic defect identification of PV panels with IR images through unmanned aircraft. IET Renewable Power Generation 17 (12), 3108–3119.

Tsanakas, J.A., Chrysostomou, D., Botsaris, P.N., Gasteratos, A., 2015. Fault diagnosis of photovoltaic modules through image processing and Canny edge detection on field thermographic measurements. International Journal of Sustainable Energy 34 (6), 351–372.

Yang, C., Sun, F., Zou, Y., Lv, Z., Xue, L., Jiang, C., et al., 2024. A survey of photovoltaic panel overlay and fault detection methods. Energies 17 (4), 837.

Zbib, B., Al Sheikh, H., 2020. Fault detection and diagnosis of photovoltaic systems through I-V curve analysis. In: 2nd International Conference on Electrical, Communication and Computer Engineering, ICECCE. Available at https://doi.org/10.1109/ICECCE49384.2020.9179390 http://ieeexplore.ieee.org/xpl/mostRecentIssue.jsp?punumber=9169751

Chapter 8

Modern maintenance strategies for photovoltaic power plants

Rokia Chekirou[1], Adel Mellit[1], and Giuseppe M. Tina[2]
[1]*Faculty of Sciences and Technology, University of Jijel, Jijel, Algeria,* [2]*Department of Electrical, Electronic and Computer Engineering, University of Catania, Catania, Italy*

Introduction

Photovoltaic (PV) power plants are becoming increasingly popular as environmentally friendly power sources that decrease greenhouse gas emissions and support the generation of renewable energy. However, PV systems face several challenges such as normal wear and tear, component failures (including defects and problems like broken modules, hotspots, inverter issues, and misaligned trackers), unfavorable weather conditions (like wind, snow, sand, and dirt accumulation), and other issues (like repairing tracker failures, tightening loose cable connections, changing fuses, and resolving SCADA errors) (Hamza et al., 2025), that result in decreased efficiency and output deficiencies frequently so it affect their performance. These challenges underscore the critical need to develop efficient maintenance plans that ensure the systems' longevity and optimal performance, prevent frequent breakdowns, and mitigate their long-term effects (Osmani et al., 2020; Pingulkar and Joshi, 2021).

Maintenance plays a pivotal role in preserving system performance, improving reliability, and minimizing economic losses associated with unplanned failures. Without a systematic maintenance plan, PV plants can experience rapid degradation, reduced power output, and increased operational costs. Consequently, maintenance activities are now regarded as a strategic function rather than a secondary operational task. Modern PV maintenance can be broadly categorized into three main strategies: corrective, preventive, and predictive maintenance (PdM). Corrective maintenance addresses failures after they occur, whereas preventive maintenance focuses on scheduled inspections and repairs to avoid breakdowns. Predictive maintenance, on the other hand, uses advanced technologies such as artificial intelligence (AI), machine learning (ML), and the Internet of Things (IoT) to

Embedded Artificial Intelligence and the Internet of Things for Photovoltaic Systems.
DOI: https://doi.org/10.1016/B978-0-443-44835-5.00012-6

predict potential failures before they happen. This proactive approach allows operators to act in advance, reducing downtime and maintenance costs while enhancing overall system efficiency. Recent advances in digital twins, remote monitoring, and data-driven analytics have revolutionized the PdM of PV power plants. These technologies enable continuous supervision of module performance, early fault detection, and precise diagnosis of degradation mechanisms. The integration of AI and IoT platforms has made PV maintenance more intelligent, automated, and cost-effective, allowing for optimal decision-making and efficient resource allocation (Sepulveda-Oviedo et al., 2023).

Overall, as PV systems become larger and more interconnected, the transition from traditional maintenance practices toward intelligent, AI-driven strategies is essential (Mellit et al., 2024). Such approaches ensure long-term reliability, higher energy yield, and better financial sustainability, key elements for the success of modern PV plants. Table 8.1 presents an overview of existing research on PV plant maintenance strategies, highlighting different approaches employed to evaluate the maintenance of PV systems.

The reviewed studies summarized in Table 8.1 emphasize the diversity and growing importance of maintenance strategies in PV plants. Various approaches have been proposed, each contributing differently to system reliability and performance optimization. While these approaches demonstrate significant progress, their effectiveness often depends on the level of system complexity, data availability, and operational conditions. The increasing diversity of PV technologies and environments highlights the need for adaptable and intelligent maintenance frameworks capable of integrating real-time monitoring, IoT, and AI-based analytics. Developing such advanced strategies will strengthen system reliability, improve operational efficiency, and ensure long-term sustainability across modern PV and renewable energy infrastructures.

This chapter presents maintenance strategies for PV power plants, encompassing traditional and advanced approaches. It examines corrective, preventive, and predictive maintenance and compares their suitability across different plant scales. Preventive maintenance is discussed through time- and condition-based methods, while predictive maintenance is explored using manual diagnostics, analytical models, and AI-driven techniques. Key inspection and cleaning methods, such as visual inspection, infrared (IR) thermography, unmanned aerial vehicle (UAV)-based monitoring, and automated cleaning, are evaluated in terms of effectiveness and feasibility. Current challenges, including data availability, cost, and model interpretability, are highlighted. Finally, emerging technologies such as Edge AI, TinyML, Artificial Intelligence of Things (AIoT), large language models (LLMs), and multimodal are presented as promising solutions, along with recommendations for enhancing PV plants' reliability and operational efficiency.

TABLE 8.1 **Overview of existing studies on PV power plants' maintenance strategies.**

Ref.	Approach used	Complexity	Findings and contributions
Villarini et al. (2017)	Failure mode and effects analysis (FMEA)	Medium	Proposed a real-data-based FMEA approach to optimize PV maintenance planning, enabling more reliable, cost-effective, and risk-prioritized maintenance strategies.
De Benedetti et al. (2018)	ML (ANN)-based PdM approach.	High	Developed an intelligent neural network–based anomaly detection system that compares predicted and actual PV outputs, using residual trend analysis (TMA) to enable accurate PdM with over 90% detection accuracy. It supports daily PdM alerts, improving reliability and reducing unexpected downtime.
Chen et al. (2018)	Random forest (RF)	Medium-high	Presented an optimized RF–based ensemble model for fault detection and diagnosis in PV arrays, using only voltage and current data to achieve high accuracy, strong generalization, and real-time applicability for intelligent maintenance.
Betti et al. (2019)	DL (vision transformer (ViT), AI, and CNN comparative study)	High	Proposed a ViT–based intelligent system for PdM of PV modules, achieving high anomaly detection accuracy (AUC = 0.96) and supporting smart city energy management through automated, web-integrated analysis.

(Continued)

TABLE 8.1 (Continued)

Ref.	Approach used	Complexity	Findings and contributions
Aremu et al. (2020)	Feature selection + ML.	Medium-high	proposed a new feature selection framework (C-RE) designed specifically for asset data in PdM. It efficiently reduces high-dimensional sensor data while preserving fault-related information, leading to better performance of ML models and more interpretable maintenance insights.
Singh and Chandra (2022)	PdM using real-time sensor data	Meduim	Proposed a PdM approach for solar PV systems, using sensor data to monitor energy yield and degradation, enabling timely maintenance to ensure long-term performance and investment efficiency.
Simon, Teixeira and da Costa (2022)	Timed Petri Net (TPN)–based simulation modeling	Medium-high	Proposed a TPN–based simulation model to predict PV system availability in advance, enabling planning, PdM, and performance optimization without requiring full-scale physical monitoring.
Toubeau et al. (2022)	ML and RF	Medium-high	Proposed an ML-assisted outage planning framework using RF to predict safe maintenance windows in renewable-rich power systems, reducing computational effort while maintaining grid security.

(*Continued*)

TABLE 8.1 (Continued)

Ref.	Approach used	Complexity	Findings and contributions
Kingsley-Amaehule et al. (2022)	IoT-based intelligent monitoring and maintenance management	High	Developed a holistic IoT-based intelligent monitoring and maintenance system for PVs, combining embedded expert algorithms with fog and cloud data collection to achieve high reliability, real-time fault detection, and effective management.
Nasrfard et al. (2023)	Petri Net (PN) modeling	Medium-high	Proposed a flexible PN–based model to optimize inspection and preventive maintenance rates, reduce total costs, and support integration with asset management tools, demonstrated on circuit breakers of the RBTS.
Joseph et al. (2023)	Data-driven fault detection	Medium	Developed a real-time, data-driven PV monitoring system that detects encapsulation failures and cell corrosion, provides user-friendly alerts, and supports remote access using a database of normal operating conditions.
Sepulveda-Oviedo et al. (2023)	RF/ML	Medium-high	Developed an RF-based method with advanced feature extraction for detecting fine PV faults, improving computational efficiency while maintaining high accuracy, and supporting preventive maintenance.

(*Continued*)

TABLE 8.1 (Continued)

Ref.	Approach used	Complexity	Findings and contributions
Abdulla et al. (2024); Hamza et al. (2024)	LSTM/K-Means clustering	High	Proposed an LSTM with K-Means clustering for PV system anomaly detection, improving accuracy by focusing on key attributes and supporting PdM with tabRUL estimation.
Kumar and Nayak (2024)	CUSUM-based detection	Medium	proposed a CUSUM-based method for rapid detection and localization of electrical faults in large-scale PV arrays, achieving detection within 5 ms and improving system safety.
Amiri et al. (2024)	ML (RF, ANN, SVM, DT, logistic regression)	Medium-high	Implemented and compared multiple ML models, including RF and ANN, for detecting partial shading and dirt faults in PV systems, achieving high accuracy while highlighting the need for system-specific training.
Al-Humairi et al. (2025)	AI and ML	Medium-high	Developed an AI-based robotic cleaning system for PV modules that predicts cleaning needs, reduces unnecessary maintenance by 30%, improves energy yield by 3%–5%, and minimizes manual intervention using optimal ML models.
Ali et al. (2025)	Optimization-based preventive maintenance	Medium-high	Proposed a reliability-driven optimization framework for PV maintenance and replacement that reduces costs, improves equipment availability, and enhances system sustainability and efficiency.

AI, Artificial intelligence; *CNN*, convolutional neural network; *LSTM*, long short-term memory; *ML*, machine learning; *PV*, photovoltaic; *PdM*, predictive maintenance; *RUL*, remaining useful life.

Maintenance strategies

Maintenance of solar PV power plants means keeping the solar system operating safely and efficiently by regularly inspecting, cleaning, and repairing its components. This includes scheduled procedures to prevent problems, fix malfunctions when they occur, and use monitoring data to predict and address problems early. The goal is to reduce energy losses, extend equipment life, and maintain consistent energy production. In order to deliver the energy expected throughout its lifetime, PV modules must be periodically revised using technologies capable of detecting different kinds of failures (Ren and Man, 2022). Maintenance of PV plants can be classified globally into two key categories: (1) traditional and (2) modern maintenance (See Fig. 8.1). Traditional maintenance methods include corrective or reactive maintenance and preventive maintenance (e.g., time-based maintenance), while modern maintenance includes preventive maintenance (e.g., condition-based maintenance) and predictive maintenance.

Corrective maintenance

Corrective maintenance is reactive in nature (Ledmaoui et al., 2025), being performed only after a fault, malfunction, or performance degradation is detected in a PV plant (See Fig. 8.2). It was started in 1980. This maintenance strategy involves interventions by maintenance or service teams to restore

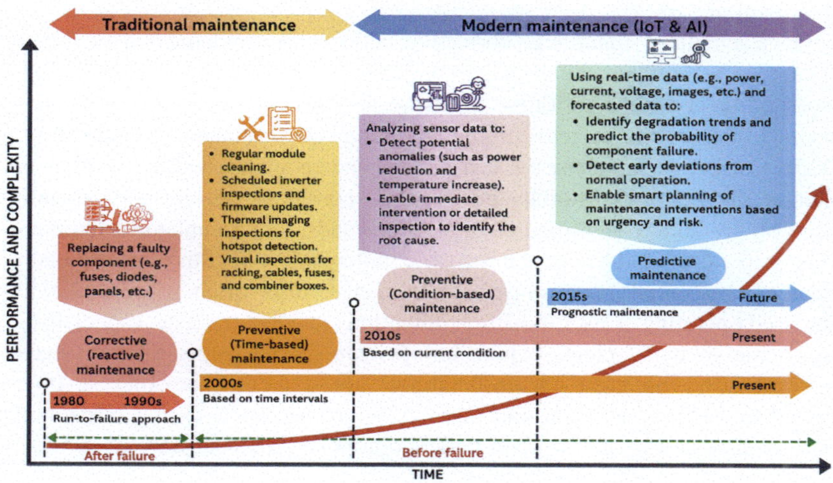

FIGURE 8.1 Timeline of maintenance methods for PV power plants. The figure shows a timeline of maintenance methods for PV power plants, illustrating how approaches have evolved over time. It highlights the progression from basic manual inspections to more advanced, technology-driven techniques (*PV*, Photovoltaics).

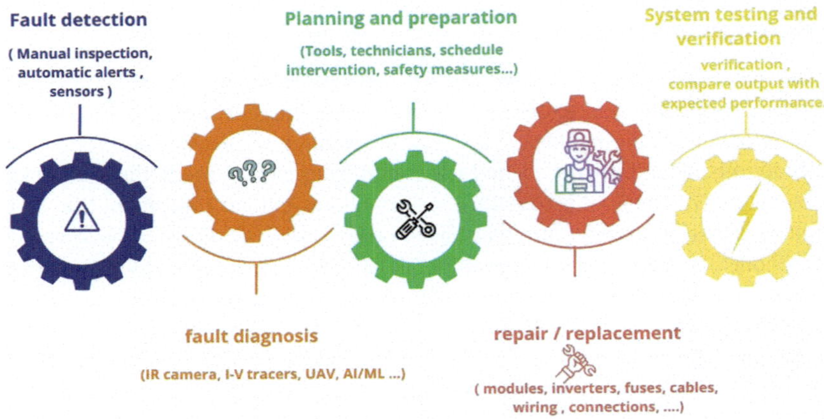

FIGURE 8.2 Basic block diagram of the corrective maintenance method. The figure shows a simple flowchart of the corrective maintenance method, illustrating how a fault is detected, diagnosed, and then repaired. It emphasizes the reactive process of fixing equipment after a failure occurs.

normal operation of a PV power plant that has been partially or completely shut down due to component or system failures. These failures can occur at various levels of the PV plant, from individual modules, strings, and inverters to data collection systems, communication interfaces, and grid connection points. The process typically begins with fault detection, either through automated monitoring systems or manual inspections (Osmani et al., 2020; Abdulla et al., 2024).

Once a fault is identified, corrective actions are implemented to repair, replace, or reconfigure the affected components. In some cases, experts may also modify control algorithms, firmware settings, or communication parameters to ensure the system returns to optimal performance (Osmani et al., 2020). Corrective maintenance can be planned or unplanned, depending on whether the fault was anticipated. Planned corrective actions are typically based on previously identified deterioration trends, while unplanned interventions respond to sudden or catastrophic failures. While this type of maintenance is essential for restoring performance, it is often associated with increased operating costs, production downtime, and lost revenue, especially in large-scale solar PV plants, where failures can spread across interconnected subsystems. In practical PV installations, corrective maintenance is often applied to address electrical and thermal failures detected during inspection. As illustrated in Fig. 8.3, several typical component defects were observed.

- The fuse holders inside the Direct Current (DC) distribution box were found to be overheated and partially melted due to excessive current flow and insufficient terminal contact. When the damage was minor, corrective actions involved cleaning the contact surfaces, tightening loose terminals,

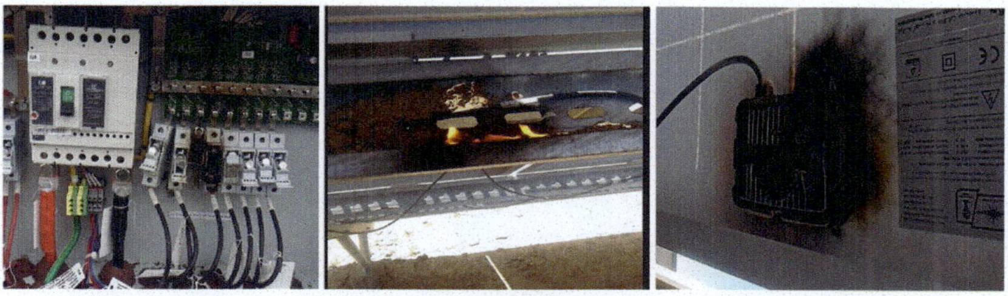

(A) Damaged fuses (B) Overheated PV cables (C) Burned junction box

FIGURE 8.3 Examples of some typical component failures requiring corrective maintenance (South of Algeria). The figure displays examples of common component failures in PV plants in southern Algeria, highlighting parts that often break or malfunction. It visually emphasizes which components typically need corrective maintenance (*PV*, Photovoltaics).

and verifying current continuity. However, in cases of severe melting or deformation, the fuses and holders were replaced to restore electrical safety.

- The PV cables exhibited insulation degradation and discoloration caused by overcurrent or improper crimping. If the insulation was only slightly affected, the damaged sections were cut and re-insulated, whereas severely burnt cables were completely replaced.
- Another recurrent fault involved burned junction boxes (J-boxes) on the rear side of PV modules, mainly resulting from poor bonding, loose internal connections, or moisture ingress, which induced internal arcing and local overheating. Corrective actions in this case consisted of resealing or replacing the J-box, followed by insulation resistance and IR testing to confirm system integrity.

Preventive maintenance

In prefailure mode, preventive maintenance is applied to mitigate potential larger damages that could result from probable future faults (Osmani et al., 2020). It involves carrying out routine inspections, repairs, and component replacements at fixed intervals (Ledmaoui et al., 2025). This strategy is performed in a planned or condition-based manner before an actual failure occurs. Fig. 8.4 illustrates the general principle of the preventive maintenance process.

Fuses and PV modules are among the main components inspected during preventive maintenance to detect early signs of malfunction and ensure reliable system operation. For fuses, preventive maintenance involves regular visual and electrical inspections to ensure proper operation and prevent potential electrical faults. As illustrated in Fig. 8.5, fuse holders within the DC distribution box are examined for discoloration, surface oxidation, or loose terminals, which may indicate overheating, poor contact, or excessive current

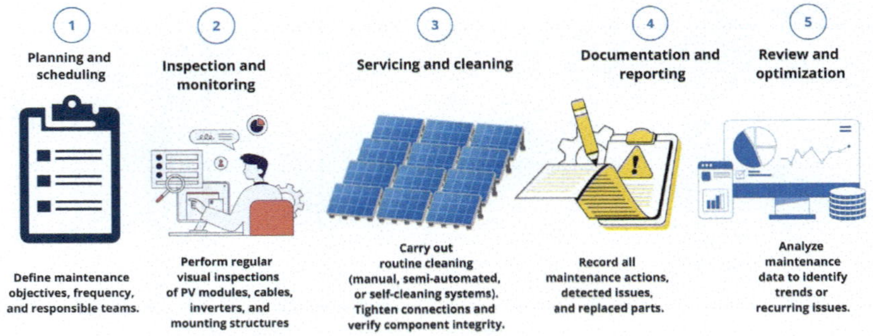

FIGURE 8.4 Basic block diagram of the preventive maintenance strategy. The figure shows a simple diagram of the preventive maintenance process, illustrating scheduled inspections and routine servicing designed to prevent equipment failures before they occur.

(A) Fuses inspection **(B) PV module inspection using I-V tracer** **(C) PV module inspection using IR thermal camera**

FIGURE 8.5 Examples of preventive inspection of a PV power plant. The figure illustrates examples of preventive inspections in PV power systems, showing routine checks on modules, wiring, and other components to detect issues before failures occur. It emphasizes proactive maintenance activities (*PV*, Photovoltaic).

flow. Minor issues, such as slight oxidation or dust accumulation, are typically corrected by cleaning and tightening the affected terminals to restore optimal conductivity (Massi Pavan et al., 2011). However, when signs of melting or deformation are observed, the fuses and holders are replaced to maintain electrical safety and system reliability. This systematic inspection approach allows early detection of degradation and prevents potential failures in the PV power distribution network.

For PV modules, preventive inspection integrates current–voltage (I–V) curve tracing and IR to assess electrical and thermal performance. As shown in Fig. 8.5, the I–V tracer allows the detection of anomalies such as current mismatch, diode failure, or degradation of cell efficiency. Meanwhile, thermographic imaging, depicted in Fig. 8.5, identifies localized hotspots and uneven temperature patterns, which are early indicators of defective cells or poor interconnections. These complementary diagnostic techniques provide a reliable, data-driven foundation for preventive maintenance, enabling timely interventions and extending the operational lifespan of the PV system.

Preventive maintenance can be classified into two groups: (1) time-based and (2) condition-based. Both aim to ensure the reliability, efficiency, and longevity of the PV system, yet they differ in their implementation strategies. While time-based maintenance relies on fixed schedules and routine inspections, condition-based maintenance depends on real-time system data and performance analysis to determine when intervention is necessary.

Time-based maintenance

Time-based or scheduled maintenance refers to routine, scheduled maintenance activities performed at specific intervals, regardless of the current condition of the PV plant components. The primary goal is to prevent performance degradation and ensure the system continues to operate within expected standards. Typical scheduled maintenance tasks include:

- Regular cleaning of modules to remove dust, dirt, and bird droppings that may reduce solar radiation and energy production.
- Scheduled inverter inspections and firmware updates to maintain operational reliability and software compatibility.
- Thermal imaging inspections to identify potential weaknesses or partial shading effects that may indicate cell incompatibility or degradation.
- Visual inspections of mounting structures, electrical connections, junction boxes, and cables to detect any physical damage, corrosion, or loose connections.

These planned interventions help reduce unexpected failures, ensure stable energy production, and extend the lifespan of the PV plant. Despite its simplicity and widespread use, this method has limitations: it can lead to unnecessary interventions (leading to increasing maintenance costs) or missed and sudden failures between inspection intervals. For this reason, many modern PV plants are gradually integrating time-based maintenance with condition-based or predictive maintenance strategies to enhance efficiency and reduce costs.

Condition-based maintenance

Condition-based maintenance focuses on monitoring the condition and performance indicators of the PV plant in real-time to detect early signs of malfunction or deterioration. Instead of following a fixed schedule, maintenance actions are triggered only when sensor data indicates potential problems, making this approach more predictive and cost-effective. This technique relies on continuously collecting data from various sensors and monitoring systems to analyze parameters such as:

- Power output fluctuations may indicate problems such as partial shading, dirt, or module deterioration.
- Abnormal temperature increases, particularly at the module or inverter level, may indicate electrical faults, hot spots, or ageing components.
- Voltage or current imbalances between wires may indicate potential wiring faults or connection problems.

In practical applications, this data can be visualized using SCADA systems or electronic monitoring platforms. The visualization method depends on the type of SCADA architecture implemented, centralized, hybrid, distributed, or cloud-based. When a fault is detected, immediate intervention or a detailed inspection can be performed to determine the root cause of the fault. Integrating AI or ML models can further enhance condition-based maintenance by enabling automatic fault detection, classification, and predictive analytics, reducing downtime and maintenance costs.

However, implementing these approaches requires continuous monitoring, reliable data acquisition, and advanced analytics infrastructure, which can increase both complexity and cost. The installation and maintenance of sensors add to the initial investment, and the interpretation of large volumes of data may sometimes result in false alarms if the underlying algorithms are not adequately trained. Additionally, the transmission and storage of monitoring data demand robust cybersecurity measures and reliable communication protocols to ensure system integrity and prevent unauthorized access. Despite these limitations, the benefits of early fault detection, optimized maintenance scheduling, and extended system lifetime make these strategies increasingly essential for modern PV plant management.

Predictive maintenance

Predictive maintenance represents a more advanced form of condition-based maintenance. It aims to anticipate equipment failures before they occur while optimizing maintenance schedules to balance service frequency, system availability, and operating costs. This approach leverages real-time and forecasted data, including power output, current, voltage, radiation, temperature, thermal and optical images (Hamza et al., 2025) (See Fig. 8.6), to:

- Identify deterioration trends and predict the likelihood of component failure.
- Early detection of deviations from normal operation.
- Enable intelligent planning of maintenance interventions based on urgency and risk.

Predictive maintenance relies on the continuous collection of data from PV system sensors and measuring devices, combined with AI and IoT technologies for advanced data analysis. These tools help determine the optimal

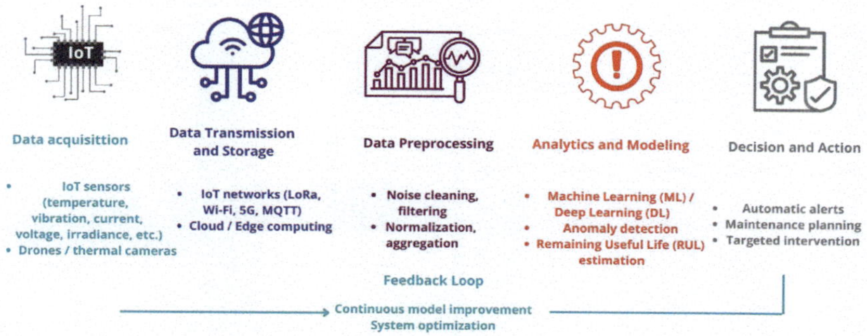

FIGURE 8.6 Basic block diagram of the predictive maintenance strategy. The figure shows a basic flowchart of the predictive maintenance process, illustrating how data is collected, analyzed, and used to anticipate and prevent equipment failures before they happen.

timing of maintenance procedures, maximizing equipment uptime, reducing operating costs, and improving overall system reliability (Ghazali et al., 2021).

Comparison between maintenance strategies

The effectiveness of PV system operation largely depends on the maintenance strategy adopted. Over time, maintenance practices have evolved from reactive interventions to advanced predictive approaches enabled by real-time monitoring and data analytics. Each method differs in terms of implementation requirements, reliability outcomes, and cost implications. To provide a structured comparison, Table 8.2 summarizes the key characteristics, advantages, limitations, and applications of the main maintenance strategies applied to PV power systems.

Based on the above Table 8.2, the following comments can be highlighted:

- Predictive maintenance is the most advanced option for an efficient maintenance strategy, which gives a company the ability to better anticipate or handle unscheduled downtime resulting from both preventable and unforeseen failures. According to a recent study, predictive maintenance enhances productivity, safety management skills, and operational cost reductions. Additionally, predictive maintenance restricts needless maintenance efforts because it has been shown to offer a better understanding of the cause of malfunction or failure (Aremu et al., 2020).
- In order to perform condition monitoring and deliver information on actual circumstances, Predictive maintenance often needs a large number of sensors. Although predictive maintenance appears to be a strong and suitable technique for maintaining solar plants, implementation gaps still exist, such as the high cost of gathering the required input data (Syamsuddin et al., 2024).

TABLE 8.2 Comparative overview of maintenance strategies applied to PV plants.

Maintenance method	Principle	Advantages	Disadvantages	Example in PV systems
Corrective maintenance (reactive)	Repair or replace a component only after it fails.	Simple to implement, no prior monitoring cost	Unexpected downtime. High repair costs. Risk of severe damage.	Replace an inverter only when it breaks down. Replace damaged fuses.
Preventive Maintenance (scheduled-based)	Interventions are scheduled at fixed intervals (time-based or usage-based).	Reduces sudden failures. More reliable than corrective maintenance.	May lead to unnecessary interventions. Higher costs if components are still in good condition.	Cleaning PV modules every 6 months, even if not required. Check every component of PV systems
Preventive maintenance (Condition-based)	Maintenance is triggered according to the actual measured condition of the equipment.	Optimizes timing of interventions. Avoids unnecessary downtime.	Requires sensors and real-time monitoring. Higher initial investment.	Checking cable insulation only when measurements indicate degradation.
Predictive maintenance	Uses data analysis (statistics and AI) to predict future failures.	Anticipates failures before they occur. Reduces corrective costs. Improves system availability and lifespan.	Requires reliable and large datasets. Higher analytical complexity. -nitial implementation cost.	Using voltage, current, and temperature data to predict module or inverter failure.

PV, Photovoltaic.

- Furthermore, the predictive maintenance architecture used in solar power generation is intricate and necessitates a substantial amount of data for the training and testing procedures. Sensor data input variability is also considerable when it comes to electrical output, which is prone to fluctuations (Chang and Han, 2024). When it comes to generalization, predictive maintenance models typically have a tendency to be location-specific, which means that they cannot be used for PV systems in different locations because the data input features vary when predictive maintenance models are constructed (Syamsuddin et al., 2024).

Current approaches and opportunities for photovoltaic predictive maintenance

As PV systems expand worldwide, ensuring their long-term reliability and cost-effectiveness requires more advanced maintenance strategies. Traditional approaches, such as corrective and preventive maintenance, are increasingly complemented or replaced by predictive techniques that rely on real-time monitoring and data analytics. Against this background, the following section examines the current approaches and emerging opportunities for evaluating PV systems for potential performance degradation including: manual diagnostics such as visual inspection, I–V curve analysis and IR thermography, as well as analytical methods like UAV detection, failure modes and effects analysis, machine learning and forecasting, deep learning and real-time sensor monitoring (Bosman et al., 2020) (Fig. 8.7).

Predictive maintenance based on manual diagnostics

Visual inspection

It remains the most fundamental method, enabling the detection of surface-level issues such as cracks, corrosion, discoloration, or physical damage to PV components (Mellit et al., 2018). It can also be used to detect defects (such as snow, fallen leaves, bird droppings, dust, or other debris) that may have accumulated on the modules (Bosman et al., 2020). Despite its simplicity and low cost, this method relies heavily on human observation, which makes it time-consuming, sometimes monotonous, and not always reliable (Abdulmawjood et al., 2018). Moreover, visual inspection is limited to detecting visible defects and cannot reveal hidden electrical or thermal faults.

Current–voltage curve tracing

Electrical testing techniques, particularly I–V curve tracing, provide a deeper assessment of PV system behavior by capturing the I–V characteristics of modules and strings. This well-established diagnostic method enables the detection of shading-related faults, mismatches, connection problems, and

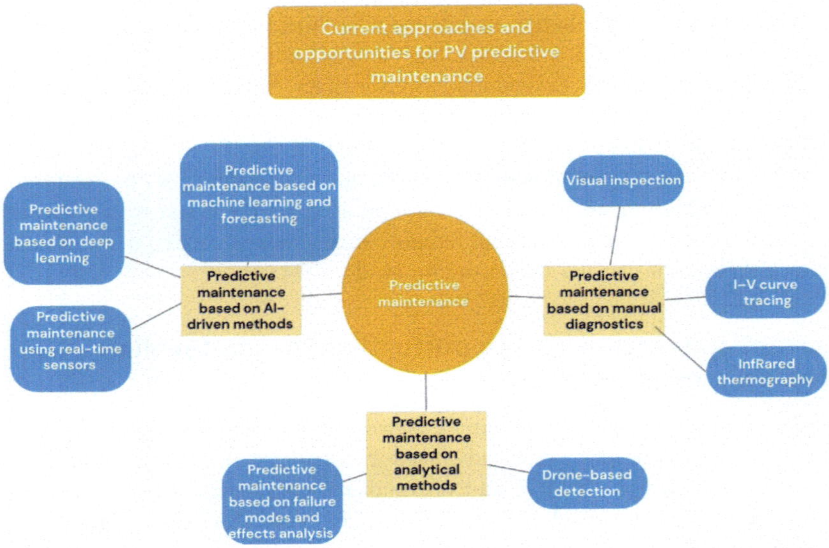

FIGURE 8.7 Current approaches and opportunities for PV predictive maintenance. The figure presents the current approaches and opportunities in PV predictive maintenance, highlighting methods such as machine learning and forecasting, deep learning, and real-time sensor monitoring, along with potential areas for improvement and optimization (*PV*, Photovoltaic).

aging effects by comparing the measured parameters with those defined under standard test conditions. Moreover, it helps obtain and verify that the electrical features of different PV module technologies correspond to their expected theoretical values (Ren and Man, 2022). Another technique is based on the comparisons among measured I-V curves and reference I-V curves in the same operating conditions generated by numerical models (e.g., Neuro-Fuzzy approach) (Bonsignore et al., 2014).

Infrared thermography

It has become a valuable noninvasive diagnostic tool, capable of detecting hotspots, defective cells, or faulty bypass diodes by visualizing temperature variations across PV modules. This method is advantageous as it can be applied to both small-scale solar cells and full-sized PV modules (Goudelis et al., 2022). In large PV plants, it is often combined with UAVs to enable faster and more accurate inspections, though at a higher cost (Bosman et al., 2020). Although effective in identifying early degradation, its accuracy can be influenced by environmental conditions such as irradiance and wind. It also faces challenges related to image quality, which depends on UAV altitude, speed, and observation angle. Another limitation is the lack of standardized pass/fail criteria for interpreting thermal images (Bosman et al., 2020).

Predictive maintenance based on analytical methods

Analytical methods leverage data-driven insights to provide a deeper understanding of system performance. Performance comparison between expected and actual generation offers a simple way to detect underperformance, though it does not identify the root cause.

Unmanned aerial vehicle-based detection

In the domain of PV predictive maintenance, the use of UAVs equipped with thermal imaging cameras has significantly transformed inspection practices. Unlike traditional ground-based methods, aerial surveys allow for rapid, large-scale, and high-resolution assessments of PV plants. By capturing thermal images from above, UAVs can highlight subtle temperature anomalies across modules, which may indicate hotspots, microcracks, defective cells, or connection failures that are often invisible to the naked eye. This capability enables maintenance teams to detect and localize faults with remarkable precision and efficiency. Moreover, UAV-based thermal inspections reduce the need for manual labor, minimize downtime, and enhance safety by limiting human exposure to hazardous conditions. When combined with advanced data analytics and ML models, the technique not only identifies existing faults but also predicts the likelihood of future failures, thus facilitating proactive interventions. Overall, UAV-enabled thermographic monitoring represents a versatile and cost-effective approach to enhancing the reliability, performance, and lifespan of PV power systems (Ledmaoui et al., 2025).

Predictive maintenance based on failure modes and effects analysis

Failure modes and effects analysis (FMEA) is a preventive, bottom-up analytical method designed to enhance system reliability by identifying and evaluating potential failure modes at the component level (Syamsuddin et al., 2024). It systematically examines each part of a system to determine possible faults, their causes, and their effects, allowing engineers to assess the criticality of each failure mode based on its severity, frequency, and detectability. Within predictive maintenance, FMEA supports the prioritization of maintenance actions to address the most critical risks, thereby minimizing downtime and improving overall system performance.

In PV applications, FMEA is used to identify, prioritize, and mitigate potential failure modes across key components such as PV modules, inverters, junction boxes, and wiring. Each failure mode is assessed through the calculation of a risk priority number, which helps rank maintenance priorities. By integrating real operational data with FMEA, PV operators can anticipate degradation mechanisms such as potential induced degradation, inverter malfunctions, or connection faults, and design cost-effective maintenance schedules focused on high-risk elements. Furthermore, the application of the FMEA methodology to optimize maintenance planning based on real-world operational data has proven effective in improving system reliability, extending service life, and supporting data-driven decision-making in PV maintenance strategies (Ali et al., 2025). When combined with reliability centered maintenance frameworks or ML algorithms, FMEA further enhances predictive capacity and enables more dynamic and evidence-based risk assessments of complex PV systems.

Predictive maintenance based on artificial intelligence-driven methods

Predictive maintenance based on machine learning and forecasting

ML and forecasting techniques have emerged as powerful tools for predictive maintenance in PV systems, enabling early fault detection and performance prediction beyond the capabilities of manual diagnostics or traditional statistical models. By leveraging large datasets collected from sensors and monitoring systems, such as irradiance, temperature, voltage, current, and power output. Major ML algorithms used in PV applications include support vector machine, linear regression, k-nearest neighbors, logistic regression, decision trees, random forest, naïve Bayes, and neural networks (Mellit and Kalogirou, 2022). These algorithms can identify hidden patterns and correlations that signal early stages of degradation or impending failures. Supervised learning approaches, such as Support Vector Machine (SVM), decision trees, and neural networks, are commonly applied to classify normal versus faulty operating states, while unsupervised methods, such as clustering and anomaly detection, help identify irregular behaviors without prior labeling.

Forecasting methods, including time-series models and DL techniques, allow operators to anticipate future performance trends and potential degradation pathways. For instance, recurrent neural networks and long short-term memory models can be used to predict PV power generation and detect deviations from expected values, serving as an early warning of faults. These predictive insights support optimized maintenance scheduling, reducing unplanned downtime and extending system lifetime.

In practice, ML and forecasting are applied to various PV components. Examples include predicting inverter failures through temperature and load data, forecasting module degradation due to soiling or potential induced degradation, and detecting anomalies in string-level performance. While these methods offer high accuracy and adaptability, challenges remain regarding the availability of high-quality training data, computational requirements, and the interpretability of complex models. To address these limitations, hybrid approaches combining ML with domain knowledge (e.g., physical models and FMEA) are increasingly explored, providing a more robust framework for predictive maintenance in PV systems.

Predictive maintenance based on deep learning

DL has become one of the most promising methods for predictive maintenance in PV power plants. Deep neural networks, such as CNNs, are capable of processing vast amounts of historical and real-time data, including temperature profiles, power generation records, irradiance, and environmental conditions. By identifying complex, nonlinear patterns and subtle anomalies that traditional analytical methods may overlook, these models enable accurate predictions of system degradation or failures (Bosman et al., 2020). In practice, DL has been successfully applied to classify defects using data from UAVs equipped with thermal and visual cameras, IoT-based monitoring platforms, and panel-level sensors. Such models not only provide early warnings of potential issues but also predict the optimal time for intervention, thus reducing downtime, minimizing energy losses, and enhancing the overall reliability and efficiency of PV plants.

Predictive maintenance using real-time sensors

Real-time sensor monitoring represents one of the most advanced approaches to predictive maintenance in PV systems. Unlike manual diagnostics or periodic inspections, this method provides continuous, high-resolution data on critical operating parameters such as irradiance, ambient and module temperature, current, voltage, power output, and in some cases, humidity, vibration, soiling, and acoustic signals. By capturing these variables in real time, operators can detect deviations from normal behavior immediately and apply predictive analytics to anticipate failures before they occur. This approach is particularly effective for identifying issues that evolve dynamically, such as inverter overheating, string mismatches, or early-stage module degradation. For example, temperature and current sensors can detect localized hotspots or abnormal electrical imbalances, while irradiance and power sensors allow for performance ratio calculations to identify efficiency losses. When integrated with the IoT platforms, these sensor networks enable remote monitoring, data aggregation, and automatic alerts, significantly reducing the response time for maintenance actions.

Applications of real-time sensor monitoring in PV predictive maintenance include early detection of potential induced degradation, real-time inverter fault diagnostics, and soiling detection for scheduling cleaning cycles. Advanced deployments may also use fiber-optic sensors or embedded microsensors within PV modules to capture highly localized conditions. Although real-time sensors provide the highest accuracy and fastest response for predictive maintenance, they come with notable challenges. High initial installation costs, system integration complexity, and the need for cybersecurity measures in IoT-based networks can limit widespread adoption. Nevertheless, as sensor costs decline and communication technologies improve, real-time monitoring is expected to become a cornerstone of predictive maintenance strategies in large-scale PV plants.

Table 8.3 presents a comparison of different predictive maintenance methods for PV systems, showing their principles, advantages, limitations,

TABLE 8.3 Comparative summary of predictive maintenance methods.

Method	Principle	Advantages	Limitations	Cost	Accuracy
Visual inspection	Direct observation of PV components	Simple, low cost	Limited to visible/ surface defects	Low	Low
I–V curve analysis	Electrical measurement of PV strings/modules	Detects mismatches, shading, and connections	Time-consuming, equipment needed	Medium	Medium
Thermal imaging (handheld)	IR camera to detect hotspots on modules	Noninvasive, easy to deploy	Small-scale, operator-dependent	Medium	Medium
UAV-based thermography	UAV-mounted IR for large PV fields	Fast, scalable, covers large plants	Weather-dependent, high equipment cost	High	High
FMEA	Systematic analysis of possible failure modes	Structured, preventive, risk prioritization	Requires expertise, qualitative bias	Medium	Medium
Real-time sensor monitoring	Sensors track current, voltage, temperature, and irradiance	Continuous monitoring, fast fault detection (provides immediate alerts)	Sensor cost, data handling challenges	Medium	High
Machine learning and forecasting	Algorithms learn from historical data	Predicts trends, preventive maintenance	Needs quality datasets, moderate computing	High	High
Deep learning	Advanced neural networks for fault prediction	Detects complex/hidden patterns	Requires big data, high computation	High	Very High

IR, Infrared; *PV*, photovoltaic; *UAV*, unmanned aerial vehicle.

cost, and accuracy. It highlights how each technique differs in complexity, performance, and suitability for various maintenance needs.

Current methods for photovoltaic module cleaning

Cleaning is one of the most important maintenance tasks for PV systems, requiring careful selection of appropriate techniques to avoid severe degradation caused by dust accumulation and its negative effects (Massi Pavan et al., 2013). Maintaining the cleanliness of PV modules is essential for maximizing operational efficiency, improving light transmission, and preventing premature degradation (Abuzaid et al., 2024). In desert regions, cleaning PV modules presents significant challenges due to the continuous accumulation of dust, limited water availability, and high operating costs (Latreche et al., 2024).

Natural cleaning

Cleaning PV modules through natural processes (rain, wind, snow, and gravity) (see Fig. 8.8) largely depends on environmental conditions and factors beyond human control, making it an unreliable standalone method. The effectiveness of natural cleaning is determined by several meteorological and design parameters,

FIGURE 8.8 Natural cleaning. The figure shows PV modules being cleaned naturally by rain without human intervention. It highlights how environmental conditions can reduce dust accumulation on the modules (*PV*, Photovoltaic).

such as wind speed, humidity, module tilt angle, and the composition of dust deposits (Abuzaid et al., 2024). However, this method is more successful in restoring its original capacity in regions with low dust accumulation, such as the US and Europe (Gupta et al., 2019). Natural cleaning methods offer several advantages, as they are cost-effective, environmentally friendly, and require no maintenance efforts (Abuzaid et al., 2024). However, one of the main drawbacks of this approach is its inability to fully clean PV modules, since the process often leaves residual dust adhered to the surface as a result of interactions with atmospheric moisture (Kazem et al., 2020).

Manual cleaning

Manual cleaning is a labor-intensive approach that employs tools such as mops or brushes to remove dust and debris from PV module surfaces (See Fig. 8.9). The frequency of cleaning depends on several factors, including the module's tilt angle, local environmental and weather conditions, and the overall system configuration. This approach is best suited for small-scale PV systems and can be carried out using either dry or wet cleaning techniques. Although dry cleaning is generally more cost-effective due to minimal water consumption, wet cleaning usually achieves better results. However, it also introduces certain risks, including potential surface scratches from worn brushes and the formation of hotspots when cleaning is conducted under high-temperature conditions (Abuzaid et al., 2024). This method is also more costly than the other ones, as it requires skilled labor (Kazem et al., 2020).

FIGURE 8.9 Manual cleaning of PV modules. The figure shows a worker manually cleaning PV modules using basic tools to remove dust and dirt. It highlights the traditional, labor-intensive approach to maintaining panel performance (*PV*, Photovoltaic).

Self-cleaning system

Self-cleaning techniques offer a passive solution for maintaining PV module cleanliness. They consist of applying a specialized coating to the front surface of the modules, functioning as an antireflective or antisoiling layer that minimizes dust accumulation and reduces the need for manual or mechanical cleaning interventions (Fig. 8.10) (Abuzaid et al., 2024). The use of such automated mechanisms helps conserve water while maintaining PV module performance at an optimal level (Gupta et al., 2019). However, their large-scale adoption remains limited due to challenges in manufacturing integration and uncertainties regarding their long-term efficiency and durability. Self-cleaning technologies for PV modules in the desert appear as a promising solution to combat dust accumulation and water scarcity, but they face technical and economic obstacles (Abuzaid et al., 2024).

Automated cleaning system

Automatic cleaning methods reduce human involvement by utilizing specialized machinery such as robotic cleaners, wipers, or UAVs to automatically remove dust and dirt from PV modules (See Fig. 8.11). These systems can be programmed to perform cleaning cycles at predetermined intervals or based on sensor inputs, operating in either wet or dry modes. They are often integrated with water supply systems to enhance cleaning efficiency. However, implementing water-based automation requires careful management of collection, storage, and disposal processes. Innovative approaches such as using recycled

FIGURE 8.10 Self-cleaning. The figure shows a PV panel equipped with a self-cleaning mechanism that removes dust automatically. It highlights a low-effort, automated approach that reduces manual maintenance (*PV*, Photovoltaic).

FIGURE 8.11 Automatic cleaning system. The figure shows an automated cleaning system moving across PV modules, using UAVs or a mechanical swing-glass mechanism to remove dust without human labor. It illustrates a self-operating solution designed for regular and consistent cleaning (*PV*, Photovoltaic; *UAV*, unmanned aerial vehicle).

air from air conditioning systems for dust removal offer sustainable alternatives but may not always achieve satisfactory results, particularly when wet cleaning is necessary. Despite their advantages in reducing labor, improving consistency, and enabling wide-area coverage with data collection through UAVs, automated cleaning systems remain costly, complex, and energy-demanding, which limits their widespread adoption (Abuzaid et al., 2024).

Semiautomated cleaning

Semiautomatic cleaning methods combine mechanical assistance with human supervision. In this approach, operators use partially automated machines such as motorized brushes or portable cleaning units to execute the cleaning process. Human intervention is still required to operate, monitor, and control the equipment, as well as to make real-time adjustments during cleaning (see Fig. 8.12). While these methods reduce manual effort compared to traditional cleaning and can be more cost-effective than fully automated systems, they still present challenges related to operational complexity, maintenance needs, and efficiency consistency. Consequently, both automated and semiautomated cleaning techniques require further optimization to ensure reliability and long-term sustainability in PV maintenance applications (Abuzaid et al., 2024).

Maintenance approaches for utility, building-integrated PV, agrivoltaic, and floating photovoltaic systems

Effective maintenance is essential for maximizing the performance, reliability, and lifespan of PV systems, not only in grid-connected utility plants but also in building-integrated PV (BIPV), agrivoltaics, and floating PV (FPV) systems. While the primary goal of all PV installations is to ensure continuous power

FIGURE 8.12 Semiautomated cleaning. The figure shows a semiautomated cleaning setup where mechanical tools or assisted devices help clean PV modules, but still require some level of human operation. It illustrates a hybrid approach that reduces manual effort while remaining less complex than fully automatic systems.

generation, each system type presents unique operational challenges that influence the choice of maintenance strategies. Utility-scale PV plants, BIPV systems, agrivoltaic installations, and floating PV systems differ in terms of accessibility, environmental sensitivity, and integration with other functions, such as agriculture or building structures. Table 8.4 summarizes common maintenance practices for these different types of PV systems, providing an overview of the strategies used to maintain optimal performance and mitigate potential problems. These strategies will be discussed in more detail later, highlighting the specific considerations for each PV system configuration.

In fact, maintenance strategies for PV systems vary considerably depending on the system type and installation environment, as well as factors such as operational complexity and associated costs. In large-scale PV installations, maintenance focuses on ensuring high operational reliability and efficiency. Common practices include periodic panel cleaning to prevent efficiency loss, regular inspections of inverters and transformers, monitoring of strings and individual modules, and vegetation control around the arrays. These large-scale installations often rely on SCADA systems for remote performance monitoring and fault detection. Maintenance in these systems is relatively simple, but costs can range from moderate to high due to the scale and complexity of the operations. For example, in BIPV systems, maintenance must consider the electrical and structural aspects of the building. Periodic inspections of the roof and supporting structures are essential to ensure safety and integrity. Panel cleaning and electrical system checks are easily

TABLE 8.4 Comparative maintenance strategies, simplicity, and cost across PV system types.

PV system type	Common maintenance strategies	Considerations	Simplicity	Cost
Utility-scale PV	Periodic cleaning of panels, Inverter and transformer checks, string and module monitoring, and vegetation control	Large-scale operations often use SCADA for remote monitoring	Moderate	Medium-High
BIPV	Roof inspection and structural integrity checks, module cleaning (as accessible), and electrical system inspection	Integration with the building envelope may limit access; esthetics and safety are important	Low	High
Agrivoltaic	Module cleaning, crop management under panels, Monitoring of module shading effects, and electrical inspections	Maintenance must balance energy generation and agricultural productivity	Low-Moderate	Medium
FPV	Cleaning of panels (may use automated floating systems), floatation and mooring inspection, electrical checks, and water quality monitoring	Exposure to water, corrosion, and biofouling are key concerns	Complex	Medium-High

BIPV, Building-integrated PV; *FPV*, floating PV; *PV*, photovoltaic.

performed, with particular attention paid to esthetics and occupant safety due to the integration of the PV modules into the building's structure.

Maintaining BIPV solar systems is more difficult (simply put, low) and tends to be relatively expensive. In the case of agripholtaic systems, where solar modules are installed on farmland, maintenance must balance energy production with crop management. In addition to standard panel cleaning and electrical checks, monitoring the effects of shading on crops and adjusting management practices is crucial for optimizing both crop yields and PV production. Maintenance is moderate to low in simplicity, and costs are generally moderate due to the dual-purpose nature of these installations. FPV systems present unique maintenance challenges due to their aquatic environment. Modules require cleaning to prevent efficiency loss from debris and biological contamination, while floating structures and moorings must be regularly inspected to ensure stability and durability. Electrical components are also monitored, and water quality is often tracked to mitigate corrosion and other environmental impacts. Maintaining FPV solar power systems is relatively simple, but can be moderately to high-cost due to specialized equipment and environmental considerations.

Overall, these comparisons highlight the need to tailor maintenance strategies to each type of PV system, taking into account not only technical requirements but also ease of operation and cost-effectiveness. Understanding these differences is crucial for improving the performance, reliability, and long-term sustainability of various PV installations. FPV systems present unique maintenance challenges due to their aquatic environment. Modules require cleaning to prevent efficiency loss from debris and biofouling, while floating structures and moorings must be regularly inspected to ensure stability and durability. Electrical components are also monitored, and water quality is often tracked to mitigate corrosion and other environmental impacts. These combined maintenance practices ensure long-term performance, reliability, and safety across various FPV system configurations.

Modern predictive maintenance based on tinyML, Artificial Intelligence of Things, large language models, and multimodal artificial intelligence

While traditional predictive maintenance methods, such as visual inspection, I–V curve analysis, thermographic imaging, and even data-driven ML/DL models, have improved the reliability of PV plants, they often face key challenges. These include the lack of interpretability, the need for large computational resources, and the time-intensive process of analyzing extensive datasets or images. To overcome these limitations, recent advances are shifting toward lightweight, embedded, and intelligent approaches (Mellit and Kalogirou, 2025) (Fig. 8.13).

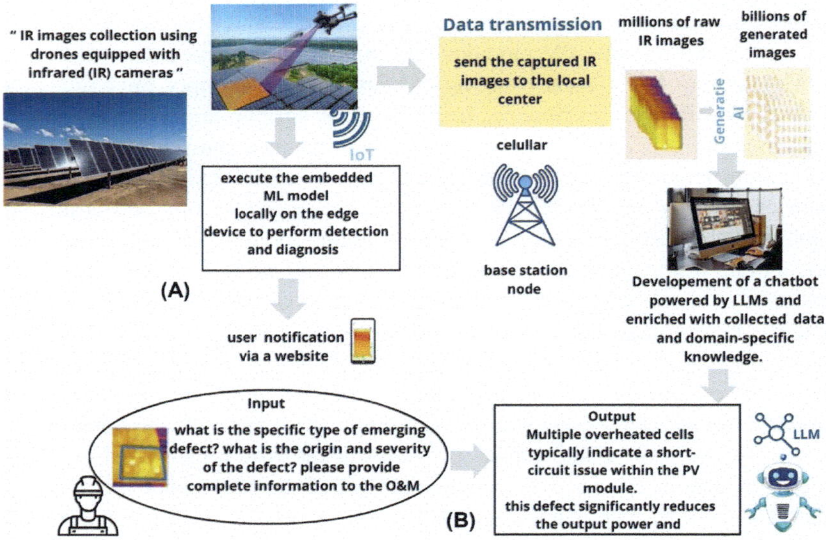

FIGURE 8.13 Modern predictive maintenance of PV plants utilizing (A) embedded ML, IoT integrated with UAVs, and (B) LLMs (small chatbot). The figure shows modern predictive maintenance for PV plants, combining embedded machine learning and IoT sensors with UAVs to monitor system health in real time. It also illustrates the use of a small LLM-based chatbot that assists operators by analyzing data and providing quick diagnostic insights (*IoT*, Internet of Things; *LLM*, large language models; *ML*, machine learning; *PV*, photovoltaic; *UAV*, unmanned aerial vehicle).

LLMs represent a transformative advancement in the evolution of predictive maintenance strategies for PV systems. Unlike conventional ML or DL models that primarily rely on statistical correlations and numerical pattern recognition, LLMs process data through semantic understanding, contextual reasoning, and knowledge-based interpretation. This unique capability enables them to generate human-readable explanations of anomalies, bridging the gap between raw sensor data and actionable operational insights (Fig. 8.13). When integrated with edge machine learning and AIoT architectures, LLMs can operate as intelligent reasoning layers within PdM frameworks. For instance, an AIoT-enabled UAV equipped with thermal and optical sensors can continuously scan PV modules and detect abnormal temperature gradients or irregular power generation patterns. The onboard TinyML model performs real-time anomaly detection, while the LLM interprets these anomalies in context, providing detailed insights (such as the temperature rise is likely due to partial shading or early-stage cell degradation). This type of interpretability enhances operator trust and accelerates maintenance decision-making.

Furthermore, LLMs play a crucial role in root-cause analysis and adaptive learning. They can correlate multiple data streams, including environmental conditions, electrical parameters, and historical maintenance records, to infer not only what fault occurred but also why it occurred. For example, by analyzing textual maintenance logs alongside sensor data, an LLM could identify patterns indicating that repeated overheating in a specific module area is linked to mounting structure misalignment rather than internal cell failure. In addition, LLMs are able to perform incremental and continuous learning, meaning they can adapt to emerging or previously unseen defect patterns without full retraining. This capability is critical in PV plants where new failure modes can arise due to evolving environmental factors, ageing components, or novel system designs. Through self-updating knowledge bases, LLMs assist O&M teams in classifying and describing new defects, even when they are not present in the original labeled dataset. From an operational standpoint, the combination of LLMs, AIoT, Edge AI, and multimodel AI offers several transformative advantages:

- LLMs convert complex, data-intensive outputs into descriptive insights, reducing the cognitive load for human operators.
- Maintenance teams receive clear recommendations, such as schedule module cleaning within 24 hours or replace bypass diode in string # x.
- LLMs can combine structured data (e.g., sensor readings) with unstructured sources (e.g., textual logs, weather reports, or operator notes).
- Intelligent, context-aware alerts enable targeted interventions, avoiding unnecessary inspections or delayed repairs.
- Embedded AIoT systems allow for distributed on-site processing, reducing dependence on cloud computing and improving resilience in remote areas.
- Multimodal AI integrates heterogeneous inputs, such as I–V curves, IR images, electrical measurements, and environmental data, to enable accurate fault detection, failure prediction, and root-cause analysis.

Intelligent AIoT devices embedded with compact LLM architectures could autonomously monitor, interpret, and respond to PV system anomalies. Such systems would not only detect and describe issues but could also recommend optimized maintenance schedules, prioritize repair actions, and communicate seamlessly with digital twins for long-term performance forecasting. This marks a major step toward self-healing, self-optimizing, and knowledge-driven PV infrastructures, aligning predictive maintenance with the broader vision of Industry 5.0 and sustainable energy intelligence.

Challenges and constraints in predictive maintenance implementation

Despite the significant progress enabled by AI-driven, IoT-enabled, and data-centric approaches, the practical deployment of predictive maintenance in PV plants still faces multiple challenges. These challenges stem from technological, economic, and operational constraints that limit scalability and effectiveness and are summarized below:

- Predictive maintenance in PV systems relies heavily on the continuous acquisition of high-quality data, including irradiance, temperature, voltage, current, and environmental parameters. However, missing, noisy, or inaccurate data can significantly compromise model performance. Issues such as sensor degradation, communication errors, or synchronization failures between monitoring devices are common in large-scale PV farms. For instance, soiling or calibration drift in an irradiance sensor may interrupt data flow, leading algorithms to misinterpret normal power fluctuations caused by shading or weather as system faults. Ensuring data accuracy and completeness, therefore requires robust cleaning, validation, and sensor maintenance strategies
- Modern predictive maintenance increasingly depends on advanced AI and DL algorithms, which demand substantial computational power and large training datasets. This is especially challenging for remote PV plants or operators with limited access to cloud-based infrastructure. Furthermore, real-time fault detection on edge devices remains a challenge due to limited processing and memory capacities, even with lightweight AI models like TinyML.
- Many PV systems were designed before the widespread adoption of IoT and AI technologies. Integrating PdM tools into such legacy systems often requires additional communication interfaces, hardware upgrades, or redesigns of existing control architectures. This lack of interoperability between old and modern systems remains a critical obstacle for industrial scalability.
- Although advanced ML and DL models have achieved impressive accuracy in fault detection and performance forecasting, they often operate as black boxes, offering predictions without clear explanations. This lack of interpretability makes it difficult for operators to trust or act upon model outputs, especially in safety-critical environments like PV power plants. For example, a DL model may identify an anomaly in a PV string, but fail to clarify whether it stems from dust accumulation, microcracks, or a bypass diode malfunction. Such ambiguity limits maintenance personnel's ability to make informed and timely decisions. To overcome this, recent research is focusing on integrating explainable AI (XAI) methods and LLMs to transform complex model predictions into human-understandable insights, thereby improving trust, transparency, and operational decision-making in predictive maintenance.
- Predictive maintenance involves high initial investments in monitoring hardware, UAVs for aerial inspection, cloud infrastructure, and specialized personnel for data analysis. While large-scale solar plants can amortize these costs over time, small and medium installations often cannot justify the expense.

- Many predictive maintenance algorithms are validated on small experimental setups or specific datasets and fail to generalize when applied to large-scale, geographically diverse PV farms. Variations in panel technology, orientation, inverter topology, and local environmental conditions often degrade model performance. Ensuring scalability requires diverse datasets, model retraining, and adaptive learning strategies capable of handling site-specific conditions.
- Environmental factors such as dust, storms, seasonal irradiance variations, shading from vegetation, and sudden temperature fluctuations introduce uncertainty into predictive models. These natural variations can mimic or mask real faults, leading to false alarms or missed detections. Robust models must therefore include environmental compensation mechanisms and adaptive thresholds for accurate fault diagnosis under varying climatic conditions.
- The integration of IoT and cloud computing introduces cybersecurity and data privacy challenges. Predictive maintenance systems often collect sensitive operational data, which, if intercepted or tampered with, could compromise the safety and performance of the PV plant. Ensuring encrypted communication, secure data storage, and resilient network protocols is thus essential for safe and reliable implementation.

Recommendations and future trends in photovoltaic maintenance strategies

As PV systems continue to grow in scale, efficiency, and complexity, maintenance strategies must evolve in parallel. Based on the recent advances and challenges discussed previously, here are the key future trends and practical recommendations to guide PV maintenance toward more.

Recommendations for advancing PV plant maintenance can be summarized as follows:

- Combine data-driven approaches with physical models or domain knowledge (e.g., physics-informed AI) to enhance interpretability and reduce false alarms.
- Apply explainability techniques such as feature importance, saliency maps, or LLM-based explanations to help operators understand model decisions.
- Begin with small-scale pilots (e.g., a single inverter string or module type) before full deployment to refine accuracy and cut costs.
- Upskill O&M teams in data science, AI/ML tools, UAV operation, and edge device management, ensuring human oversight in interpreting AI results.
- Evaluate performance beyond detection accuracy, including cost savings, downtime, maintenance intervals, and safety improvements, using feedback loops for refinement.

- Promote shared fault datasets, open sensor logs, and standardized benchmarks to strengthen comparisons and advance the field of intelligent, reliable, and cost-efficient solutions.
- Incorporate XAI techniques to interpret multimodal model outputs, providing operators with actionable insights and building trust in AI-driven decisions.

Some future directions can be summarized as follows:

- Using lightweight ML models embedded directly in sensors or UAVs will reduce latency, data transfer costs, and dependency on cloud infrastructure. These models can perform initial anomaly detection on-site, complementing more powerful cloud/LLM-based reasoning layers.
- Combining AIoT with LLMs for interpretability will become more widespread. LLMs can help translate predictive signals, logs, and image data into actionable and understandable guidance for operators.
- Because operators require more than just accurate faults, they want to know why a fault is occurring. Research will increasingly focus on interpretable models, hybrid approaches that incorporate physical models or domain knowledge, and methods that can provide confidence levels, decision rationales, or root-cause explanations.
- Digital twins (virtual replicas of PV plants) will offer simulation environments, where maintenance strategies can be tested, fault scenarios can be predicted, and long-term degradation patterns can be forecast. This helps in planning, optimization, and risk assessment.
- Smart systems using UAVs, sensors, AI, and LLMs will increasingly support autonomous or semiautonomous operations: real-time anomaly detection, automatic prioritization, fault reporting, and possibly even automated minor corrective actions (e.g., cleaning, adjusting trackers) in remote or large-scale PV plants.

Conclusion

Maintenance plays a crucial role in enhancing the reliability, safety, and energy production of PV power plants. This chapter presented and compared traditional maintenance strategies, including corrective, preventive, and predictive, highlighting their strengths, limitations, and areas of application. While preventive maintenance remains essential for routine scheduled maintenance, predictive maintenance, supported by advanced diagnostic technologies and AI-based analytics, emerges as the most promising approach for minimizing downtime and operating costs. The integration of UAV inspections, imaging, and real-time sensor data has significantly improved early fault detection accuracy. Furthermore, innovative cleaning technologies, from automated systems to self-cleaning solutions, contribute to maintaining optimal performance. Despite these advancements, challenges remain, such as high implementation costs, limited data availability, and the interpretability of AI models.

In the near future, the adoption of Edge AI, AIoT, LLMs, multimodal AI, and XAI will enable advanced predictive maintenance frameworks for PV power plants, allowing early fault detection, transparent decision-making, and autonomous maintenance actions, thereby significantly improving operational efficiency, reliability, and long-term sustainability.

AI Disclosure

During the preparation of this work, the author(s) used ChatGPT for proof-reading. After using this tool/service, the author(s) reviewed and edited the content as needed and take full responsibility for the content of the publication.

References

Abdulla, H., Sleptchenko, A., Nayfeh, A., 2024. Photovoltaic systems operation and maintenance: a review and future directions. Renewable and Sustainable Energy Reviews 195, 114342. Available at https://doi.org/10.1016/j.rser.2024.114342.

Abdulmawjood, K., Refaat, S.S., Morsi, W.G., 2018. Detection and prediction of faults in photovoltaic arrays: a review. Institute of Electrical and Electronics Engineers Inc., Canada Proceedings - 2018 IEEE 12th International Conference on Compatibility, Power Electronics and Power Engineering, CPE-POWERENG 2018, pp. 1–8, Available from: http://ieeexplore.ieee.org/xpl/mostRecentIssue.jsp?punumber=8362839, https://doi.org/10.1109/CPE.2018.8372609.

Abuzaid, H., Awad, M., Shamayleh, A., 2024. Photovoltaic modules' cleaning method selection for the MENA region. Sustainability 16 (21), 9331. Available at https://doi.org/10.3390/su16219331.

Al-Humairi, A., Khalis, E., Al-Hemyari, Z.A., Jung, P., 2025. Machine learning-based predictive maintenance for photovoltaic systems. AI 6 (7), 133. Available at https://doi.org/10.3390/ai6070133.

Ali, B.M., Al-Musawi, T.J., Mohammed, A., Fakhruldeen, H.F., Hanoon, T.M., Khurramov, A., et al., 2025. Sustainable strategies for preventive maintenance and replacement in photovoltaic power systems: enhancing reliability, efficiency, and system economy. Unconventional Resources 6. Available at https://doi.org/10.1016/j.uncres.2025.100170, https://www.sciencedirect.com/science/journal/26665190.

Amiri, A.F., Oudira, H., Chouder, A., Kichou, S., 2024. Faults detection and diagnosis of PV systems based on machine learning approach using random forest classifier. Energy Conversion and Management 301, 118076. Available at https://doi.org/10.1016/j.enconman.2024.118076.

Aremu, O.O., Cody, R.A., Hyland-Wood, D., McAree, P.R., 2020. A relative entropy based feature selection framework for asset data in predictive maintenance. Computers & Industrial Engineering 145. Available at https://doi.org/10.1016/j.cie.2020.106536, https://www.journals.elsevier.com/computers-and-industrial-engineering.

Betti, A., Trovato, M.L., Leotta, G., Ruffini, F., Leonardi, F.S., Lanzetta, C., 2019. Predictive maintenance in photovoltaic plants with a big data approach. arXiv, Italy arXiv. 23318422. Available from: https://arxiv.org. doi: 10.48550/arxiv.1901.10855.

Bonsignore, L., Davarifar, M., Rabhi, A., Tina, G.M., Elhajjaji, A., 2014. Neuro-Fuzzy fault detection method for photovoltaic systems. Energy Procedia 62, 431–441. Available at http://doi.org/10.1016/j.egypro.2014.12.405.

Bosman, L.B., Leon-Salas, W.D., Hutzel, W., Soto, E.A., 2020. PV system predictive maintenance: challenges, current approaches, and opportunities. Energies 16 (3). Available at https://doi.org/10.3390/en13061398, https://www.mdpi.com/1996-1073/13/6/1398.

Chang, Z., Han, T., 2024. Prognostics and health management of photovoltaic systems based on deep learning: a state-of-the-art review and future perspectives. Renewable and Sustainable Energy Reviews 205, 114861. Available at https://doi.org/10.1016/j.rser.2024.114861.

Chen, Z., Han, F., Wu, L., Yu, J., Cheng, S., Lin, P., et al., 2018. Random forest based intelligent fault diagnosis for PV arrays using array voltage and string currents. Energy Conversion and Management 178, 250–264. Available at https://doi.org/10.1016/j.enconman.2018.10.040.

De Benedetti, M., Leonardi, F., Messina, F., Santoro, C., Vasilakos, A., 2018. Anomaly detection and predictive maintenance for photovoltaic systems. Neurocomputing 310, 59–68. Available at https://doi.org/10.1016/j.neucom.2018.05.017.

Ghazali, S.N.A.M., Sujod, M.Z., Jadin, M.S., 2021. Forensic of solar PV: a review of potential faults and maintenance strategies. In: 7th International Conference on Engineering and Emerging Technologies, ICEET 2021. Institute of Electrical and Electronics Engineers Inc., Malaysia. https://doi.org/10.1109/ICEET53442.2021.9659624, http://ieeexplore.ieee.org/xpl/mostRecentIssue.jsp?punumber=9659488

Goudelis, G., Lazaridis, P.I., Dhimish, M., 2022. A review of models for photovoltaic crack and hotspot prediction. Energies 15 (12). Available at https://doi.org/10.3390/en15124303, https://www.mdpi.com/1996-1073/15/12/4303/pdf?version=1655016782.

Gupta, V., Sharma, M., Pachauri, R.K., Dinesh Babu, K.N., 2019. Comprehensive review on effect of dust on solar photovoltaic system and mitigation techniques. Solar Energy 191, 596–622. Available at https://doi.org/10.1016/j.solener.2019.08.079, www.elsevier.com/inca/publications/store/3/2/9/index.htt.

Hamza, A., Ali, Z., Dudley, S., Saleem, K., 2024. Optimizing PV array performance: A2 LSTM for anomaly detection and predictive maintenance based on machine learning. In: IEEE Energy Conversion Congress and Exposition, ECCE 2024 - Proceedings, pp. 1681–1688 Institute of Electrical and Electronics Engineers Inc., United Kingdom. https://doi.org/10.1109/ECCE55643.2024.10861733, http://ieeexplore.ieee.org/xpl/mostRecentIssue.jsp?punumber=10860271

Hamza, A., Ali, Z., Dudley, S., Saleem, K., Uneeb, M., Christofides, N., 2025. A multi-stage review framework for AI-driven predictive maintenance and fault diagnosis in photovoltaic systems. Applied Energy 393, 126108. Available at https://doi.org/10.1016/j.apenergy.2025.126108.

Joseph, E., Vijaya Kumar, P.M., Mahinder Singh, B.S., Ching, D.L.C., 2023. Performance monitoring algorithm for detection of encapsulation failures and cell corrosion in PV modules. Energies 16 (8). Available at https://doi.org/10.3390/en16083391, http://www.mdpi.com/journal/energies/.

Kazem, H.A., Chaichan, M.T., Al-Waeli, A.H.A., Sopian, K., 2020. A review of dust accumulation and cleaning methods for solar photovoltaic systems. Journal of Cleaner Production 276. Available at https://doi.org/10.1016/j.jclepro.2020.123187, https://www.journals.elsevier.com/journal-of-cleaner-production.

Kingsley-Amaehule, M., Uhunmwangho, R., Nwazor, N., Okedu, K.E., 2022. Smart intelligent monitoring and maintenance management of photo-voltaic systems. International Journal of

Smart Grid 6 (4), 110–122. Available at https://doi.org/10.20508/ijsmartgrid.v6i4.260.g246, https://www.ijsmartgrid.org/index.php/ijsmartgridnew/article/view/260.

Kumar, S., Nayak, P.K., 2024. An effective method for detection and location estimation of faults in large-scale solar PV arrays. Solar Energy 277. Available at https://doi.org/10.1016/j.solener.2024.112727, https://www.sciencedirect.com/science/journal/0038092X.

Latreche, M., Medjahed, B., Mellit, A., Rabhi, A., Hajji, B., 2024. Design of a dust cleaning prototype for photovoltaic generators. In: International Conference on Electronic Engineering and Renewable Energy Systems, pp. 227–235.

Ledmaoui, Y., El Maghraoui, A., El Aroussi, M., Saadane, R., 2025. Review of recent advances in predictive maintenance and cybersecurity for solar plants. Sensors 25 (1), 206. Available at https://doi.org/10.3390/s25010206.

Massi Pavan, A., Mellit, A., De Pieri, D., 2011. The effect of soiling on energy production for large-scale photovoltaic plants. Solar Energy 85 (5), 1128–1136.

Massi Pavan, A., Mellit, A., De Pieri, D., Kalogirou, S., 2013. A comparison between BNN and regression polynomial methods for the evaluation of the effect of soiling in large scale photovoltaic plants. Applied Energy 108.

Mellit, A., Chourouk, M., Zennaro, M., 2024. Predictive method for cleaning photovoltaic plants in desert areas using TinyML and IoT technique. In: 3rd International Conference on Advanced Electrical Engineering (ICAEE) IEEE, pp. 1–6.

Mellit, A., Kalogirou, S., 2025. Recent advances in the application of infrared thermographic imaging and embedded artificial intelligence for fault diagnosis and predictive maintenance of photovoltaic plants: challenges and future directions. Renewable and Sustainable Energy Reviews 223, 116057. Available at https://doi.org/10.1016/j.rser.2025.116057.

Mellit, A., Kalogirou, S.A., 2022. Machine learning and deep learning for photovoltaic applications. Artificial Intelligence in Smart Photovoltaic Technology 1.

Mellit, A., Tina, G.M., Kalogirou, S.A., 2018. Fault detection and diagnosis methods for photovoltaic systems: a review. Renewable and Sustainable Energy Reviews 91, 1–17. Available at https://doi.org/10.1016/j.rser.2018.03.062.

Nasrfard, F., Mohammadi, M., Karimi, M., 2023. A Petri net model for optimization of inspection and preventive maintenance rates. Electric Power Systems Research 216, 109003. Available at https://doi.org/10.1016/j.epsr.2022.109003.

Osmani, K., Haddad, A., Lemenand, T., Castanier, B., Ramadan, M., 2020. A review on maintenance strategies for PV systems. Science of the Total Environment 746, 141753. Available at https://doi.org/10.1016/j.scitotenv.2020.141753.

Pingulkar, C., Joshi, A., 2021. AI and data analytics for predictive maintenance in solar power plants. International Journal of Progressive Research in Engineering, Management and Science 1, 52–69.

Ren, J., Man, Y., 2022. Artificial Intelligence for Smart Photovoltaic Technologies. AIP Publishing. Available at http://doi.org/10.1063/9780735424999.

Sepulveda-Oviedo, E.H., Trave-Massuyes, L., Subias, A., Pavlov, M., Alonso, C., 2023. Detection and classification of faults aimed at preventive maintenance of PV systems. arXiv.

Simon, D.F., Teixeira, M., da Costa, J.P., 2022. Availability estimation in photovoltaic generation systems using Timed Petri Net simulation models. International Journal of Electrical Power & Energy Systems 137. Available at https://doi.org/10.1016/j.ijepes.2021.106897, https://www.journals.elsevier.com/international-journal-of-electrical-power-and-energy-systems.

Singh, U.P., Chandra, S., 2022. Control Applications in Modern Power Systems: Select Proceedings of EPREC 2021A predictive maintenance scheme for solar PV system Control Applications in Modern Power Systems: Select Proceedings of EPREC 2021.

Syamsuddin, A., Adhi, A.C., Kusumawardhani, A., Prahasto, T., Widodo, A., 2024. Predictive maintenance based on anomaly detection in photovoltaic system using SCADA data and machine learning. Results in Engineering 24, 103589. Available at https://doi.org/10.1016/j.rineng.2024.103589.

Toubeau, J.-F., Pardoen, L., Hubert, L., Marenne, N., Sprooten, J., De Grève, Z., et al., 2022. Machine learning-assisted outage planning for maintenance activities in power systems with renewables. Energy 238, 121993. Available at https://doi.org/10.1016/j.energy.2021.121993.

Villarini, M., Cesarotti, V., Alfonsi, L., Introna, V., 2017. Optimization of photovoltaic maintenance plan by means of a FMEA approach based on real data. Energy Conversion and Management 152, 1–12. Available at https://doi.org/10.1016/j.enconman.2017.08.090, https://www.journals.elsevier.com/energy-conversion-and-management.

Chapter 9

Cyber-physical security in photovoltaic power plants

Nassiha Boutana[1], Adel Mellit[2], and Abdesselam Bouzerdoum[3]
[1]*Department of Electronics, USTHB University, Algiers, Algeria,* [2]*Faculty of Sciences and Technology, University of Jijel, Jijel, Algeria,* [3]*Hamad Bin Khalifa University, Doha, Qatar*

Introduction

Conventional energy systems are being transformed into highly integrated and intelligent structures due to the rapid advancement in communication and information technology. Internet of Things (IoT) integration into power systems improves control, monitoring, and service delivery by facilitating effective two-way communication between utilities and end users (Diaba et al., 2024). As a crucial part of this digital revolution, photovoltaic (PV) power plants are being included in smart grids more and more, taking advantage of data-driven optimization, automation, and remote management. However, this growing reliance on IoT and digital infrastructure also introduces significant cybersecurity challenges. Like other grid components, PV systems are susceptible to both cyber and physical attacks that might compromise the availability, confidentiality, and integrity of vital data (Tatipatri and Arun, 2024). IoT devices, supervisory control and data acquisition (SCADA) systems, remote monitoring, intelligent inverters, and cloud-based platforms used in energy production, storage, and distribution processes are all potential points of attack for these systems. Ransomware, spear phishing, and the use of unpatched or weakly authenticated devices, data integrity attacks (DIA), and Denial of Service (DoS) attacks are among the cybersecurity threats to PV plants that can result in lost productivity, income, and operations (Boutana et al., 2025). Insurance policies and best practices, such as network isolation, role-based access and authentication enforcement, encryption, software patching, backups, network monitoring, physical security, and supply chain risk management, are recommended to mitigate such risks (Cybersecurity in photovoltaic plant operations, 2021).

A review of the cybersecurity risks that face PV systems is given in Harrou et al. (2023), shedding light on the evolving cybersecurity landscape for PV

Embedded Artificial Intelligence and the Internet of Things for Photovoltaic Systems.
DOI: https://doi.org/10.1016/B978-0-443-44835-5.00008-4

systems. Sandia National Laboratories (Carter et al., 2017) performed cybersecurity assessments of a communications-enabled PV inverter and a remote grid monitoring gateway of two residential-scale interoperable distributed energy resource devices. Cybersecurity in power systems has therefore become a vital and dynamic field of study and research (Presekal et al., 2025; Singh et al., 2025). The power system anomaly's deviation from normal behavior could be a sign of equipment malfunctions, ineffective operations, or cyberattacks (Gupta et al., 2022; Pan et al., 2020). Good detection of cyber-attacks is crucial for grid-connected PV (GCPV) plant availability and stability to maintain a continuous power supply.

AI techniques have been particularly effective in detecting cyber-attacks in PV systems and enhancing the reliability and efficacy of PV operations. Various machine learning (ML) and deep learning (DL) techniques have been proposed for PV system threat identification and mitigation. For example, a deep sequence learning-based diagnostic method for data integrity attacks on PV systems in smart grids, including DC–DC and DC–AC converters, is proposed in (Li et al., 2021). A study in Hassan et al. (2025) assesses the contribution of three DL methods: long short-term memory (LSTM), artificial neural networks (ANN), and convolutional neural networks (CNN) to PV cyber-attacks detection. A CNN technique is suggested in Zhang et al. (2023) for detecting cyber-attacks in PV plants by point-of-common coupling sensors alone. Authors in Ahmadzadeh et al. (2024) proposed a deep CNN-based approach to detect false data injection (FDI) attacks on PV-integrated distribution systems (IDS). The performance of the trained framework has been compared with other supervised ML and DL-based techniques. Moreover, the study in Aflaki et al. (2021) uses an unsupervised learning technique called hierarchical clustering to develop an artificial sensor that can identify two distinct cyber-attacks (e.g., FDI and DoS) during the power system's dynamic behavior.

This chapter presents the latest developments in the security of GCPV systems. As PV power plants become increasingly digitized and integrated with smart grids, they are exposed to a wider range of cyber and physical vulnerabilities. The chapter highlights the interconnection between cyber and physical security risks, focusing on how disruptions in one domain can interact with the other, impacting overall system reliability, safety, and performance. It also explores potential threats targeting various PV plant components, from sensors, inverters, and data acquisition units to communication networks and cloud-based supervisory control systems. In response to these challenges, the chapter reviews a range of detection and mitigation strategies, including traditional model-based approaches and emerging data-driven technologies powered by artificial intelligence and machine learning. These intelligent methods have demonstrated strong potential in identifying abnormal patterns, predicting failures, and enabling proactive defense mechanisms, enhancing the resilience and reliability of modern grid-connected PV infrastructures.

Cyberattacks on grids and renewable energy plants

Numerous cyberattacks have targeted the global control system for the electric power in recent years. Cyberattacks in Ukraine were the first ones that were known to cause power outages directly (Roadmap for photovoltaic cyber security, 2017; Rajkumar et al., 2023). A brief timeline of cyber-attacks involving power systems over the previous nine to 10 years is shown in Fig. 9.1.

Starting from 2015, a cybersecurity attack in Ukraine caused approximately 225.000 customers to lose power (an outage of 200 MW) across various areas. Sophisticated malware was carried out a year later, which led to a disruption in the distribution network. In 2017, cyberattacks known as Dragonfly/Energetic Bear No. 2 targeted a nation's critical infrastructure, including its water supply and electrical grids (Hemsley and Fisher, 2018). In 2019, a DoS attack was carried out in the United States of America, which was the first cyberattack in the field of wind and solar energy (Yohanandhan et al., 2020). Attacks on Venezuela's electrical grid in 2020 resulted in a power outage in a number of states (Du et al., 2023).

In 2021, a ransomware attack by the DarkSide group led to a shutdown of the largest US fuel pipeline. Although not an electricity grid, it disrupted fuel supply for several days (Duo et al., 2022). In 2022, a new variant of the industrial malware (Industroyer2) was used in an attempted power grid disruption in Ukraine. The largest-ever coordinated cyberattack on the energy sector occurred in Denmark in 2023 when hackers targeted 22 energy companies, including electricity distribution operators, by exploiting a Zyxel firewall vulnerability (Diaba et al., 2024). This forced several utilities to disconnect from the internet to protect their SCADA systems. In 2024, hackers installed backdoors by taking advantage of vulnerabilities in about 800 SolarView Compact remote monitoring devices at PV plants in Japan. These devices were used to siphon bank funds through illicit online transactions (Addressing the growing cybersecurity concerns within solar, 2025).

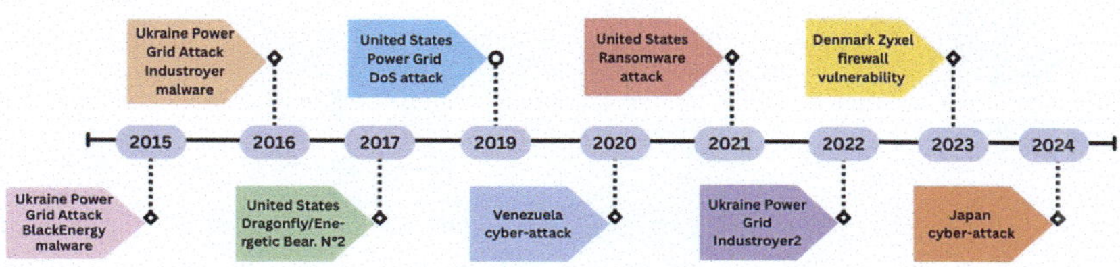

FIGURE 9.1 Timeline of cyberattacks in power systems from 2015 to 2024. This figure illustrates a brief timeline of the well-known cyberattacks targeted at the electric power systems starting from 2015 to 2024.

Cyber-physical security in grid-connected photovoltaic power plants

Definition

The framework of a GCPV power plant is illustrated in Fig. 9.2 along with its main components, communication lines, and possible attack points. It demonstrates how cyber systems and physical infrastructure are interdependent, highlighting vulnerabilities that hackers could exploit.

Generally, the system includes a PV array with sensors, DC–DC converters, DC–AC inverters, a monitoring and diagnostics unit that collects real-time data from sensors and other components such as PV module temperature, irradiance, currents, and voltages measurements, and then sends operational data via communication links to the control unit. The grid and consumers receive power from electrical connections, while the control unit manages system operations, monitors performance, and enables remote access, data exchange, and control via cloud services, servers, and access points via the Internet connections. As shown in the figure, any interface with the Internet: monitoring systems, control units' infrastructure, cloud servers, and access points are vulnerable to cyber-attacks that can manipulate operational data, disrupt control commands, or disable plant functionality, while on-site components like sensors, PV modules, and inverters remain exposed to direct tampering or physical damage. Table 9.1 illustrates a comparison between physical and cyberattacks.

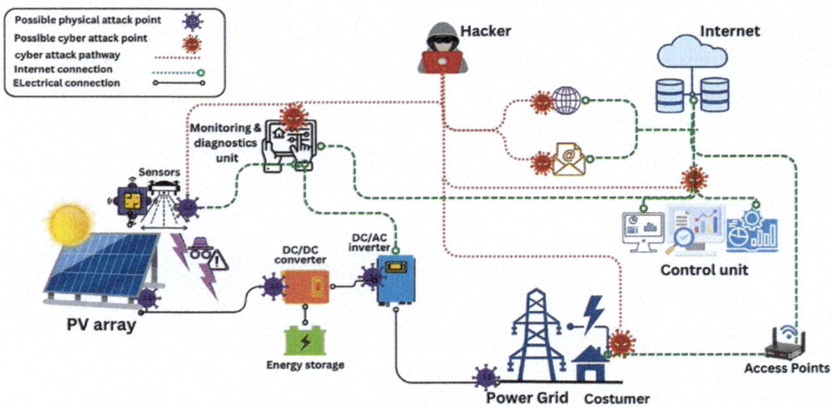

FIGURE 9.2 **The framework of the GCPV power plant with possible cyber physical attacks points.** This figure illustrates the framework of a grid-connected PV plant, showing the main components from the PV array to the grid interface. It highlights potential cyberphysical attack points along the communication and control pathways. *GCPV*, grid-connected PV; *PV*, photovoltaic.

TABLE 9.1 **A comparison between physical attacks and cyber attacks in PV plants.**

	Physical attacks	Cyber attacks
Main targets	Attacks target the PV plant's physical components as well as its computer systems: PV modules, inverters, combiner boxes, wiring, sensors, physical access control, and control/monitoring systems.	Attacks only target information technology (IT) and digital systems without making direct physical contact: SCADA systems, monitoring platforms, data loggers, communication networks, user accounts, servers, routers, and cloud-based dashboards.
Attack methods	• Cutting cables or damaging equipment • Tampering with inverters and combiner boxes • Physically altering sensors • Stealing PV panels or batteries	• Phishing • Malware • Man in the middle • Denial of service • Data integrity attack
Consequences	Equipment damage, power outages, and high repair/replacement costs	Disruption in operations, loss or manipulation of data, loss of monitoring/control access, Financial deception

PV, photovoltaic; *SCADA*, supervisory control and data acquisition.

From the table above, it is evident that physical attacks target the various physical components of the PV infrastructure, such as, but not limited to, the modules, inverters, junction box, cabling, and mounting structures, and these can be caused by acts of theft, sabotage, or even natural events. The potential consequences of physical attacks are location-based, but often occur in the immediate term, resulting in equipment damage, loss of generated power, increased costs associated with repair and recovery, and loss of revenue in the form of lost operating time. In any large-scale PV power plant, any single physical failure can generate substantial declines in energy production and operational efficiency, and without timely repairs and action, potentially result in asset deterioration in the long term. On the other hand, cyber-attacks compromise the digital and communications layers of the photovoltaic system, resulting in indirect effects. These attacks affect control systems, data acquisition units, IoT devices, and cloud services used for monitoring and automated control.

An attacker can alter system behavior, disrupt real-time control, or inject false data into supervisory control and data acquisition (SCADA) systems by changing or disrupting data flows. Cyber breaches can lead to unsafe operational conditions, degraded grid stability, and financial fraud through false production or performance reports. In extreme cases, simultaneous actions can spread across interconnected power grids, which can create additional risk factors for areas beyond the local PV plant. In general, while physical attacks result in visible, immediate losses, cyber-attacks can pose systemic threats to integrity, trustworthiness, and operational security and may do so quietly over time, which is why it is important to develop and implement defense strategies that deal with both dimensions simultaneously.

Potential attacks in grid-connected photovoltaic power plants

The most common types of attacks targeting PV systems are categorized in Fig. 9.3.

Man-in-the-Middle

Man-in-the-Middle (MitM) cyberattacks are classified according to the adversary's network position. By positioning themselves between two or more communicating hosts, attackers can secretly monitor and alter the traffic that is transmitted. Potential MitM attacks on PV plants' infrastructures fall into different categories:

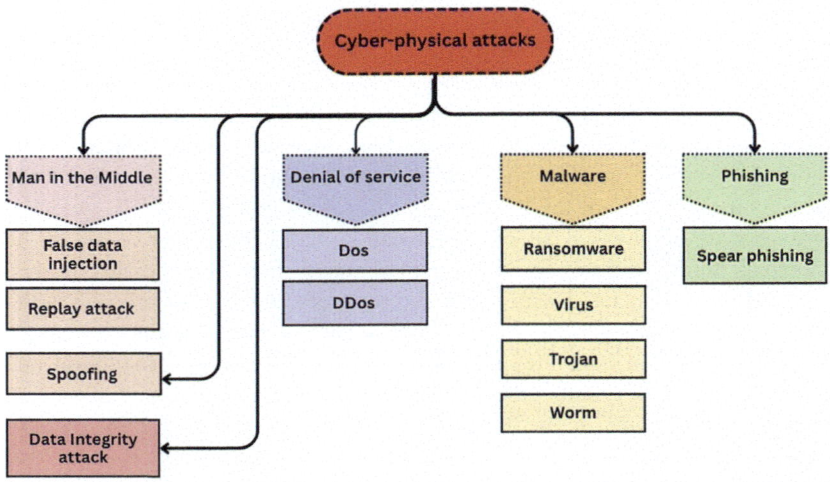

FIGURE 9.3 Potential cyber-physical attacks in GCPV power plants. This figure shows potential cyberphysical attack points in a grid-connected PV power plant. It highlights where attackers could target communication, control, or monitoring systems. *GCPV*, grid-connected PV; *PV*, photovoltaic.

False data injection

In FDI, attackers inject malicious data into communication channels between inverters and control systems, leading to incorrect operational decisions (Zhang et al., 2022). These attacks can target both measurement data and control commands. Systems with a high PV penetration rate are more vulnerable to FDI attacks. An attacker, who is familiar with the parameters and structure of the grid (particularly how PV plants operate), may regularly generate false data to alter power flow measurements. This might be used to manipulate grid pricing (Xie et al., 2020), disrupt the economic dispatch of solar energy, or even cause physical damage by imposing line overloads due to inaccurate solar generation estimations (Hug and Giampapa, 2012). An example of an FDI attack is shown in Fig. 9.4.

Replay attacks

The replay attack is a type of MitM attack, which performs in two phases: the monitoring phase and the replay phase. First, the goal is to track and log a real sequence of data measurements, and then use these previously recorded values to replace the actual measurements during the invasion. Until the attack is completed, the captured data are continuously replayed. An example of this attack in the GCPV power plant is shown in Fig. 9.5; in this scenario, attackers might record an accurate instruction to shut down a solar inverter. They later

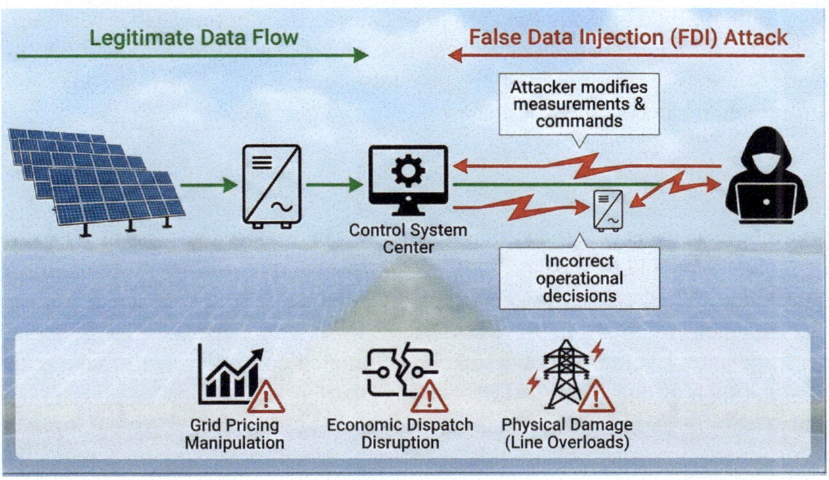

FIGURE 9.4 **Example of an FDI attack in a GCPV power plant.** This figure illustrates an example of a false data injection attack. It shows an attacker intercepting and modifying data between solar panels and a control center, leading to incorrect decisions and consequences such as grid pricing manipulation, economic dispatch disruption, and physical damage. *FDI*, false data injection; *GCPV*, grid-connected photovoltaic.

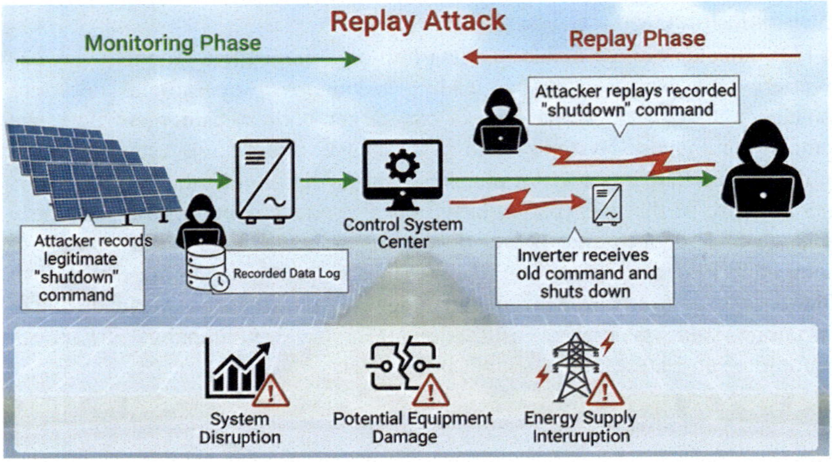

FIGURE 9.5 **Example of a replay attack scenario in a GCPV power plant.** This figure shows an example of a replay attack in a PV system, when attackers record a legitimate shutdown command during an initial monitoring phase. In the subsequent replay phase, the attacker retransmits this recorded data to the inverter, tricking the system into executing the outdated command. *GCPV*, grid-connected PV; *PV*, photovoltaic.

replay the identical command, which causes the inverter to suddenly shut down, potentially damaging the system or causing disruptions (Ribas Monteiro et al., 2023). The cyber-physical system's sensors, actuators, controllers, or inverters are typically targeted by this false replay attack, which does not require extensive system expertise.

Spoofing

Spoofing is a type of attack that occurs when real data is replaced with manipulated or falsified data in the control or monitoring systems of a photovoltaic plant. Spoofing can occur on sensors, GPS, or the communications channels to mislead the SCADA of PV plants regarding their actual operating state. Eventually, operators or automated controllers will make poor decisions such as changing grid connections, battery operations, or inverter set points (e.g., by using manipulated data) (Presekal et al., 2025). An example is demonstrated in Fig. 9.6. In this scenario, sensor data spoofing typically involves falsifying readings like irradiance, voltage, current, or temperature, which might lead to losses in efficiency, destabilizing of the grid interface, or equipment damage (Ibrahim et al., 2025). Additionally, by spoofing sensor data, an attacker might manipulate and falsify a PV converter's power output. A study in Barua and Al Faruque (2020) demonstrates how an intelligent attacker can use a noninvasive magnetic spoofing technique to insert faulty

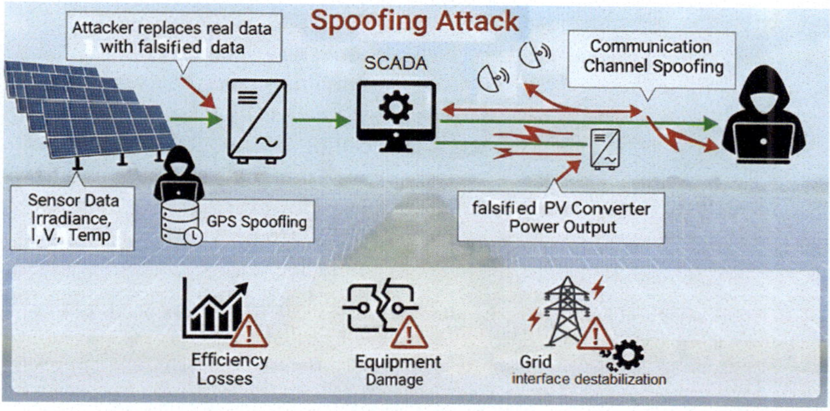

FIGURE 9.6 Example of a spoofing attack targeting the SCADA system of a GCPV power plant. This figure depicts a spoofing attack where attackers manipulate communication channels and inject falsified sensor or GPS data into the SCADA system. This causes the system to operate on false information, leading to efficiency losses, hardware damage, and grid interface destabilization. *GCPV*, grid-connected photovoltaic; *SCADA*, supervisory control and data acquisition.

measurements into an inverter's Hall sensors. From the damaged Hall sensor, the injected errors may spread to the inverter's internal controllers and ultimately damage the inverter itself.

Data integrity attacks

Data integrity attacks can be conducted as part of a MitM attack; however, they can also be carried out via compromised systems, malware within the system, or direct access. An example of this attack is illustrated in Fig. 9.7. To change the observed PV system state, attackers try to compromise the system's stability by providing it with false information, such as injecting fake data, manipulating sensor measurements, and changing energy production values (Ye et al., 2022).

Denial of service attack

DoS attacks attempt to saturate networks and shut down communication channels so that authorized users cannot access critical services. As illustrated in Fig. 9.8, these types of attacks are carried out by flooding the control channel with messages to saturate the network and interfere with regular communication. DoS attacks can be executed by using a network of infected devices, or bots, to perform DoS attacks indirectly or directly from home computers. In GCPV plants, these threats can block sensors' data communication (e.g., irradiance, current, and voltage measurements), disrupt inverter

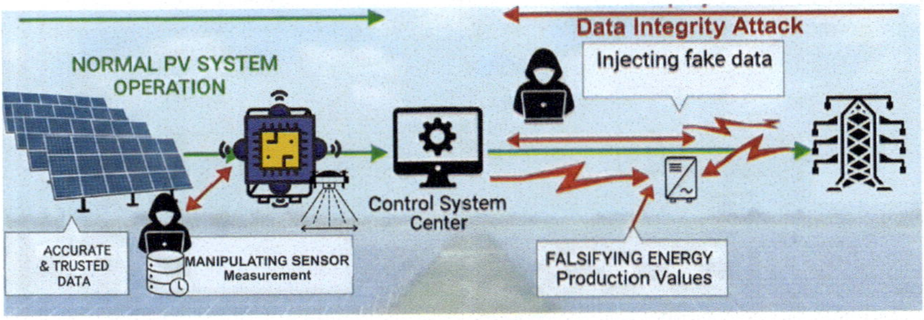

FIGURE 9.7 Example of a data integrity attack scenario in a GCPV power plant. This figure demonstrates how attackers compromise system stability by injecting fake data and manipulating sensor measurements to alter the observed system state. The attack compromises the control center's decision-making and disrupts normal photovoltaic operations. *GCPV*, grid-connected photovoltaic.

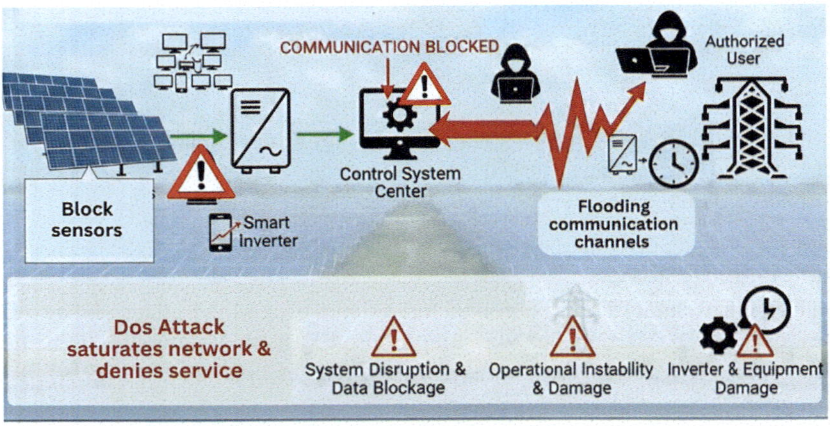

FIGURE 9.8 Denial of service attack in a GCPV power plant. This figure illustrates a denial of service attack where attackers saturate communication channels and block sensor data to prevent authorized users from accessing critical PV services. The resulting network flooding leads to severe system disruption, operational instability, and potential physical damage to inverters and related grid equipment. *GCPV*, grid-connected PV; *PV*, photovoltaic.

commands, or delay timely grid coordination signals (Ye et al., 2022). An even more dangerous variety is the distributed denial-of-service attack, whereby an attacker commands several bots to launch an attack against a system at once, making network resources totally inaccessible (Biju et al., 2019; de Neira et al., 2023).

Malware

Malware refers to malicious code composed to take control of software vulnerabilities in PV plant infrastructure, targeting particularly SCADA systems, inverters, and communication networks. In PV power systems, malware is introduced as viruses, Trojans, worms, ransomware, or botnets that disrupt operation, steal sensitive data, or damage equipment (Humayed et al., 2017). Advanced malware, such as worms and botnets, is especially dangerous in its ability to propagate independently from device to device and coordinate large-scale attacks (Khan et al., 2016). Malware in PV plants can infect inverter firmware, manipulate measurement and control data, or lock up SCADA communications, leading to loss of monitoring, power outage, or even grid instability. Fig. 9.9 shows how malware attacks target and affect the GCPV power plant. The rise of sophisticated industrial malware like Industroyer demonstrates how renewable energy infrastructures, including PV plants, are under greater threat from cyber-physical attacks (Achaal et al., 2024).

Phishing

Phishing in PV plants is a cyberattack in which attackers pose as reliable organizations (such as utility companies, equipment suppliers, or internal staff) in an attempt to confuse their victims into opening malicious attachments,

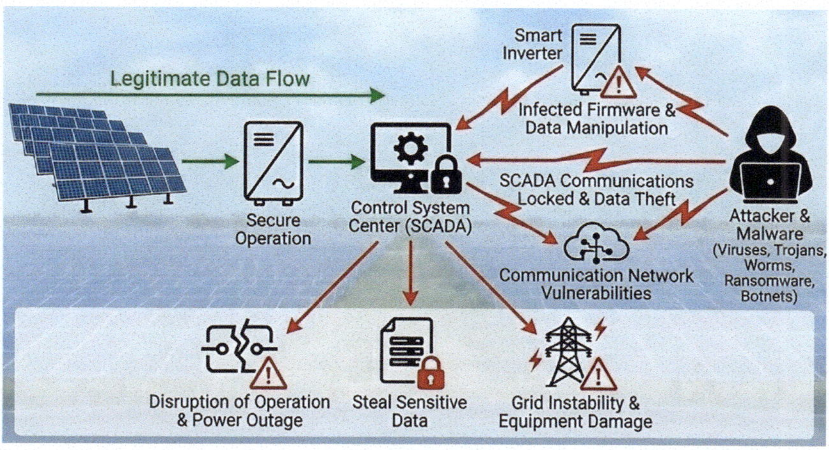

FIGURE 9.9 **Example of attacks targeting the GCPV power plant by introducing malware into different elements.** This figure illustrates a malware attack where viruses, Trojans, or ransomware infect GCPV power plant infrastructure, specifically targeting SCADA systems and inverter firmware. The attack results in significant operational disruption, sensitive data theft, and grid instability through the manipulation of control data and equipment damage. *GCPV*, grid-connected photovoltaic; *SCADA*, supervisory control and data acquisition.

clicking on infected links, or providing private information. Attackers can obtain unauthorized access to corporate IT networks, control systems, and monitoring platforms that facilitate PV operations through phishing. This access can then be used to spread malware throughout the system, modify performance data, or interfere with plant operations (Biju et al., 2019). An example of a phishing attack targeting the GCPV power plant is shown in Fig. 9.10.

Spear phishing is a more focused kind of phishing, where Attackers target specific users or organizations that conduct PV operations. Attackers first gather specific information about their targets, such as the positions of suppliers, technical staff, or plant operators, and then create very convincing emails. For example, attackers might act as grid operators or inverter suppliers and send particular emails with malicious attachments. Because the content appears authentic, spear phishing has a considerably higher success rate. It is commonly used as a basis for more complex attacks on photovoltaic plants, including data theft, malware installation, and supervisory control system tampering (Presekal et al., 2025).

The diagram block in Fig. 9.11 summarizes the different cyber-attacks targeting each element in the GCPV power plant.

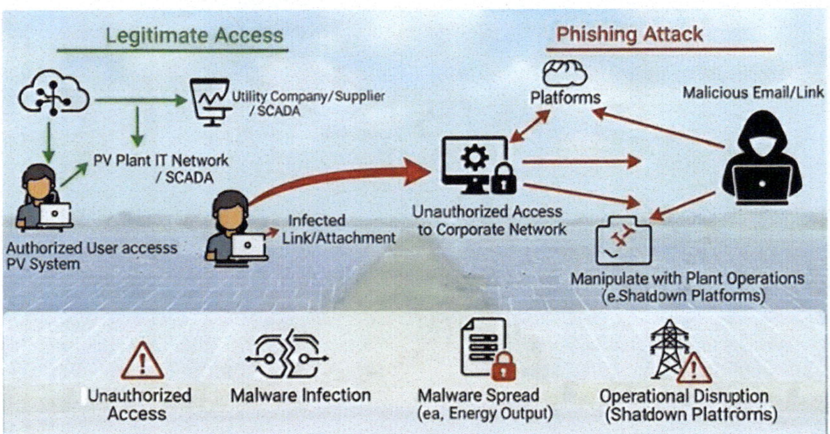

FIGURE 9.10 Phishing attack scenario targeting the GCPV power plant. The figure illustrates a phishing attack where an attacker uses malicious emails or links to trick authorized users into providing access to the GCPV plant's IT network. This unauthorized access allows the attacker to spread malware throughout the system and manipulate plant operations, leading to severe disruptions such as platform shutdowns. *GCPV*, grid-connected photovoltaic.

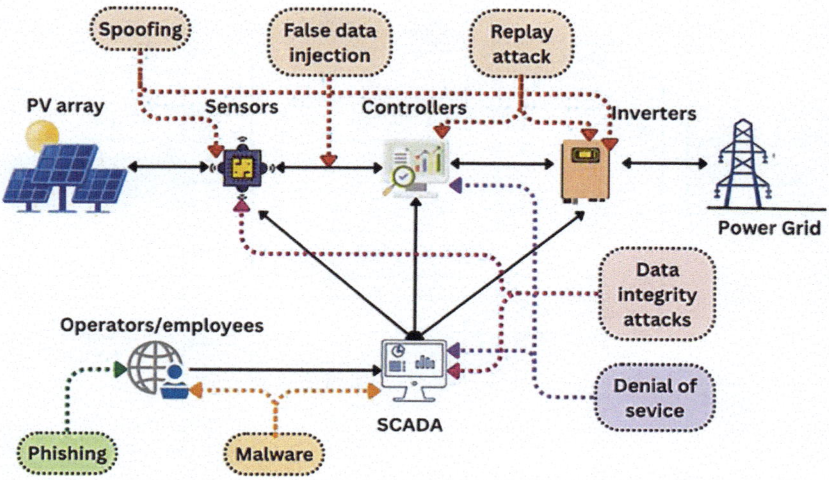

FIGURE 9.11 **Potential cyber-attacks targeting each element in the GCPV power plant.** This figure illustrates the possible cyberattacks that can target individual components of a GCPV power plant. It shows how each element, such as sensors, controllers, and communication links, can be exploited by attackers. *GCPV*, grid-connected photovoltaic.

Cyberattacks detection methods in grid-connected photovoltaic power plants

Understanding the nature of threats and vulnerability types is key to effectively detecting cyberphysical attacks and threats in GCPV plants. Various methods and strategies for hardware and software have been proposed in the literature, aiming to detect these cyberattacks. Fig. 9.12 depicts the cyber-attack detection methods highlighted in this chapter.

- Model-based methods
- Data-driven methods: ML and DL
- Intrusion detection systems

Model-based detection methods

Model-based methods detect threats and unusual activity by using analytical models, which are frequently designed to simulate systems under no-fault situations. This discrepancy is assessed in relation to a residual threshold (Ye et al., 2022). The most recent advancements in these methods are Kalman Filter (KF)-based state-estimation techniques, which use iterative mathematical algorithms to predict and update system states (such as voltage, current, frequency, or power output) under noisy and uncertain conditions. These techniques create residuals (differences) by real-time comparing the measured

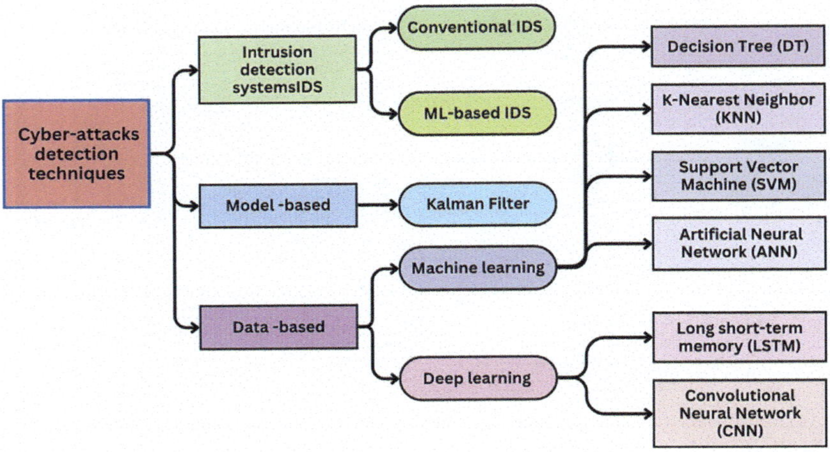

FIGURE 9.12 **Cyber-attacks detection methods.** This figure presents different methods used to detect cyber-attacks in GCPV power plants. It categorizes the detection approaches based on how anomalies or malicious activities are identified. *GCPV*, grid-connected photovoltaic.

signals from PV inverters, controllers, and sensors with the KF-estimated states.

Considerable or continuous deviations in these residuals reveal the possibility of cyber-attacks, such as replay attacks, false data injection (Ye et al., 2022). For example, authors in Teymouri et al. (2018) develop a cyber-physical model for a grid-connected PV system that employs an attack detection algorithm to look for irregularities in grid voltage measurements. DIA controller vulnerabilities are used in Zhang et al. (2020). Furthermore, authors in Larkin et al. (2020) propose deploying a data diode to detect DIA in the PV system. KF methods are often combined with other anomaly detection approaches (e.g., dynamic watermarking, observers, ML) to increase robustness against different types of attacks.

Data-driven methods

GCPV plants generate a huge volume of data: voltage, current, frequency, power, communication logs, and user access, etc. Data-driven methods can work with data and find hidden patterns. Consequently, they can proactively identify, mitigate, or recover from cyber-attacks on those systems.

Machine learning-based detection methods

Such attacks on the GCPV plant's edge devices may go undetected by model-based and rule-based methods. Therefore there is a need to employ ML-based methods to analyze and detect attacks (Sourav et al., 2022). The ML-based detection methods do not require mathematical models of the physical system, in contrast to the model-based detection methods that were previously discussed. Instead, they depend entirely on the system's historical data. ML is used to extract knowledge and patterns from data. It can be applied to both classification and regression tasks: Regression focuses on numerical prediction, and it is commonly used in applications like forecasting power system load. Classification entails giving predicted values to particular groups, which makes it ideal for detecting cyberattacks. A ML-based classifier, for instance, can be trained using historical data to identify possible cyber-physical attacks by detecting unusual changes in the system data.

Decision tree

A decision tree is a classification and prediction algorithm based on a tree structure where internal nodes correspond to decisions made at a certain level based on input features of the system (e.g., sensor readings, communication routines, inverter logs), branches represent the conditions that are responsible for linking features, and the leaf nodes correspond to the end state of the decision-making process that can be outcome categories such as normal operation versus detection of a cyber-attack. In the context of a GCPV security system, decision trees are used to identify and classify potential attacks, such as false data injection, spoofing, or DoS, by learning a sequence of conditions through training data. Its benefits include fast inference and no assumptions regarding the structure of the data. However, it can very easily become overfit (Li et al., 2020).

K-nearest neighbor

The K-nearest neighbor (KNN) algorithm is a classification approach that assumes similar system behaviors occur close to each other in feature space. This is useful for surveillance of an unclassified event (system parameters such as an unusual inverter reading, sensor data, or a SCADA communications pattern) in a PV system; the occurrence of an event is established based on the classes of the k nearest events previously observed. Effectively, this method can be used to detect and classify cyber threats by referring new data to previous data known to be an attack or normal system operation patterns, such as false data injection, spoofing, or denial of service. KNN methods are intuitive, easy to implement, and scale to multiclass problems in the cybersecurity of GCPV power plants (e.g., normal vs different types of attacks on a GCPV plant). However, as the dataset continues to grow for PV operational logs, KNN's effectiveness decreases. In dealing with biased datasets where attack data is much smaller than the normal operation data, KNN can also have difficulties. Finally, KNN's effectiveness could depend highly on identifying the number of neighbors being considered, which has an immediate effect on detection accuracy (Li et al., 2020).

Support vector machine

A support vector machine (SVM) is a supervised learning algorithm used to classify a dataset by finding the optimal boundary, also called a hyperplane, between the two different classes. Data points that are far from the boundary are classified with higher confidence. When GCPV plant data are not linearly separable (which is often the case), SVM implements the kernel trick, which maps the data into a higher-dimensional space where a separating hyperplane can be identified. SVM is mathematically rooted and easy to interpret; it is effective for binary or multiclass classification problems. However, the use of SVM can be computationally expensive for very large datasets. Through the analysis of system attributes (such as power generation deviations, unusual inverter switching patterns, or Abnormal SCADA login frequency), SVM can be trained to differentiate between malicious behavior, such as FDI, and normal operation (Larkin et al., 2020; Li et al., 2020).

Artificial neural network

Inspired by biological neural structures, an ANN is a multilayered computational model consisting of interconnected layers of artificial neurons that are capable of learning complex nonlinear mappings between inputs and outputs.

- Intrusion detection and anomaly recognition: Differentiating between malicious or abnormal behavior, such as FDI or control signal manipulation, and normal PV system operation (such as voltage, current, and irradiance characteristics).
- Classification and prediction: Using operational data channels to predict potential attacks or classify cyber threats (such as DOS, spoofing, or malware).
- Adaptive learning: continuously adding updated operational and threat data to models to help them become resilient if attack scenarios change.

Deep learning methods

Long short-term memory

LSTM is a variant of recurrent neural networks (RNNs), particularly effective for modeling time-series data because it maintains long-term dependencies and mitigates the vanishing (or exploding) gradient problem that standard RNNs face. In the context of cybersecurity in photovoltaic plants. LSTMs are used to analyze sequential data streams from the physical layer (such as sensor time-series: voltage, current, irradiance, and power output) and cyber layer (such as SCADA/ communications links and inverter control commands). LSTMs can detect attacks (such as spoofing, FDI, and DoS) that diverge from predicted dynamic profiles by learning the temporal patterns of typical PV plant behavior. In addition to increasing the model capacity, deep or stacked LSTMs (multiple LSTM layers) allow for the collection of more complex temporal correlations, such as those between seasons, daily cycles, or physical signals (Luo et al., 2022). For example, a study in Hassan et al. (2025) employs multilayer LSTMs to improve cyber threat detection by capturing time-dependent patterns in PV farm datasets. A study in Li et al. (2020) shows that LSTM still achieved a very good performance for detecting cyber-attacks in GCPV plants.

Convolutional neural network

CNNs were first created for object recognition and image classification and can be used successfully for cybersecurity in PV plants because of their tolerance to noise and high capacity to extract spatial and hierarchical data. In this context, time-series cyber-physical data (such as network traffic features or sensor measurements, such as voltage, current, power output, and irradiance) can be converted into matrix formats. The CNN then processes these inputs via convolutional layers that learn local, short-range spatial patterns (e.g., relationships among neighboring time steps or among correlated sensors) and pooling layers to reduce dimensionality, followed by fully connected layers that decide whether data is normal or under attack (false-data injection, DoS, replay, etc.). CNNs have been demonstrated to identify cyber-attacks in GCPV plants or PV-integrated distribution systems with great accuracy, especially against FDI attacks on voltage regulation, when used alone or in hybrid models (e.g., CNN + LSTM) (Li et al., 2020).

For instance, the authors in Ahmadzadeh et al. (2024) demonstrate better detection performance by using a CNN architecture to differentiate between normal and FDI-attack conditions in actually observed PV datasets that include seasonal and load variations. In addition, a study in Mao et al. (2022) proposes a data-driven approach to detect and classify time-series data that identifies cyber-attacks, such as FDI, DoS, and delays, using LSTM and CNN-LSTM. The simulation results for grid-connected PV inverters show stable and accurate detection based on optimized feature selection and system measurements.

Comparison between machine learning and deep learning methods in cyber-attack detection

Data-driven methods, including ML and DL can train models based on prior data to distinguish the model based on combinations of attributes between normal activity and suspicious activities. Table 9.2 demonstrates a comparison of ML and DL methods for cyber-attack detection in PV power plants.

TABLE 9.2 Comparison of ML and DL methods for cyber-attack detection in GCPV power plants.

Ref.	Method	Robustness	Accuracy and precision	Training time	Computational efficiency	Data requirement
Li et al. (2020)	DT	Moderate; sensitive to noise and adversarial inputs	80%–90% (good for structured attack datasets)	Very fast (seconds)	High efficiency; suitable for edge devices	Small datasets sufficient
Li et al. (2020)	K-NN	Low robustness, performance drops with irrelevant/noisy features	78%–88%; precision depends on k value & distance metric	Moderate (slower with large datasets)	Poor efficiency for large-scale inference	Requires moderately sized datasets
Ozay et al. (2016)	SVM	Good robustness	85%–95%; strong precision for binary attacks	Moderate,	Moderate efficiency; scales poorly with huge data	Works with small-to-medium datasets
Li et al. (2020)	ANN	Robust against moderate attack variations	90%–95%; effective for intrusion detection	Long	feasible for PV controllers	Needs larger datasets
Mao et al. (2022)	LSTM	High robustness against sequential/temporal attacks	92%–98%; excels at detecting FDI and DoS attacks.	High (minutes–hours)	Lower efficiency;	Requires large sequential datasets
Mao et al. (2022), Zhang et al. (2023)	CNN	Very robust; strong generalization for unseen attacks	94%–99%; high precision & recall	High (minutes–hours)	Requiring calculation	Requires large labeled datasets

ANN, artificial neural network; *CNN*, convolutional neural network; *DL*, deep learning; *DT*, decision tree; *K-NN*, K-nearest neighbor; *LSTM*, long short term memory; *ML*, machine learning; *SVM*, support vector machine.

Intrusion detection systems

Definition

An intrusion detection system (IDS) is usually software (sometimes embedded in hardware) that operates to control or monitor the network of a GCPV plant. Fig. 9.13 shows how an intrusion detection system works as a bridge between cyber and physical GCPV plants to detect potential cyber-attacks.

IDSs are frequently categorized into two different types. Any deviation from typical behavior is used by anomaly-based intrusion detection systems to identify threats. When the traffic flow and known attack patterns match, signature-based IDS highlight a system danger concern. An IDS may be host-based, indicating that software programs are installed on specific client computers, or network-based, which indicates that it is positioned throughout the network as hardware sensors or installed system software that analyzes data packet flow (Chandola et al., 2009). Conventional IDS monitors communication traffic and control commands using signature and anomaly detection techniques, which are effective against popular attack patterns but are less adaptive when it comes to emerging threats (Sahani et al., 2023). ML-based intrusion detection systems, on the other hand, use measurement data

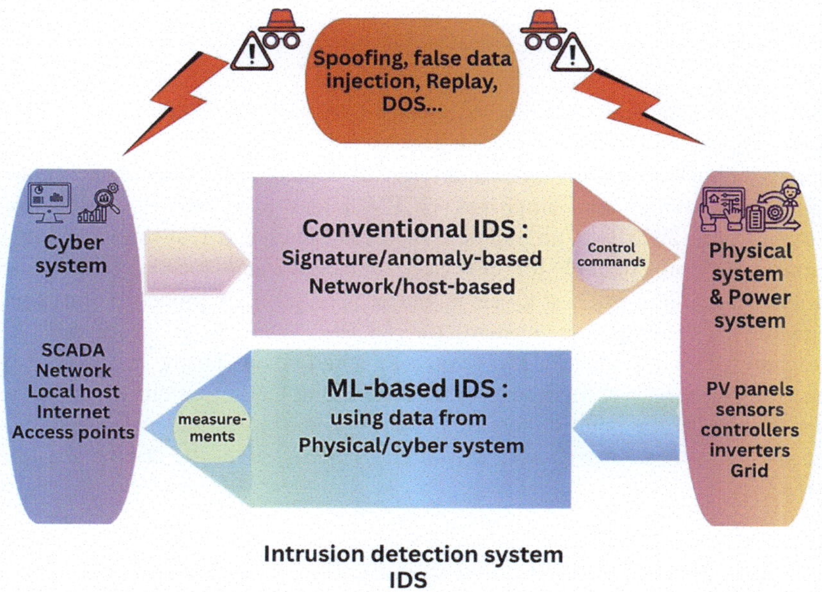

FIGURE 9.13 **Intrusion detection system for a GCPV power plant.** This figure shows an intrusion detection system designed for a grid-connected PV plant. It illustrates how data flow is monitored to detect abnormal or malicious activities. *GCPV*, grid-connected PV; *PV*, photovoltaic.

TABLE 9.3 Advantages and limits of IDS systems.

Advantages	Limits
Identifies unauthorized access, malware, and data manipulation before they affect PV performance or grid stability.	Normal system variations (e.g., due to weather or load changes) may be misinterpreted as attacks.
Prevents cyber-induced disturbances (e.g., inverter control manipulation, data spoofing) that could destabilize the grid.	Continuous monitoring and analysis increase processing and bandwidth requirements in embedded GCPV plants.
Provides continuous observation of inverter communication, sensors, and SCADA links.	Lack of large, labeled datasets for GCPV-specific attacks reduces AI-based IDS accuracy.
ML-based IDS can evolve with new attack signatures and network behaviors.	Signature-based IDS cannot identify novel or zero-day threats.
Aids in meeting standards like IEC 62443 and NIST guidelines for smart grid cybersecurity.	Difficult to retrofit IDS into legacy inverters and control systems without protocol support.

AI, artificial intelligence; *GCPV*, grid-connected PV; *IDS*, intrusion detection system; *ML*, machine learning; *PV*, photovoltaic; *SCADA*, supervisory control and data acquisition.

from physical and cyber layers to learn normal system behavior and detect anomalies caused by new or hidden attacks.

By establishing a multilayered defense, IDS guarantees early detection of malicious activities and improves GCPV plants' cybersecurity resilience (Sahani et al., 2023). As an example of recent studies, a real-time programmable detection method for precisely detecting and locating malicious intrusions in micro-grids is given in Ma et al. (2022). A proactive intrusion detection and mitigation system for GCPV inverters is developed in Mohammadi et al. (2023). Moreover, a technique in Zhang et al. (2023) combines device-level model-based detection with CNN (system-level waveform features at the point of common coupling) in a fusion scheme in order to improve robustness. Advantages and limits of IDS systems are briefly listed in Table 9.3.

Intrusion detection system types and their implementation

Network-based intrusion detection system

Network-based IDS observes network traffic passively in order to detect malicious activity. It reviews packets across the network, looking for signatures of known attacks or deviations from established baseline behavior.

It detects, for example, improperly formatted packets in industrial protocols (MODBUS, SunSpec), bad commands (e.g., a STOP command or a SET_POWER command from an unknown IP address), network reconnaissance scan, DoS floods targeting an inverter's IP, or MitM attack.

It can be implemented in:

- *Between field and supervisory networks:* This is the most critical location. A network-based IDS placed in this location monitors all communication between the SCADA system and the field devices (inverters, sensors). It can see every suspicious command.
- *At the network edge*: Placed just inside the firewall connecting the supervisory network to the corporate IT network or the internet. This detects intrusion attempts from the corporate side.
- *Inside the field network*: On a span port of a switch connecting a group of inverters to monitor lateral movement between field devices.

-Host-based intrusion detection system

A host intrusion detection system (HIDS) is installed directly on a critical host, allowing it to monitor its internal state. It uses agent software to check the host's file integrity, running processes, system logs, and network connections for signs of compromise. It can detect: unauthorized changes to SCADA software or configuration files, Malware execution or suspicious processes, unusual login attempts or privilege escalation on the SCADA server, or unauthorized access to process critical set point files or control logic.

It can be implemented in:

1. *SCADA:* The first target for attackers, Host-based IDS is indispensable.
2. *Data historian server:* Protects the integrity of operational data, which is crucial for detection algorithms and performance monitoring.
3. *Engineering workstation:* Engineering workstations are often highly privileged and used for configuration. They are a primary entry point for malware.

-Machine learning-based intrusion detection system

This approach employs ML to establish a model of normal behavior and then highlights major variations from that normal behavior. A baseline of normal operation is established in a system (e.g., normal traffic patterns, normal power curves, and standard sequences of commands), followed by statistical or ML models (like Isolation Forests or Auto-encoders) that identify which observations appear unusual. This is often a function of specific modern operational technology IDS platforms rather than a stand-alone sensor. It can detect:

- *FDI*: small deviations, in other words, slow and small drifts in sensor data (such as irradiance, voltage, etc.) that evade FDI attack detection based on a threshold.

- *Novel and zero-day attacks:* Data alterations or misrepresentation of information that do not have a specific signature; these attacks are meaningful enough to deviate from normal behavior.

  It might be implemented as:

- *Network anomaly detector:* for analyzing traffic flows and protocol sequences, this approach can only be used as a network anomaly detector.
- *As a process anomaly detector:* This can share attributes with the aforementioned detector; moreover, it correlates network data with physical process data, as in the following example: Inverter A is reporting high power, but the irradiance sensor is reporting zero.

General process of machine learning-based intrusion detection system

An ML-based IDS, a type of intelligent security system, learns normal and abnormal behaviors of the institutional systems or networks. By using historical data collected from different sources, including network traffic, device logs, and process parameters, it trains predictive models to differentiate between legitimate and malicious activity. During the training process, the ML algorithm identifies statistical relationships and behavioral signatures that define the normal state of a system (e.g., typical inverter connection rates, voltage levels, or packet structures). When the system is implemented, the ML-IDS continuously monitors new data in real time, then compares it to the learned patterns. If the observed data varies appreciably from what is expected, it will be classified as anomalous or suspiciously intrusive. This allows early detection of cyber-attacks such as data spoofing, DoS, or FDI. ML-based IDS options increase situational awareness in real-time by automating the learning and detection processes, lowering the need for manual monitoring, and can adapt over time as system operational conditions evolve. The performance of ML-based IDS in GCPV power systems involves several important steps (See Fig. 9.14):

- **Dataset collection:** involves gathering data from the GCPV plant's physical and cyber layers. For instance, SCADA traffic under protocols such as Modbus may be included in communication datasets, whereas inverter outputs, voltage, current, and irradiance values represent physical measurements. Normal operational conditions as well as simulated cyber-attacks, such as replay, denial of service, and false data injection, must be recorded in these datasets.
- **Preprocessing**: include data classifying as normal or malicious and normalizing feature values to scale them, which increases model efficiency.
- **Feature extraction:** The most instructive characteristics are then selected from the collected measurements and network traffic using feature

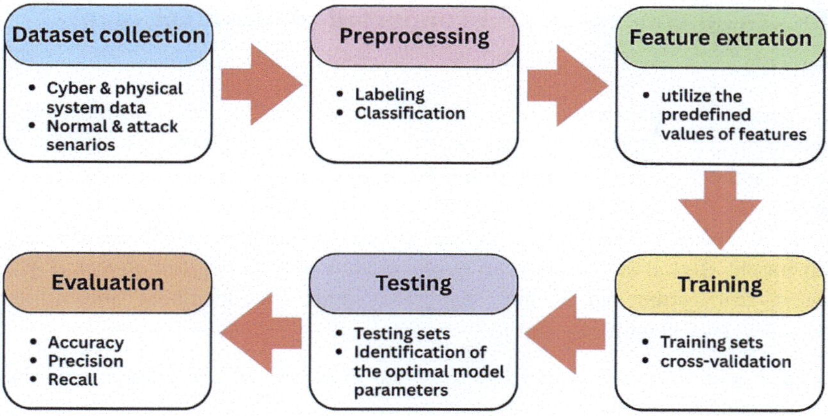

FIGURE 9.14 **Process of ML-based IDS in GCPV power plants.** This figure illustrates the process of a machine learning-based intrusion detection system in grid-connected PV plants. It shows how data is collected, processed, and analyzed to identify potential cyber threats. *GCPV*, grid-connected PV; *IDS*, intrusion detection system; *ML*, machine learning; *PV*, photovoltaic.

extraction, frequently using dimensionality reduction methods such as principal component analysis (Hall, 1999).

- **Training phase:** Once features are defined, the dataset is divided into training and testing subsets at the beginning of the training phase, and models are iteratively adjusted using cross-validation to achieve optimal performance.
- **Testing**: Following fitting and tuning to identify the optimal model parameters for each algorithm, the performance of the selected model is evaluated using the testing sets.
- **Evaluation**: To evaluate the IDS's consistency, resilience, and efficacy in detecting cyberattacks, evaluation metrics including accuracy, precision, and recall can be employed.

ML-IDS present considerable advantages for improving the cybersecurity of GCPV plants, including their ability to automatically identify abnormal behavior and detect cyber-attacks in real-time. ML-IDS can learn complex patterns in both communication and operational data, adapt to new threats, and can minimize manual intervention, consequently, enhancing both system reliability and grid stability. However, there are drawbacks to using ML-IDS, including the need for large, high-quality labeled datasets to train a competent model, susceptibility levels for alarm generation due to normal fluctuations of PV energy generation (in relation to weather conditions), and high computational capacity to reach the timeliness needed for real-time processing. In addition, ML-IDS implementation within preexisting PV infrastructures can be challenging, particularly legacy systems that are limited in processing or communication capability.

Blockchain concept in grid-connected photovoltaic plants

Definition

Blockchain is a distributed ledger technology that enables the safe recording, sharing, and verification of data across a decentralized network without a central authority. Data is stored in blocks, with each one linked cumulatively to the previous by a cryptographic hash, creating a continuous and immutable chain of records (Ye et al., 2022). Blockchain serves as a secure and transparent digital layer integrating all elements in the grid-connected PV system from solar panels and storage systems to consumers and utility operators. Blockchain can support:

- Safe and secure energy data transfer between GCPV plants and the power grid
- Automated energy trading using intelligent contracts
- Authentication and traceability of delivered and consumed power
- Tamper-proof record of all transactions and control operations

Moreover, blockchain serves as a common trust platform for producers, consumers, and grid operators. Every energy transaction (i.e., energy and data injection, consumption, or storage of power) is recorded as a transaction of the blockchain. The distributed ledger offers real-time audits of energy flows while monitoring for fraudulent reporting. Automation of supply and demand balancing, billing, and renewable energy certificates to producers established through smart contracts. For example, a householder who has PV panels delivers excess energy to the grid; the blockchain records would include: the amount of energy supplied, time, and identity of the seller/producer, and the transaction that would deliver a payment to the producer. Blockchain's advantages and limitations are summarized in Table 9.4. The main role of blockchain is to prevent and mitigate the impact of and possibly cyber-attacks by creating a more secure and trustworthy solution through:

- Data integrity: a locked file cannot be altered or deleted once it is recorded.
- Decentralization: There is no single point of failure that can be exploited.
- Authentication: Every device or user has its own recognizable cryptographic identity.
- Transparency: All activities are visible and traceable across the distributed network.

Blockchain-based grid-connected photovoltaic power plant architecture

Cyberphysical GCPV power system architecture is illustrated in Fig. 9.15 where blockchain operates as a trusted communication and control layer to all PV components. It guarantees that data integrity, authentication, and

TABLE 9.4 Advantages and limits of blockchain.

Advantages	Limits
• Lessens the potential for unauthorized access, tampering, or spoofing of data. • Noneditable or tamper-proof data enhances the precision of monitoring and reporting. • No longer need for intermediaries (energy brokers) to manage the transactions. • End users can trade electricity directly, peer-to-peer to each other. • Smart contracts automate control processes and payment transfers.	• High computational cost for consensus algorithms (specifically Proof of Work). • Scalability barriers regarding the number of IoT-connected PV devices. • Latency on recording real-time energy transactions. • Integration of legacy grid control systems. • Regulatory and interoperability challenges in national energy markets.

IoT, Internet of Things; *PV*, photovoltaic.

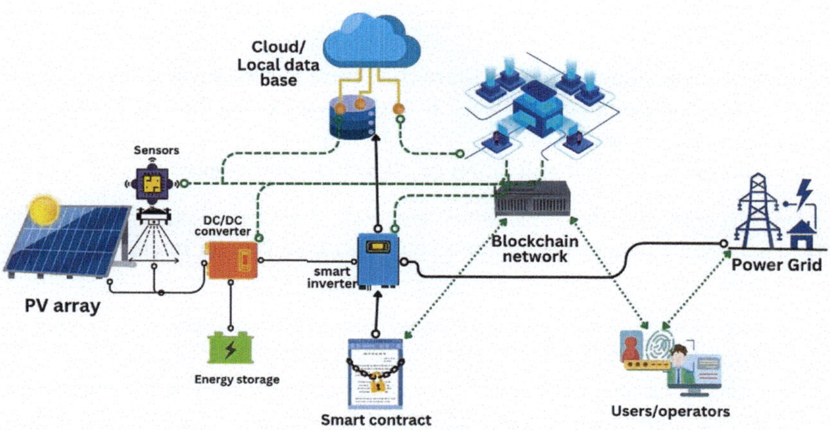

FIGURE 9.15 **Cyberphysical GCPV plant architecture.** This figure depicts the architecture of a cyber-physical grid-connected PV plant. It shows the integration of the blockchain, physical components, communication networks, and control systems. *GCPV*, grid-connected PV; IDS, intrusion detection system; *PV*, photovoltaic.

transparency are maintained from the smallest IoT sensor to the grid operator, thereby establishing a resilient, cyber-secure photovoltaic system. Each part of the system is described as follows:

PV sensors (IoT Nodes): These are the data collection elements of the GCPV plant. Each sensor has its own unique cryptographic identity (public/

private key pair), and the data being collected is digitally signed before transmission.

Security role:

- Prevent the falsification of data or the insertion of fake data (e.g., FDI).
- Each reading is connected and traceable to the authentic source.

Smart inverter controller: Controls the flow of power and synchronizes with the grid, communicates with the blockchain to record control actions and authenticates blockchain orders using digital signatures.

Security role:

- Prevents unauthorized control actions, such as spoofing
- Ensures that only authenticated blockchain transactions can alter inverter settings.

Blockchain network (Decentralized Ledger Layer): This component acts as the security backbone of the overall system. It consists of multiple nodes (computers or gateways) distributed across the grid or local area network. Each node: stores a copy of the ledger; engages in agreement to validate transactions and verifies blocks of sensor data and inverter actions, before adding them to the chain.

Security role:

- Removes any single point of failure (i.e., there is no central server to hack)
- Provides data immutability; once it has been added to the chain, that data cannot be changed
- Facilitates tracing and validation of all system activities

Smart contracts: these elements are coding programs that run on the blockchain and execute rules autonomously without human intervention.

Security role:

- Regulate energy trading transactions between consumers and producers (peer-to-peer markets).
- Automatic maintenance reminders when outputs drop below threshold levels (e.g., reminder for electrical maintenance services when output falls under predetermined parameters).
- Enforce payment and transaction remittance when a certain amount of energy is delivered. For example, if node A delivers 10 kWh to node B, then it notifies the system to transfer 1.5 USD from node B to node A.
- Disallow human interference or manipulation in trading or billing for energy-related services.
- Provide transparency and fairness through an automated agreement.

Cloud/local database: while blockchain stores transactional records, it becomes indefensible for large, raw data (like continuous sensor streams). Thus a hybrid architecture could be executed where hashes, identities, and verified summaries are stored in the blockchain, and the detailed sensor datasets are stored in the cloud or local database.

The relation between both ensures that the blockchain hash can verify that no data in the database has been tampered with.

Security role:

- Efficient and scalable storage
- Cryptographically linked to the blockchain for verification of integrity

User/operator dashboard: This is the interface in which engineers or grid operators would monitor and control the PV system. It allows them to monitor energy generation and consumption in real-time, track the verified blockchain transactions, receive alerts from smart contracts, and approve or deny action in the system.

Security role:

- Access is controlled by blockchain-authenticated access.
- Every action taken by the operator is immutably recorded in the blockchain.

Complementary roles of blockchain and machine learning/intrusion detection system for stronger security

Blockchain ensures trust, immutability, and secure data sharing; it's a preventive cybersecurity foundation. However, ML/IDS provides intelligence to detect and respond to threats; it's a reactive and adaptive layer. Together, they create a resilient, autonomous, and transparent security framework capable of protecting GCPV power plants from both known and emerging cyber threats (See Table 9.5).

When integrated, blockchain technology and (ML/IDS) create a powerful, multilayered defense that greatly enhances the cybersecurity of GCPV plants and the grid. Constant data exchange is fundamental, requiring collaborative communication between solar panels, inverters, control units, and grid operators, which makes them vulnerable to cyber threats such as FDI, unauthorized access, and DoS attacks. The integration of blockchain and ML/IDS will provide both preventative protection and intelligent detection for attackers, resulting in greater reliability, transparency, and trust between all network elements. Blockchain serves as the foundation of security and enables the secure storage of all transactions and data exchanged in the GCPV plant network in a decentralized and tamper-proof ledger. Each participant or system device is assigned a unique cryptographic identity that permits only validated members to exchange data.

On the other hand, LM-based IDS serves as the intelligent monitoring layer of the GCPV network with a continuous analysis of the operational data that the system is generating to identify any events that would indicate either a

TABLE 9.5 Comparison between blockchain and ML/IDS.

Aspect	Blockchain	ML / IDS
Purpose	Prevent attacks and ensure data integrity	Detect, classify, and respond to attacks
Type	Preventive/protective measure	Analytical/reactive system
Data handling	Stores data immutably in a decentralized ledger.	Analyzes data to identify anomalies
Architecture	Distributed (peer-to-peer nodes)	Centralized or hybrid (local or cloud analysis)
Key techniques	Cryptography, hashing, consensus algorithms, smart contracts	Supervised, unsupervised, ML or DL
Reaction to attack	Makes tampering almost impossible	Identifies and alerts about ongoing or past attacks
Weakness	High computational cost, limited real-time detection	Needs high-quality training data, can produce false alarms
Example function	Verify the authenticity of PV data	Detect false data injection or DoS attack

DL, deep learning; *ML*, machine learning; *IDS*, intrusion detection system; *PV*, photovoltaic.

malfunction or a cyber-attack. By combining Blockchain and ML/IDS, they form a feedback loop security mechanism that prevents, detects, and responds to threats instantaneously. Blockchain assures that all information coming into the network is authentic and traceable, while the ML/IDS mechanism confirms device and data flow behavior during operation. When an intrusion is detected, corresponding response actions (such as event logging, notifying administrators, and triggering automatic smart contracts) are initiated and immutably recorded on the blockchain for future auditing and analysis. Together, these capabilities create secure and transparent operations. The PV system and grid will safely operate within a trusted distributed arrangement with the ability to intelligently react to the future emergence of cyber-attacks.

Challenges and recommendations

As GCPV power plants evolve into connected, data-driven infrastructures, they generate vast amounts of data from sensors, inverters, meters, weather stations, and SCADA systems. While this digitization can improve visibility of operations and efficiency, it also increases the attack surface, making cybersecurity a key concern in the world of big data. In addition to improvements in the digitization of the PV industry, the progress of IoT has enhanced the performance and monitoring of solar PV power plants, but it has also raised new challenges for cyber-physical security. Components of GCPV plants with internet connectivity, such as inverters, controllers, and data loggers, often lack authentication standards, use old firmware, or have poor default security settings, making them vulnerable to remote exploitation by attackers. The discovery of undocumented communication modules in some imported inverters has raised concerns about hidden vulnerabilities and data leakage. Furthermore, the lack of standardized cybersecurity frameworks and limited regulatory coverage for distributed PV systems hinders consistent protection practices across the sector (Cybersecurity: European solar power plants under pressure, 2025). Human and operational factors further exacerbate these problems: lack of employee training, poor update management, and limited real-time visibility can all delay the detection and response to cyber incidents.

To address these vulnerabilities, a multilayered and coordinated security approach is essential. The implementation of cybersecurity standards for PV systems is recommended by experts, which includes essential secure firmware, encryption, and strong authentication by default. Reducing the attack surface and isolating vulnerable devices is possible by Adopting network segmentation, zero-trust architecture, and encrypted protocols. Additionally, the integration of ML-IDS will enhance the detection of anomalies and may enable early detection of emerging threats (Ledmaoui et al., 2025). Furthermore, it is best practice for operators to periodically conduct vulnerability assessments; validate supply chain (particularly cybersecurity) transparency; and provide regular cybersecurity training for personnel at plant sites to develop a culture of security awareness. Collectively, these approaches contribute to increasing the resilience and operational reliability of modern PV power plants in the face of evolving cyber-physical threats. Dealing with big data in large-scale PV systems is not just a matter of technology or a technical challenge; it is a priority in cybersecurity. The combination of AI-driven analytics, blockchain-based integrity assurance, and digital twin validation will enable a secure, transparent, and resilient solar energy ecosystem.

Conclusion

This chapter gave a comprehensive overview of the cyber-physical security of GCPV plants, including their structure, security framework, potential vulnerabilities, and detection methods. We focused on the various cyber threats that can target PV infrastructure, such as FDI, DoS, and replay attacks, as well as the potential consequences for grid stability and energy reliability. Antiviruses, firewalls, network segmentation, and intrusion detection systems are the most prevalent security mechanisms used in power grids today. Unfortunately, even these security control measures may be outdated or insufficient. As a result, cyberattacks on GCPV plants using diverse threat vectors can have a terrible impact on system operation. To address these concerns, both model-based and data-driven approaches were highlighted, with a focus on current advancements in ML and DL. Based on recent literature studies, AI-driven methods demonstrate promising results and show great potential for anomaly detection and real-time resilience improvement in GCPV power plants. Their effectiveness is strongly reliant on the availability of high-quality datasets and efficient feature selection processes. Furthermore, intrusion detection systems were offered as promising ways for increasing resilience. These solutions emphasize the necessity for proactive and adaptive security to defend GCPV plants from cyberattack scenarios.

The future roadmap for cybersecurity in GCPV power plants is advancing toward a smart, integrated, and autonomous defense system based on AI, digital twins, and blockchain. AI will act as an important enabler for predictive threat detection and adaptive response by analyzing real-time data, enabling organizations to anticipate and respond to a cyberattack before it interrupts operations. Digital twin technology will expand upon AI capabilities by creating virtual replicas of PV assets that track performance, simulate attack scenarios, and verify system integrity in real-time, enhancing recovery speed and resilience. At the same time, the use of blockchain will contribute to data integrity, authentication, and trust through decentralized, unchangeable ledgers, supporting secure communication and transparency of transactions among the parties involved. These technologies provide a clear vision to self-aware PV infrastructures that are resilient to cyber vulnerabilities, and autonomously assess, identify, and mitigate cyber-physical vulnerabilities, marking the onset of a new area of safe, smart, and sustainable solar energy systems.

AI disclosure

During the preparation of this work, the author(s) used ChatGPT for proofreading. After using this tool/service, the author(s) reviewed and edited the content as needed and take(s) full responsibility for the content of the publication.

References

Achaal, B., Adda, M., Berger, M., Ibrahim, H., Awde, A., 2024. Study of smart grid cybersecurity, examining architectures, communication networks, cyber-attacks, countermeasure techniques, and challenges. Cybersecurity 7 (1), 2523–3246. Available at https://doi.org/10.1186/s42400-023-00200-w.

Addressing the growing cybersecurity concerns within solar, 2025. Available at https://www.pv-tech.org/addressing-the-growing-cybersecurity-concerns-within-solar.

Aflaki, A., Gitizadeh, M., Razavi-Far, R., Palade, V., Ghasemi, A.A., 2021). A hybrid framework for detecting and eliminating cyber-attacks in power grids. Energies 14 (18), 5823. Available at https://doi.org/10.3390/en14185823.

Ahmadzadeh, M., Abazari, A., Ghafouri, M., Ameli, A., Muyeen, S.M., 2024. A deep convolutional neural network-based approach to detect false data injection attacks on PV-integrated distribution systems. IEEE Access 12, 2803–2816. Available at https://doi.org/10.1109/access.2023.3348549.

Barua, A., & Al Faruque, M.A., 2020. Hall spoofing: a non-invasive DoS attack on grid-tied solar inverter. 29th USENIX Security Symposium (USENIX Security 20). (2020).

Biju, J.M., Gopal, N., Prakash, A.J., 2019. Cyber attacks and its different types. International Research Journal of Engineering and Technology 6 (3).

Boutana, N., Mellit, A., Chekirou, R., Bettayeb, M., 2025. Security of Grid-Connected Photovoltaic Plants: A Mini-Review and Future Directions. 2025 International Conference on Artificial Intelligence, Embedded Systems, and Renewable Energy (AIESRE).

Carter, C., Onunkwo, I., Cordeiro, I.P., Johnson, J., 2017. Cyber security assessment of distributed energy resources. 2017 IEEE 44th Photovoltaic Specialist Conference (PVSC).

Chandola, V., Banerjee, A., Kumar, V., 2009. Anomaly detection: a survey. ACM Computing Surveys 41 (3). Available at https://doi.org/10.1145/1541880.1541882.

Cybersecurity: European solar power plants under pressure, 2025. https://www.pv-magazine.com/2025/05/09/cybersecurity-european-solar-power-plants-under-pressure/.

Cybersecurity in photovoltaic plant operations, 2021. https://docs.nrel.gov/docs/fy21osti/78755.pdf.

Diaba, S.Y., Shafie-khah, M., Elmusrati, M., 2024. Cyber-physical attack and the future energy systems: a review. Energy Reports 12, 2914–2932. Available at https://doi.org/10.1016/j.egyr.2024.08.060, https://www.sciencedirect.com/science/journal/23524847.

de Neira, A.B., Kantarci, B., Nogueira, M., 2023. Distributed denial of service attack prediction: challenges, open issues and opportunities. Computer Networks 222, 109553. Available at https://doi.org/10.1016/j.comnet.2022.109553.

Du, D., Zhu, M., Li, X., Fei, M., Bu, S., Wu, L., et al., 2023. A review on cybersecurity analysis, attack detection, and attack defense methods in cyber-physical power systems. Journal of Modern Power Systems and Clean Energy 11 (3), 727–743. Available at https://doi.org/10.35833/mpce.2021.000604.

Duo, W., Zhou, M.C., Abusorrah, A., 2022. A survey of cyber attacks on cyber physical systems: recent advances and challenges. IEEE/CAA Journal of Automatica Sinica 9 (5), 784–800. Available at https://doi.org/10.1109/jas.2022.105548.

Gupta, K., Sahoo, S., Mohanty, R., Panigrahi, B.K., Blaabjerg, F., 2022. Distinguishing between cyber attacks and faults in power electronic systems—a noninvasive approach. IEEE Journal of Emerging and Selected Topics in Power Electronics 11 (2), 1578–1588.

Hall, M.A., 1999. Feature selection for machine learning: comparing a correlation-based filter approach to the wrapper. Proceedings of the twelfth international Florida artificial intelligence research society conference.

Harrou, F., Taghezouit, B., Bouyeddou, B., Sun, Y., 2023. Cybersecurity of photovoltaic systems: challenges, threats, and mitigation strategies: a short survey. Frontiers in Energy Research 11. Available at https://doi.org/10.3389/fenrg.2023.1274451.

Hassan, G.F., Ahmed, O.A., Sallal, M., 2025. Evaluation of deep learning techniques in PV farm cyber attacks detection. Electronics 14 (3), 546. Available at https://doi.org/10.3390/electronics14030546.

Hemsley, K., Fisher, R., 2018. A history of cyber incidents and threats involving industrial control systems. International Conference on Critical Infrastructure Protection. Springer.

Hug, G., Giampapa, J.A., 2012. Vulnerability assessment of AC state estimation with respect to false data injection cyber-attacks. IEEE Transactions on Smart Grid 3 (3), 1362–1370. Available at https://doi.org/10.1109/TSG.2012.2195338.

Humayed, A., Lin, J., Li, F., Luo, B., 2017. Cyber-physical systems security—a survey. IEEE Internet of Things Journal 4 (6), 1802–1831. Available at https://doi.org/10.1109/jiot.2017.2703172.

Ibrahim, H., Kim, J., Huang, P.H., Raval, V., Enjeti, P. 2025. Methods to enhance cybersecurity of multiple inverters in large grid connected PV/battery energy storage systems. IEEE Applied Power Electronics Conference and Exposition (APEC). pp. 727–32.

Khan, R., Maynard, P., McLaughlin, K., Laverty, D., Sezer, S. 2016. Threat analysis of blackenergy malware for synchrophasor based real-time control and monitoring in smart grid. 4th International Symposium for ICS & SCADA Cyber Security Research.pp. 53–63.

Larkin, R.D., Wagner, T. J., Mullins, B. E. 2020. Securing photovoltaic system deployments with data diodes. 2020 47th IEEE Photovoltaic Specialists Conference (PVSC).

Ledmaoui, Y., El Maghraoui, A., El Aroussi, M., Saadane, R., 2025. Review of recent advances in predictive maintenance and cybersecurity for solar plants. Sensors 25 (1), 206. Available at https://doi.org/10.3390/s25010206.

Li, F., Li, Q., Zhang, J., Kou, J., Ye, J., Song, W.Z., et al., 2021. Detection and diagnosis of data integrity attacks in solar farms based on multilayer long short-term memory network. IEEE Transactions on Power Electronics 36 (3), 2495–2498. Available at https://doi.org/10.1109/tpel.2020.3017935.

Li, Q., Li, F., Zhang, J., Ye, J., Song, W., Mantooth, A., 2020. Data-driven cyberattack detection for photovoltaic (PV) systems through analyzing micro-PMU data. ECCE 2020 - IEEE Energy Conversion Congress and Exposition 431–436. Available at https://doi.org/10.1109/ECCE44975.2020.9236274, http://ieeexplore.ieee.org/xpl/mostRecentIssue.jsp?punumber=9235288.

Luo, Y., Xiao, Y., Cheng, L., Peng, G., Yao, D.D., 2022. Deep learning-based anomaly detection in cyber-physical systems: progress and opportunities. ACM Computing Surveys 54 (5). Available at https://doi.org/10.1145/3453155, http://dl.acm.org/citation.cfm?id=J204.

Ma, S., Li, Y., Du, L., Wu, J., Zhou, Y., Zhang, Y., et al., 2022. Programmable intrusion detection for distributed energy resources in cyber–physical networked microgrids. Applied Energy 306, 118056. Available at https://doi.org/10.1016/j.apenergy.2021.118056.

Mao, J., Zhang, M., Xu, Q., 2022. CNN and LSTM based data-driven cyberattack detection for grid-connected PV inverter. Conference CNN and LSTM based Data-driven Cyberattack Detection for Grid-connected PV Inverter. IEEE., pp. 704–709.

Mohammadi, F., Bok, R., Saif, M., 2023. A proactive intrusion detection and mitigation system for grid-connected photovoltaic inverters. IEEE Transactions on Industrial Cyber-Physical Systems 1, 273–286. Available at https://doi.org/10.1109/ticps.2023.3326773.

Ozay, M., Esnaola, I., Yarman Vural, F.T., Kulkarni, S.R., Poor, H.V., 2016. Machine learning methods for attack detection in the smart grid. IEEE Transactions on Neural Networks and Learning Systems 27 (8), 1773–1786. Available at https://doi.org/10.1109/TNNLS.2015.2404803, http://ieeexplore.ieee.org/xpl/RecentIssue.jsp?punumber=5962385.

Pan, K., Palensky, P., Esfahani, P.M., 2020. From static to dynamic anomaly detection with application to power system cyber security. IEEE Transactions on Power Systems 35 (2), 1584–1596. Available at https://doi.org/10.1109/tpwrs.2019.2943304.

Presekal, A., Rajkumar, V.S., Ştefanov, A., Pan, K., Palensky, P., 2025. Cyberattacks on power systems. Smart Cyber-Physical Power Systems: Fundamental Concepts, Challenges, and Solutions. wiley, Netherlands, pp. 365–403. Available at https://onlinelibrary.wiley.com/doi/book/10.1002/9781394191529, http://doi.org/10.1002/9781394191529.ch15.

Rajkumar, V.S., Ştefanov, A., Presekal, A., Palensky, P., Torres, J.L.R., 2023. Cyber attacks on power grids: causes and propagation of cascading failures. IEEE Access 11, 103154–103176. Available at https://doi.org/10.1109/access.2023.3317695.

Ribas Monteiro, L.F., Rodrigues, Y.R., Zambroni de Souza, A.C., 2023. Cybersecurity in cyber–physical power systems. Energies 16 (12), 4556. Available at https://doi.org/10.3390/en16124556.

Roadmap for photovoltaic cyber security, 2017. https://www.sandia.gov/research/publications/details/roadmap-for-photovoltaic-cyber-security-2017-12-01/.

Sahani, N., Zhu, R., Cho, J.H., Liu, C.C., 2023. Machine learning-based intrusion detection for smart grid computing: a survey. ACM Transactions on Cyber-Physical Systems 7 (2). Available at https://doi.org/10.1145/3578366, https://dl.acm.org/pub.cfm?id=J1536&CFID=818762994&CFTOKEN=30473978.

Singh, J., Shah, O.A., Arora, S., 2025. Smart grid cybersecurity: anomaly detection in solar power systems using deep learning. Energy Storage and Saving. Available at https://doi.org/10.1016/j.enss.2025.04.001.

Sourav, S., Biswas, P.P., Chen, B., Mashima, D. 2022. IEEE International Conference on Communications, Control, and Computing Technologies for Smart Grids, SmartGridComm. Institute of Electrical and Electronics Engineers Inc. Singapore Detecting Hidden Attackers in Photovoltaic Systems Using Machine Learning http://ieeexplore.ieee.org/xpl/mostRecentIssue.jsp?punumber=9960863.

Tatipatri, N., Arun, S.L., 2024. A comprehensive review on cyber-attacks in power systems: impact analysis, detection, and cyber security. IEEE Access 12, 18147–18167. Available at https://doi.org/10.1109/access.2024.3361039.

Teymouri, A., Mehrizi-Sani, A., Liu, C.-C. 2018. Cyber security risk assessment of solar PV units with reactive power capability. IECON 2018-44th Annual Conference of the IEEE Industrial Electronics Society.

Xie, L., Mo, Y., Sinopoli, B. 2020. False data injection attacks in electricity markets. 2010 first IEEE international conference on smart grid communications. pp. 226–31.

Ye, J., Giani, A., Elasser, A., Mazumder, S.K., Farnell, C., Mantooth, H.A., et al., 2022. A review of cyber-physical security for photovoltaic systems. IEEE Journal of Emerging and Selected Topics in Power Electronics 10 (4), 4879–4901. Available at https://doi.org/10.1109/JESTPE.2021.3111728, http://ieeexplore.ieee.org/xpl/RecentIssue.jsp?punumber=6245517.

Yohanandhan, R.V., Elavarasan, R.M., Manoharan, P., Mihet-Popa, L., 2020. Cyber-physical power system (CPPS): a review on modeling, simulation, and analysis with cyber security applications. IEEE Access 8, 151019–151064. Available at https://doi.org/10.1109/ACCESS.2020.3016826, http://ieeexplore.ieee.org/xpl/RecentIssue.jsp?punumber=6287639.

Zhang, J., Ye, J., Song, W., Lian, J., Zhao, D., Yang, H., 2023. Hybrid cyber-attack detection in photovoltaic farms. 2023 IEEE Energy Conversion Congress and Exposition, ECCE 2023 6295–6300. Available at https://doi.org/10.1109/ECCE53617.2023.10362667, http://ieeexplore.ieee.org/xpl/mostRecentIssue.jsp?punumber=10361838.

Zhang, J., Li, Q., Ye, J., L, 2020. Cyber-physical security framework for photovoltaic farms. 2020 IEEE CyberPELS (CyberPELS) 1–7.

Zhang, J., Guo, L., Ye, J., Giani, A., Elasser, A., Song, W., et al., 2023. Machine learning-based cyber-attack detection in photovoltaic farms. IEEE Open Journal of Power Electronics 4 (4), 658–673. Available at https://doi.org/10.1109/ojpel.2023.3309897.

Zhang, J., Guo, L., Ye, J., 2022. Cyber-attack detection for photovoltaic farms based on power-electronics-enabled harmonic state space modeling. IEEE Transactions on Smart Grid 13 (5), 3929–3942.

Index

Note: Page numbers followed by "*f*" and "*t*" refer to figures and tables, respectively.

D

E

F

G

Printed and bound by CPI Group (UK) Ltd, Croydon, CR0 4YY
17/09/2026
02218522-0001